INDIGENOUS KNOWLEDGE FOR CLIMATE CHANGE ASSESSMENT AND ADAPTATION

Many indigenous peoples and marginalized populations live in environments that are highly exposed to climate change impacts, such as arid zones, small islands, high-altitude regions, and the Arctic. As a result of this heightened exposure and their natural resource-based livelihoods, these societies are already observing and responding to changes exacerbated by climate change. Local and indigenous knowledge is therefore a source of invaluable information for climate change assessment and adaptation. This unique transdisciplinary publication is the result of collaboration between UNESCO's Local and Indigenous Knowledge Systems (LINKS) programme, the United Nations University's Traditional Knowledge Initiative, the Intergovernmental Panel on Climate Change (IPCC), and other organizations. Chapters written by indigenous peoples, scientists and development experts provide insight into how diverse societies observe and adapt to changing environments. A broad range of case studies illustrate how these societies, building upon traditional knowledge handed down through generations, are already developing their own solutions for dealing with a rapidly changing climate and how this might be useful on a global scale. Of interest to policymakers, social and natural scientists, and indigenous peoples and experts, this book provides an indispensable reference for those interested in climate science, policy and adaptation. This publication is the second in the "Local & Indigenous Knowledge" series published by UNESCO.

DOUGLAS NAKASHIMA is Director *ad interim* of UNESCO's Division of Science Policy and Capacity-building, and former Chief of the Small Islands and Indigenous Knowledge section. He created UNESCO's global programme on Local and Indigenous Knowledge Systems (LINKS) in 2002 that addresses the role of indigenous knowledge in environmental management, including in response to climate change, and reinforces its intergenerational transmission. Dr. Nakashima has been working within the field of indigenous knowledge for over 35 years, with his initial research focussing on Inuit and Cree First Nations in Arctic and Subarctic Canada. He recently led UNESCO's work with the IPCC to highlight, in the Fifth Assessment Report, the importance of indigenous knowledge for climate change assessment and adaptation.

IGOR KRUPNIK is Curator of Arctic Ethnology collections at the National Museum of Natural History, Smithsonian Institution in Washington, DC. Trained as a cultural anthropologist and ecologist, Dr. Krupnik has worked among the Yupik, Chukchi, Aleut, Nenets, and Inupiaq peoples, primarily in Alaska and the Russian Arctic region. His area of expertise lies in modern cultures, indigenous ecological knowledge, and the impact of modern environmental and social change on human life in the North. He has published more than 20 books, catalogues, and edited collections, including several "sourcebooks" on indigenous ecological and historical knowledge produced jointly with local partners for community use. He led major efforts in the documentation of indigenous knowledge of sea ice in the changing Arctic during the International Polar Year 2007–2008, and in 2012, he was awarded a medal from the International Arctic Science Committee for building bridges among social and natural scientists and polar indigenous peoples.

JENNIFER T. RUBIS is a Programme Specialist and Coordinator of UNESCO's Climate Frontlines project, focussing on indigenous knowledge in relation to climate change. She is a native Dayak from Sarawak in Borneo, and is descended from a line of Jagoi shamans and priestesses. She is a strong advocate of community organizing and the inclusion of indigenous perspectives in decision-making. She has worked for over ten years on forest and environmental issues within United Nations agencies and in civil society organizations at the international, national and community level.

INDIGENOUS KNOWLEDGE FOR CLIMATE CHANGE ASSESSMENT AND ADAPTATION

Edited by

DOUGLAS NAKASHIMA
UNESCO

IGOR KRUPNIK
Smithsonian Institution

JENNIFER T. RUBIS
UNESCO

CAMBRIDGE
UNIVERSITY PRESS

This book should be cited as

Nakashima, D., Krupnik, I. and Rubis, J.T. 2018. *Indigenous Knowledge for Climate Change Assessment and Adaptation.* Local & Indigenous Knowledge 2. Cambridge University Press and UNESCO: Cambridge and Paris.

University Printing House, Cambridge CB2 8BS, United Kingdom

One Liberty Plaza, 20th Floor, New York, NY 10006, USA

477 Williamstown Road, Port Melbourne, VIC 3207, Australia

314–321, 3rd Floor, Plot 3, Splendor Forum, Jasola District Centre, New Delhi – 110025, India

79 Anson Road, #06-04/06, Singapore 079906

Cambridge University Press is part of the University of Cambridge.

It furthers the University's mission by disseminating knowledge in the pursuit of education, learning, and research at the highest international levels of excellence.

Published jointly by the United Nations Educational, Scientific and Cultural Organization (UNESCO), 7, Place de Fontenoy, 75007 Paris, France, and Cambridge University Press, University Printing House, Shaftesbury Road, Cambridge CB2 8BS, United Kingdom.

www.cambridge.org
Information on this title: www.cambridge.org/9781107137882
DOI: 10.1017/9781316481066

© UNESCO 2018

First published 2018

Printed in the United Kingdom by TJ International Ltd. Padstow Cornwall

A catalogue record for this publication is available from the British Library.

Library of Congress Cataloging-in-Publication Data
Names: Nakashima, Douglas, editor. | Krupnik, Igor, editor. | Rubis, Jennifer T., editor.
Title: Indigenous knowledge for climate change assessment and adaptation / edited by Douglas Nakashima, Igor Krupnik, and Jennifer T. Rubis.
Description: Cambridge, United Kingdom; New York, NY: Cambridge University Press, 2018. | Includes bibliographical references.
Identifiers: LCCN 2017037842 | ISBN 9781107137882 (hardback)
Subjects: LCSH: Indigenous peoples – Ecology – Case studies. | Human beings – Effect of climate on – Case studies. | Traditional ecological knowledge – Case studies. | Climatic changes – Case studies. | BISAC: SCIENCE / Earth Sciences / Meteorology & Climatology.
Classification: LCC G50.I53 2018 | DDC 305.8–dc23
LC record available at https://lccn.loc.gov/2017037842

ISBN Cambridge 978-1-107-13788-2 Hardback
ISBN UNESCO 9789231002762

Cambridge University Press has no responsibility for the persistence or accuracy of URLs for external or third-party internet websites referred to in this publication and does not guarantee that any content on such websites is, or will remain, accurate or appropriate.

One year following first publication of this book, electronic files of the content will be available under the terms of a Creative Commons Attribution-Non-Commercial-ShareAlike 3.0 IGO (CC-BY-NC-SA 3.0 IGO) license (http://creativecommons.org/licenses/by-nc-sa/3.0/igo/) from http://www.unesco.org/new/en/unesco/resources/publications/unesdoc-database/

The designations employed and the presentation of material throughout this publication do not imply the expression of any opinion whatsoever on the part of UNESCO concerning the legal status of any country, territory, city or area or of its authorities, or the delimitation of its frontiers or boundaries.

The authors are responsible for the choice and the presentation of the facts contained in this book and for the opinions expressed therein, which are not necessarily those of UNESCO and do not commit the Organization.
Unless otherwise indicated, copyright of the illustrations belongs to the respective authors.

Contents

List of Contributors	*page* ix
Foreword	xiii
Acknowledgements	xvii
List of Abbreviations	xix

1 Indigenous Knowledge for Climate Change Assessment and
   Adaptation: Introduction 1
   DOUGLAS NAKASHIMA, JENNIFER T. RUBIS AND IGOR KRUPNIK

Part I Knowing our Weather and Climate

2 Forest, Reef and Sea-Level Rise in North Vanuatu: Seasonal Environmental
  Practices and Climate Fluctuations in Island Melanesia 23
  CARLOS MONDRAGÓN

3 Annual Cycles in Indigenous North-Western Amazon: A Collaborative
  Research Towards Climate Change Monitoring 41
  ALOISIO CABALZAR

4 Indigenous Knowledge in the Time of Climate Change (with Reference to
  Chuuk, Federated States of Micronesia) 58
  ROSITA HENRY AND CHRISTINE PAM

5 Local Responses to Variability and Climate Change by Zoque Indigenous
  Communities in Chiapas, Mexico 75
  MARÍA SILVIA SÁNCHEZ-CORTÉS AND ELENA LAZOS CHAVERO

6 Climate Knowledge of Ch'ol Farmers in Chiapas, Mexico 84
  FERNANDO BRIONES

Part II Our Changing Homelands

7   Indigenous Forest Management as a Means for Climate Change Adaptation and Mitigation — 93
    WILFREDO V. ALANGUI, VICTORIA TAULI-CORPUZ, KIMAREN OLE RIAMIT, DENNIS MAIRENA, EDDA MORENO, WALDO MULLER, FRANS LAKON, PAULUS UNJING, VITALIS ANDI, ELIAS NGIUK, SUJARNI ALLOY AND BENYAMIN EFRAIM

8   Indigenous Knowledge, History and Environmental Change as Seen by Yolngu People of Blue Mud Bay, Northern Australia — 106
    MARCUS BARBER

9   Coping with Climate: Innovation and Adaptation in Tibetan Land Use and Agriculture — 123
    JAN SALICK, ANJA BYG, KATIE KONCHAR AND ROBBIE HART

10 Seasonal Environmental Practices and Climate Fluctuations in Island Melanesia: Transformations in a Regional System in Eastern Papua New Guinea — 142
    FREDERICK H. DAMON

11 Traditional Knowledge and Crop Varieties as Adaptation to Climate Change in South-West China, the Bolivian Andes and Coastal Kenya — 152
    KRYSTYNA SWIDERSKA, HANNAH REID, YICHING SONG, DORIS MUTTA, PAUL ONGUGO, MOHAMED PAKIA, ROLANDO OROS AND SANDRA BARRIGA

Part III Confronting Extreme Events

12 Accounts from Tribal Elders: Increasing Vulnerability of the Navajo People to Drought and Climate Change in the Southwestern United States — 171
    MARGARET HIZA REDSTEER, KLARA B. KELLEY, HARRIS FRANCIS AND DEBRA BLOCK

13 The Spirits Are Leaving: Adaptation and the Indigenous Peoples of the Caribbean Coast of Nicaragua — 188
    MIRNA CUNNINGHAM KAIN

14 Indigenous Reindeer Herding and Adaptation to New Hazards in the Arctic — 198
    SVEIN D. MATHIESEN, MATHIS P. BONGO, PHILIP BURGESS, ROBERT W. CORELL, ANNA DEGTEVA, INGER MARIE G. EIRA, INGER HANSSEN-BAUER, ALVARO IVANOFF, OLE HENRIK MAGGA, NANCY G. MAYNARD, ANDERS OSKAL, MIKHAIL POGODAEV, MIKKEL N. SARA, ELLEN INGA TURI AND DAGRUN VIKHAMAR-SCHULER

15 'Everything That Is Happening Now Is Beyond Our Capacity' – Nyangatom Livelihoods Under Threat 214
SABINE TROEGER

Part IV Sources of Indigenous Strength and Resilience

16 'Normal' Catastrophes or Harbinger of Climate Change? Reindeer-herding Sami Facing Dire Winters in Northern Sweden 229
MARIE ROUÉ

17 Canaries of Civilization: Small Island Vulnerability, Past Adaptations and Sea-Level Rise 247
MARJORIE V. C. FALANRUW

18 Peasants of the Amazonian-Andes and their Conversations with Climate Change in the San Martín Region 254
RIDER PANDURO

19 Climate Change, Whaling Tradition and Cultural Survival Among the Iñupiat of Arctic Alaska 265
CHIE SAKAKIBARA

20 Indigenous Knowledge for Climate Change Assessment and Adaptation: Epilogue 280
IGOR KRUPNIK, JENNIFER T. RUBIS AND DOUGLAS NAKASHIMA

Index 291

Colour plate section to be found between pp. 140 and 141

Contributors

Foreword. Minnie Degawan, Kankanaey Igorot from Sagada, Philippines, and Indigenous and Traditional Peoples Program, Conservation International, Washington, DC, USA.

Chapter 1. Douglas Nakashima, Local and Indigenous Knowledge Systems (LINKS) programme, Natural Sciences Sector, UNESCO, Paris, France; **Jennifer T. Rubis**, LINKS programme, Natural Sciences Sector, UNESCO, Paris, France; and **Igor Krupnik**, Arctic Studies Center, Smithsonian Institution, Washington, DC, USA.

Chapter 2. Carlos Mondragón, Centro de Estudios de Asia y Africa, El Colegio de México, Mexico City, México.

Chapter 3. Aloisio Cabalzar, Rio Negro Program, Instituto Socioambiental, São Paolo, Brazil, and Graduate Institute of International and Development Studies, Geneva, Switzerland.

Chapter 4. Rosita Henry and **Christine Pam**, College of Arts, Society and Education, James Cook University, Australia.

Chapter 5. María Silvia Sánchez-Cortés, Instituto de Ciencias Biológicas, Universidad de Ciencias y Artes de Chiapas; and **Elena Lazos Chavero**, Instituto de Investigaciones Sociales, Universidad Nacional Autónoma de México, Mexico City, México.

Chapter 6. Fernando Briones, Consortium for Capacity Building, INSTAAR, and Center for Science and Technology Policy Research, CIRES, University of Colorado, Boulder, Colorado, USA.

Chapter 7. Wilfredo V. Alangui, College of Science, University of the Philippines Baguio, Baguio City, Philippines; **Victoria Tauli-Corpuz**, Tebtebba Foundation, Baguio City, Philippines; **Kimaren Ole Riamit**, Indigenous Livelihoods Enhancement Partners (ILEPA), Kenya; **Dennis Mairena**, **Edda Moreno** and **Waldo Muller**, Centro para la Autonomía y Desarrollo de los Pueblos Indígenas, Nicaragua; **Frans Lakon, Paulus Unjing, Vitalis Andi, Elias Ngiuk, Sujarni Alloy** and **Benyamin Efraim**, Institut Dayakologi and Aliansi Masyarakat Adat Nusantara – Wilayah Kalimantan Barat, Pontianak, Indonesia.

Chapter 8. Marcus Barber, Social and Economic Sciences Program, CSIRO, Brisbane, Australia.

Chapter 9. Jan Salick, Missouri Botanical Garden, St. Louis, Missouri, USA; **Anja Byg**, Social, Economic and Geographical Sciences, The James Hutton Institute, Dundee, Scotland, UK; **Katie Konchar** and **Robbie Hart,** Missouri Botanical Garden, St. Louis, Missouri, USA.

Chapter 10. Frederick H. Damon, Department of Anthropology, University of Virginia, Charlottesville, Virginia, USA.

Chapter 11. Krystyna Swiderska and **Hannah Reid**, International Institute for Environment and Development, London, UK; **Yiching Song**, Center for Chinese Agricultural Policy, China; **Doris Mutta** and **Paul Ongugo**, Kenya Forestry Research Institute, Kenya; **Mohamed Pakia**, Pwani University College, Kilifi, Kenya; **Rolando Oros**, Fundación para la Promoción e Investigación de Productos Andinos, Cochabamba, Bolivia; and **Sandra Barriga**, Asociación TARIY, Cochabamba, Bolivia.

Chapter 12. Margaret Hiza Redsteer, University of Washington Bothell, School of Interdisciplinary Arts and Sciences, Bothell, WA; **Klara B. Kelley,** Navajo Nation, Black Hat, New Mexico; **Harris Francis**, St Michaels, Navajo Nation; and **Debra Block**, US Geological Survey, Flagstaff, Arizona, USA.

Chapter 13. Mirna Cunningham Kain, Centro para la Autonomía y Desarollo de los Pueblos Indígenas, Nicaragua. **Box 13.1: Nadezhda Fenly,** International Indigenous Women's Forum (FIMI), Nicaragua.

Chapter 14. Svein D. Mathiesen, UArctic Institute for Circumpolar Reindeer Husbandry, International Centre for Reindeer Husbandry, Saami University of Applied Science, Guovdageaidnu/Kautokeino, Norway, and North-Eastern Federal University, Republic of Sakha, Yakutia, Russia; **Mathis P. Bongo**, Sami University College, Guovdageaidnu/ Kautokeino, Norway; **Philip Burgess**, International Centre for Reindeer Husbandry, Guovdageaidnu/ Kautokeino, Norway; **Robert W. Corell**, Global Environment and Technology Foundation, Center for Energy and Climate Solutions, Arlington, Virginia, USA; **Anna Degteva**, St Petersburg State University, St Petersburg, Russia; **Inger Marie G. Eira**, Sami University College, Guovdageaidnu/Kautokeino, Norway; **Inger Hanssen-Bauer**, Telemark University College and Norwegian Meteorological Institute, Norway; **Alvaro Ivanoff**, NASA Goddard Space Flight Center, Greenbelt, MD, USA; **Ole Henrik Magga**, Sami University College, Guovdageaidnu/Kautokeino, Norway; **Nancy G. Maynard**, Cooperative Institute for Marine and Atmospheric Studies, Rosenstiel School of Marine and Atmospheric Sciences, University of Miami, Miami, Florida, USA; **Anders Oskal**, International Centre for Reindeer Husbandry, Guovdageaidnu/ Kautokeino, Norway; **Mikhail Pogodaev**, Association of World Reindeer Herders, St Petersburg, Russia; **Mikkel N. Sara**, Sami University College, Guovdageaidnu/Kautokeino, Norway; **Ellen Inga Turi**, Sami University College, Guovdageaidnu/Kautokeino, Norway; and **Dagrun Vikhamar-Schuler**, Norwegian Meteorological Institute, Oslo, Norway.

Chapter 15. Sabine Troeger, Horn of Africa Regional Environment Centre, Addis Ababa, Ethiopia.

Chapter 16. Marie Roué, National Centre for Scientific Research and the National Museum of Natural History, Paris, France.

Chapter 17. Marjorie V. C. Falanruw, Yap Institute of Natural Science, Yap, Federated States of Micronesia.

Chapter 18. Rider Panduro, Asociación Rural Amazónica Andina Choba-Choba, San Martín, Peru.

Chapter 19. Chie Sakakibara, Environmental Studies Program, Oberlin College, Oberlin, Ohio, USA.

Chapter 20. Igor Krupnik, Arctic Studies Center, Smithsonian Institution, Washington, DC, USA; **Jennifer T. Rubis** and **Douglas Nakashima**, LINKS programme, Natural Sciences Sector, UNESCO, Paris, France.

Foreword

More and more the issue of climate change is becoming an everyday reality, both in the part of the world where I live and in many places I have visited over the years. Our mountainous communities are experiencing unexpected prolonged dry seasons or too much rain. Our rivers are drying up when they should be full, and we are being exposed to unusually cold or very hot weather. Climate change is a fact – there is no more denying it.

Climate change and its impacts have become a unifying phenomenon for the entire human community. The impacts of climate change are such that governments, academic institutions and non-governmental organizations (NGOs) are spending huge amounts of money to study it and to come up with strategies and solutions. Countless studies and meetings have been done to discuss climate change and possible remedies for its adverse effects.

However, the ongoing discourse on climate change often misses a basic component – that of indigenous knowledge. What is indigenous knowledge? This is the knowledge that a community has accumulated through years of interacting with the land, the waters, the air … in short, with its environment. It includes knowledge of what is the best seed to plant during which times of the year and on which part of the field; even who should best do the planting. It includes knowledge on when the sun or the moon will be at its highest point, as well as the lowest, and how this will affect people's food production. It includes knowing when certain animals are best hunted. It includes knowing what the next season is based on and what types of birds or plants will be abundant. It includes the knowledge of what type of rainfall is expected based on the movement of the clouds, plants and temperature changes. In short, it is the knowledge that has sustained us, indigenous peoples, for generations. Indigenous knowledge, like any other type of human knowledge, is never static. It evolves as times and seasons change.

Much of this knowledge remains at the level of our communities, where it is passed from one generation to the next via the traditional ways – through actual experience, continuous practice, rituals and storytelling. Most often this knowledge is not written down and has never been. The knowledge is passed on, but it is always enriched from the actual experience of the current practitioner. It is based on keen observations by each individual and by the entire community of many things that surround us in our daily life. While such knowledge is not written down and has been in existence for innumerable years, it continues to flourish primarily because it is viable and practical, and because it can always be verified through practice.

However, it is also a reality that climate change, together with other factors, is eroding indigenous knowledge systems. The loss of such knowledge will be felt by all and will have a long-lasting impact, not just on indigenous peoples but on all people on the planet. It is very necessary and urgent that indigenous knowledge be brought to the fore in climate change discourse and assessment. While indigenous knowledge does not hold the key to unlocking the solution to the many adverse impacts of climate change, it holds valuable lessons about monitoring/assessment and examples of successful adaptation. Indigenous knowledge has helped sustain indigenous communities for generations; it is highly possible that it can help sustain us in the future.

While efforts have been made to recognize indigenous peoples' rights in climate change mitigation and adaptation programmes, much remains to be done. Recognizing and ensuring that such rights are respected are important building blocks in making climate programmes viable and sustainable; but, there is also an urgent need to understand and harness the knowledge that indigenous peoples have in relation to climate. For example, in the current efforts to design REDD+ initiatives, there is much focus on safeguards and rights recognition. But if the knowledge of the forest-dependent indigenous communities is not harnessed in the development of forest resource management plans, the possibility of success lessens. There has to be an acceptance of the fact that communities have been there long before any managers, scientists or rangers were in place and consequently that this local knowledge is far more complex. Thus, harnessing indigenous knowledge to climate change adaptation and mitigation efforts will provide much benefit.

The publication of this book is an important step towards this effort of mainstreaming indigenous knowledge in the climate change discussion. The documentation of the many cases covering different regions is already a major step in ensuring that such knowledge will not be lost. Additionally, the chapters show that indigenous peoples, despite lacking the technology of modern management and science, have been able to observe and put in place measures to prepare their communities for change.

It is hoped that the publication of this book will foster more interest towards establishing productive dialogues between the holders of indigenous knowledge and mainstream climate scientists. Such a dialogue will bring about a better understanding of our environment which will surely be beneficial not just for one group, but for all of us.

UNESCO must be commended for embarking on this very relevant and timely initiative. Due to this undertaking, indigenous peoples and the development practitioners and scholars who work with them have made an effort to write down the knowledge that communities have developed and practised. This publication under the logo of UNESCO and with Cambridge University Press will give such papers more credibility in the eyes of those who speak at political sessions, design climate policies and distribute funds to cope with climate change. It is hoped that indigenous peoples will find such partnerships useful in their continuing struggle for recognition. For scientists, the challenge is there to accept: different knowledge systems exist and they remain in practice in many parts of the world. This publication prepared by a small group of people may be one of many small steps towards bridging the gap as we search for solutions. The next steps are in the hands of the knowledge holders and the decision makers.

Minnie Degawan

Acknowledgements

First and foremost, we would like to express our deep appreciation to all of the indigenous knowledge-holders whose insights, understandings and interpretations are at the heart of every chapter in this book.

Sincere thanks to the following experts who peer reviewed one or more book chapters: Peter Bates (The Firelight Group), Gillian Cambers (Sandwatch Foundation), Minnie Degawan (Conservation International, USA), Kirsty Galloway-Mclean, John E. Hay (University of the South Pacific, Fiji), Sam Johnston, Ameyali Ramos-Castillo, Carla Roncoli (Emory University, USA), Ursula Oswald Spring (National University of Mexico, Mexico) and Hans Thulstrup (UNESCO).

We are indebted to many persons who contributed to the preparation of this book. Special thanks to: Antoine Bateman, Julia Cheftel, Veronica Gonzalez Gonzalez, Trupthi Narayan, Max Ooft, Amaury Parelle, Tanara Renard–Truong Van Nga, Mao Takeuchi, Lena Taub, Hans Thulstrup and Donara Sydeeva-Blanc.

Abbreviations

ACIMET	Associação das Comunidades Indígenas do Médio Tiquié
AEITY	Associação Escola Indígena Tukano Yupuri
ANU	Australia National University
AR3	IPCC Third Assessment Report
AR4	IPCC Fourth Assessment Report
AR5	IPCC Fifth Assessment Report
ARAA/CHOBA CHOBA	Asociación Rural Amazónica Andina Choba Choba
CBD	Convention on Biological Diversity
CCAP	Centre for Chinese Agricultural Policy
CCIAV	Climate Change Impact, Adaptation and Vulnerability
COP	Conference of the Parties
CSIRO	Commonwealth Scientific and Industrial Research Organisation, Australia
EALÁT	Reindeer Herders Vulnerability Networks Study
ECMWF	European Centre for Medium-Range Weather Forecasts
ENSO	El Niño Southern Oscillation
FAO	Food and Agriculture Organization of the United Nations
FOIRN	Regional Federation of Indigenous Organizations, Brazil
FSM	Federated States of Micronesia
ICC	Inuit Circumpolar Council
IIED	International Institute for Environment and Development
IIPFCC	International Indigenous Peoples Forum on Climate Change
IPCC	Intergovernmental Panel on Climate Change
IPRs	Intellectual Property Rights
IPY	International Polar Year
ISA	Institute Socioambiental, Brazil
KEFRI	Kenya Forestry Research Institute
LINKS	Local and Indigenous Knowledge Systems programme, UNESCO
LKAB	Luossavaara-Kiirunavaara AktieBolag

NASA	National Aeronautics and Space Administration, USA
NGO	Non-Governmental Organization
PPB	Participatory Plant Breeding
PRA	Participatory Rural Appraisal
REDD+	Reducing emissions from deforestation and forest degradation, and the role of conservation, sustainable management of forests, and enhancement of forest carbon stocks in developing countries
ROS	Rain-on-Snow
SEK	Swedish Krona
SIDS	Small Island Developing States
SPCZ	South Pacific Convergence Zone
SPREP	Secretariat of the Pacific Regional Environment Programme
TAP	Tibetan Autonomous Prefecture, China
TAR	Tibetan Autonomous Region, China
TK	Traditional Knowledge
UN	United Nations
UNEP	United Nations Environmental Programme
UNESCO	United Nations Educational, Scientific and Cultural Organization
UNFCCC	United Nations Framework Convention on Climate Change
UNPFII	United Nations Permanent Forum on Indigenous Issues
UNU	United Nations University
UNU-TKI	United Nations University-Traditional Knowledge Initiative

1

Indigenous Knowledge for Climate Change Assessment and Adaptation: Introduction

Douglas Nakashima, Jennifer T. Rubis and Igor Krupnik

Over the last decades, climate variability and, more specifically, global climate change have entered the mainstream of international discourse, reflection and concern. One recent outcome of this global preoccupation with climate has been a growing interest in how weather, climate variability and climate change might be experienced, understood and interpreted by societies and cultures around the world, including those of indigenous peoples. In these diverse ecological, social and cultural settings, what changes are people observing and what responses might be the most appropriate and effective? This in turn raises the issue of what policies and what actions are required to guide adjustments to actual or expected future climate and its effects (IPCC, 2014: 5). To ensure that climate change decision-making recognizes and supports local priorities and needs, it is critical to be aware of what is already being experienced on the ground. Without this understanding, decisions may not only fail to provide assistance to those most in need, but may inadvertently undermine local resilience and increase vulnerability.

This volume presents a selection of case studies that illustrate how knowledge and practice rooted in indigenous communities may inform our understandings of climate change processes, and how indigenous coping strategies provide a crucial foundation for community-based adaptation. It also confronts some recurrent but misleading assertions about climate change impacts and responses with actual accounts from indigenous communities around the globe. It therefore contributes to a newly emerging field that builds synergies among a wide range of disciplines, from both the natural and social sciences, to address climate change assessment and adaptation in accordance with the observations, practices, knowledge and priorities of indigenous peoples.

The Emergence of Indigenous Knowledge in the Global Climate Change Arena

Global climate change was identified in the 1980s as one of humanity's most daunting challenges (see early history in UNFCCC, 1992; Maslin, 2014: 16–19). This recognition led to the establishment of the Intergovernmental Panel on Climate Change (IPCC) in 1988, and the United Nations Conference on Environment and Development in 1992 (Earth Summit, Rio de Janeiro, Brazil) that established the United Nations Framework

Convention on Climate Change (UNFCCC). Since that time, world attention roused by the global climate change debate has largely concentrated on particular areas of the planet – low-lying tropical islands and coastlines, high-altitude zones, tropical forests and the polar regions (Orlove et al., 2014). It so happens that most of these areas, except ice-covered Antarctica, are home to indigenous peoples. Despite the world's growing interest in indigenous peoples and their homelands as harbingers of the impacts of planetary climate change, much work needs to be done to understand their concerns and appreciate their knowledge (see Box 1.1).

Interest in the perspectives and knowledge of indigenous peoples (see Box 1.2) first emerged in certain world regions and on certain themes. In the Arctic, the Earth's northern polar region, indigenous voices were heard loud and clear already by the year 2000, perhaps because the shift in its climate regime has been one of the most dramatic and pronounced on the planet (see Larsen et al., 2014). Owing to these and other factors, polar peoples' observations and concerns about transitions triggered by the changing climate and weather were rapidly reported and were widely circulated in scholarly and political circles (McDonald et al., 1997; Weller et al., 1999; IISD, 2000; Huntington, 2000; Krupnik and Jolly, 2002; Herlander and Mustonen, 2004; Nickels, 2005).

In addition, since 1996, the Arctic has a unique political body called the Arctic Council. It includes eight Arctic nation states as its founding members, several 'observers' (both other states and organizations) and also six 'permanent participants' representing polar indigenous peoples – the Inuit, Sami, Aleut, Gwich'in, other Dene/Athabaskan groups and indigenous nations of northern Russia. The Arctic Council has been very active on issues related to environmental change and it has historically welcomed indigenous peoples' interests and voices, as seen in its seminal *Arctic Climate Impact Assessment* (ACIA, 2005) and several other studies it initiated. The Arctic may be a rare example where a strong consensus has developed among scientists, indigenous peoples, politicians and local governments about the threats brought by the changing environment *and* the need to engage polar residents in common actions – observation, research, assessment, adaptation and mitigation.

A similar combination of local factors favoured a more proactive response in another area critically affected by rapid climate change, low-lying tropical islands (Lazrus, 2012; Orlove et al., 2014). Here the push for local people's voices and knowledge in climate change assessment had the additional advantage of a direct state presence in intergovernmental forums. Small island developing states (SIDS), whether from the Caribbean or the Atlantic, Indian or Pacific Oceans, are United Nations Member States and are themselves Parties to the UNFCCC. In the Pacific, SIDS are 'indigenous states' with majority populations of indigenous peoples with their distinctive languages, institutions, cultures and histories. The New York-based intergovernmental Alliance of Small Island States also provided a platform to mobilize shared concerns and demands. From the early 2000s, island voices could be heard directly in global arenas and through networks of non-governmental players, anthropological and other scientific studies, independent environmentalists and filmmakers (see Lazrus, 2012; Rudiak-Gould, 2013).

Box 1.1 Basic Concepts and Definitions: Understanding Indigenous Knowledge

Recognition of indigenous knowledge is a recent development in the climate sciences in general and in the understanding of global climate change in particular. For this reason, it may be useful to introduce some basic concepts and definitions for readers whose encounter with indigenous knowledge may be relatively recent.

The term 'indigenous knowledge' makes reference to knowledge and know-how that have been accumulated across generations and which guide human societies in their innumerable interactions with their surrounding environment. Such traditional ecological knowledge is defined as: 'a cumulative body of knowledge, practice and belief, evolving by adaptive processes and handed down through generations by cultural transmission, about the relationship of living beings (including humans) with one another and with their environment' (Berkes, 2012: 7).

These knowledge systems are transmitted and renewed by each succeeding generation. They ensure the well-being of people around the globe by providing food security from hunting, fishing, gathering, pastoralism or small-scale agriculture, as well as health care, clothing, shelter and strategies for coping with environmental fluctuations and external forces of change (Warren et al., 1995; Nakashima and Roué, 2002; Sillitoe et al., 2002; Sillitoe, 2007).

An abundance of labels for this knowledge coexist in the literature. Common terms include but are not limited to indigenous knowledge, traditional knowledge, traditional ecological knowledge, local knowledge, farmers' knowledge, folk knowledge and indigenous science. Although each term may have somewhat different connotations and reference groups, they often share sufficient meaning to be utilized interchangeably (Nakashima and Roué, 2002). In this publication, the term indigenous knowledge will be used most frequently, as many of the examples put forward relate to knowledge developed and maintained by indigenous peoples. However, it should be recalled that important sets of local knowledge of relevance for climate change assessment and adaptation is also held by non-indigenous rural societies (Grabherr, 2009; Lawrence, 2009).

It is also important to keep in mind that much indigenous knowledge is gendered (Berkes, 2012). While men and women share a great deal of knowledge, they also hold distinct knowledge sets relating to differing and complementary roles that they may fulfil in society and in production. Rocheleau (1991) comments that 'half or more of indigenous ecological science has been obscured by the prevailing "invisibility" of women, their work, their interests and especially their knowledge'.

In this publication, the term 'knowledge' is used in its broadest sense. Though knowledge (in particular scientific knowledge) is often opposed to practice (science vs technology) and the rational is distinguished from the spiritual (science vs religion), in indigenous worldviews these diverse elements are often combined. In a holistic understanding of human interactions with their surrounding milieu, indigenous knowledge encompasses not only empirical understandings and deductive thought, but also community know-how, practices and technology; social organization and institutions; and spirituality, rituals, rites and cosmologies (Nakashima and Roué, 2002).

> **Box 1.2 Basic Concepts and Definitions: Identifying Indigenous Peoples**
>
> Indigenous peoples live in all regions of the world and own, occupy or use up to 22 per cent of the world's land, which in turn harbours 80 per cent of the world's biological diversity (UNDP, 2011: 54). They are estimated to number at least 370 million, and represent the greater part of the world's cultural diversity (UNPFII, n.d.), including the major share of the world's almost 7,000 languages (Harrison, 2007).
>
> In view of the enormous cultural diversity of indigenous peoples, their many histories of contact and interaction with other societies, and the broad spectrum of political contexts in which they live, there is no single universally accepted definition of 'indigenous peoples'. Most operational definitions converge around a set of core criteria that generally include:
>
> - maintenance of social and cultural traits distinct from those of mainstream or dominant society (which may include distinct languages, production systems, social organization, political and legal systems, spirituality and worldviews)
> - unique ties to ancestral territories and to the natural resources of these places
> - self-identification and recognition by others as being part of a distinct cultural group (Cobo, 1986)
> - in many instances, a historical or continuing experience with subjugation, dispossession and marginalization.
>
> Terms used to designate indigenous peoples vary considerably with place, social context and historical moment. Native, aboriginal or tribal peoples, ethnic minorities, hill tribes, scheduled tribes, sea gypsies, bushmen, Indians or First Nations are only a few of the many terms that may be applied to indigenous peoples.
>
> Many groups that self-identify as indigenous peoples are not recognized as such by nation states. Some members of indigenous groups feel the need to hide their identity due to the negative connotations of the 'indigenous label' (Montenegro and Stephens, 2006). Indigenous homelands often extend across national borders, and in some cases a single people may find themselves divided among several countries (UNPFII, n.d.).

Similarly, indigenous peoples have actively engaged in the UNFCCC forest-related discussions, particularly on the issue of REDD+ (UNFCCC, 2009; Schroeder, 2010). In the last fifteen years, as the dams of ignorance, political neglect and sidelining have been broken, research initiatives, international conferences, scholarly papers, dissertations, special journal issues and books on indigenous peoples and climate change have proliferated. Recent literature overviews cite current publications on these topics by many dozen, often by the hundreds (i.e. Crate and Nuttall, 2009; Roncoli et al., 2009; Crate, 2011; Ford et al., 2012; Lazrus, 2012; Nakashima et al., 2012; Maldonado et al., 2013; Bennett et al., 2014; McDowell et al., 2014 Orlove et al., 2014). The flow even triggered criticism by those who call it 'a thriving industry' of studying 'The Endangered Other' (Hall and Sanders, 2015) or point to the growing 'climate fatigue' in some indigenous communities overwhelmed by visiting researchers and journalists eager to talk about the impacts of changing climate (Marino and Schweitzer, 2009).

In the past ten to fifteen years the amount of available information has changed from a trickle to a steady stream, though it still constitutes but a minute fraction of the overall climate change publication 'flood', as witnessed by the assessment reports of the IPCC. Furthermore, it is still unevenly distributed with much attention continuing to be focused on a small number of emblematic peoples and places, and a dearth of information elsewhere. A number of important gaps and shortcomings still need to be addressed and overcome.

IPCC assessment reports, issued every five to seven years since 1990, provide a convenient measure of the growth in attention to indigenous peoples and indigenous knowledge during the last twenty-five years. The First IPCC Assessment Report in 1990 and the second in 1995 include no more than a handful of entries on indigenous peoples/populations/cultures, as well as indigenous livelihoods (IPCC, 1990, 1995). No reference is made to indigenous peoples as holders of knowledge about their environment or about climate change. By 2001 and the third assessment report (AR3), however, references to indigenous peoples in the Working Group II report on Impacts, Adaptation and Vulnerability increase by an order of magnitude, and for the first time a scattering of specific references appear to indigenous or traditional knowledge (IPCC, 2001). Particularly detailed accounts addressing the nature of indigenous populations, traditional livelihoods, specific vulnerabilities, factors of resilience, indigenous knowledge and indigenous resource management are included in the regional chapters concerning Australia/New Zealand and the polar regions, reflecting ongoing research in these locations.

The Fourth IPCC Assessment Report (AR4) in 2007 continues this overall trend, but is distinctive because of a pronounced surge in the number of entries on indigenous knowledge. This includes not only the regional chapters on polar regions and Australia/New Zealand, but also substantial content addressing indigenous perspectives in Africa, North America and in small islands. Particularly noteworthy is an entry in the Africa chapter on 'Indigenous knowledge in weather forecasting' (IPCC, 2007a: 456). Important references also appear in thematic chapters, such as Chapter 2 on 'New Assessment Methods' that reports: 'Traditional knowledge of local communities represents an important, yet currently largely under-used resource for CCIAV assessment (Huntington and Fox, 2005)' (ibid.: 138) and a box in Chapter 20 regarding the 'Role of local and indigenous knowledge in adaptation and sustainability research' (ibid.: 833). The AR4 also includes cross-chapter case studies with one featuring 'Indigenous knowledge for adaptation to climate change' (ibid.: 865). Moreover, for the first time, indigenous knowledge is cited in itself as an information source for understanding the nature of environmental impacts due to climate change:

Traditional ecological knowledge from Canada has recorded current ecosystem changes such as poor vegetation growth in eastern regions associated with warmer and drier summers; increased plant biomass and growth in western regions associated with warmer, wetter and longer summers;

the spreading of some existing species, and new sightings of a few southern species; and changing grazing behaviours of musk oxen and caribou as the availability of forage increases in some areas.

(IPCC, 2007a: 666)

Responding to this growing momentum in the AR4, UNESCO worked with IPCC and other organizations to support and advance global understanding of the links between indigenous knowledge and efforts to adapt to global climate change. UNESCO's Local and Indigenous Knowledge Systems (LINKS) programme and the United Nations University Traditional Knowledge Initiative (UNU-TKI), together with Vicente Barros, co-chair, and Edwin Castellanos and Roger Pulwarty, authors of the IPCC WG II of the Fifth Assessment Report (AR5), convened an international meeting in Mexico City in 2011 to bring together knowledge holders from indigenous peoples and local communities, indigenous knowledge experts and developing country scientists. UNESCO and UNU also produced the publication 'Weathering Uncertainty: Traditional knowledge for climate change assessment and adaptation' (Nakashima et al., 2012), which provided the authors of the AR5 with a review on this theme of over 300 publications from the scientific and grey literature.

In the AR5 published in 2014, both indigenous peoples and indigenous knowledge receive broad and systematic attention in the Working Group II report on Impacts, Adaptation and Vulnerability, with specific subsections dedicated to 'Indigenous Peoples' and to 'Local and Traditional Forms of Knowledge' in Chapter 12 on Human Security. It is in the AR5 that indigenous knowledge is given explicit recognition for the first time in the all-important Summary for Policymakers of the Synthesis Report.

Indigenous, local, and traditional knowledge systems and practices, including indigenous peoples' holistic view of community and environment, are a major resource for adapting to climate change, but these have not been used consistently in existing adaptation efforts. Integrating such forms of knowledge with existing practices increases the effectiveness of adaptation.

(IPCC, 2014: 27)

This growing attention by IPCC authors to indigenous peoples, as well as to indigenous knowledge, is also confirmed by a quantitative analysis of key terms (following the method of Ford et al., 2016). The occurrence of keywords referencing 'indigenous peoples', 'indigenous knowledge' and related terms shows a dramatic increase across the 25-year span of IPCC assessment reports (Figure 1.1). From a mere handful in the first two reports, the number of occurrences suddenly jumps to well over 100 in the AR3 in 2001, followed by a steady increase in the AR4 to almost 400 in the AR5 in 2014. The keyword count shows that the jump in occurrences in the AR3 is almost entirely due to references to 'indigenous peoples or communities'. A surge in the number of occurrences of 'indigenous knowledge' only appears in the AR4 in 2007 with over 80 occurrences, which almost doubles by the time of the AR5. Tabulating the occurrence of the same keywords by region reveals identical upward trends in regional chapters across the third (the first to include regional chapters), fourth and fifth assessment reports (Figure 1.2).

In summary, IPCC authors, reviewing the current state of knowledge in five assessment reports between 1990 and 2014, collectively bear witness to an expanding global awareness

Indigenous Knowledge for Assessment and Adaptation 7

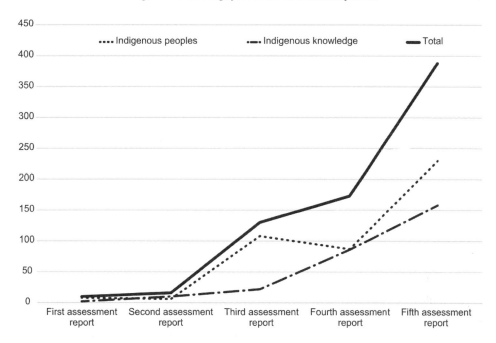

Figure 1.1 References to indigenous peoples and knowledge in IPCC assessment reports. © Tanara Renard–Truong Van Nga.

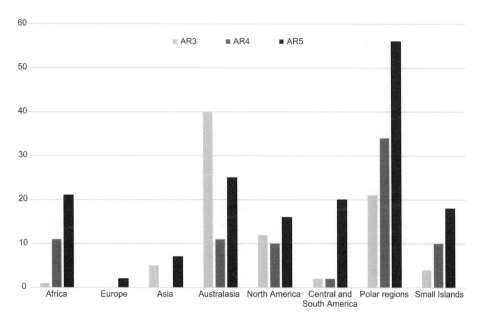

Figure 1.2 References to indigenous peoples and knowledge in the regional chapters of IPCC assessment reports. © Tanara Renard–Truong Van Nga.

of indigenous peoples' issues, in general, and the role to be played, in particular, by indigenous knowledge of climate change observation, adaptation and mitigation.

Paralleling this trend within the IPCC, indigenous issues have also been the subject of increasing attention within the UNFCCC and its subsidiary bodies. The role of indigenous knowledge has emerged in recent discussions on adaptation. The 2010 Cancun Adaptation Framework included indigenous knowledge among its principles for enhancing adaptation action (UNFCCC, 2010: para. 7). In 2013, the Nairobi Work Programme of the UNFCCC released a technical paper addressing indigenous and traditional knowledge for adaptation (UNFCCC, 2013), followed by an expert workshop (UNFCCC, 2014). These advances were then consolidated at the 21st Conference of the Parties (COP 21) of the UNFCCC through adoption of the Paris Agreement that exhorts nations to undertake adaptation action 'based on and guided by the best available science and, as appropriate, traditional knowledge, knowledge of indigenous peoples and local knowledge systems' (UNFCCC, 2015: Article 7). Just as significantly, the COP 21 decision recognizes the need to strengthen indigenous knowledge and 'establishes a platform for the exchange of experiences and sharing of best practices on mitigation and adaptation in a holistic and integrated manner' (UNFCCC, 2015: para. 135).

Indigenous peoples have been influential throughout this evolution. Involved in the COP since 1998 and acknowledged as a separate observer constituency since 2001, indigenous peoples and their representative organizations have engaged in the annual UNFCCC COP meetings to highlight the need for climate action and the desire to be able to actively participate in decision-making (Macchi et al., 2008). Through the International Indigenous Peoples Forum on Climate Change (IIPFCC), a self-organized, inclusive caucus for indigenous peoples at the UNFCCC, they have argued for, among other things, recognition of rights, full and effective participation and traditional knowledge (UNFCCC, 2004; IIPFCC, 2015).

Questioning 'Received Wisdom' About Indigenous Peoples and Climate Change

The growth of attention to indigenous peoples and their knowledge in climate change debates has been accompanied by a proliferation of assertions that, with repetition, are becoming a sort of 'received wisdom'. Increasingly widespread, these claims about indigenous peoples and their knowledge are at best misleading, and in some cases dangerously erroneous. Two of these oft-encountered statements are examined in greater depth here: the reputed vulnerability of indigenous peoples and the announced premature demise of indigenous knowledge.

Indigenous Peoples: Reconsidering Vulnerability

It has become common currency to present indigenous peoples as the first victims of global climate change, underlining their heightened vulnerability in the face of climate change impacts (IPCC 2007a, 2014; Ford et al., 2016). While not false, such a simplified formulation warrants more careful consideration. For sure, impacts on indigenous communities and their territories are anticipated to be both early and severe due to their location in vulnerable environments, including small islands, high-altitude zones, desert margins and the circumpolar Arctic. Furthermore, climate change poses a direct threat to many indigenous societies due to their continuing reliance upon resource-based livelihoods. Heightened exposure to negative impacts, however, is not the only reason for specific attention and concern. As many indigenous populations are socially and culturally distinct from mainstream society, the decisions, policies and actions undertaken by the major group, even when well-intended, may prove inadequate, ill-adapted and even inappropriate. IPCC points out that 'those in the weakest economic position are often the most vulnerable to climate change and are frequently the most susceptible to climate-related damages, especially when they face multiple stresses' (IPCC, 2007b: 65). In this respect, the IPCC AR4 makes specific reference to indigenous peoples and traditional ways of living, particularly in polar regions and small island states.

It would be a mistake, however, to only view indigenous peoples as potential victims of global climate change. Indeed, indigenous peoples rarely represent themselves as helpless or unable to cope in the face of change (Salick and Byg, 2007; Salick and Ross, 2009; Berkes and Armitage, 2010). They commonly emphasize that their environment has always changed and is continually changing (Fienup-Riordan and Rearden, 2010). Even though they often express grave concerns about climate change impacts on their homelands, they also systematically express confidence in their ability to adapt to whatever circumstances climate change may bring (Cochran, 2008; see also in this volume Barber, Chapter 8; Mondragón, Chapter 2). Retter (2009) contrasts the resilience of the diversified and ecosystem-based fishing economies of the indigenous coastal Sami, with the vulnerability of Norwegian commercial fisheries that rely primarily on cod, a species that soon may move out of the Norwegian economic zone as ocean waters continue to warm.

While indigenous peoples make their own detailed observations of dramatic changes in weather and ecological responses, they do not always consider this as a reason for alarm. For example, nomadic Nenets reindeer herders of the Russian Arctic, whose annual migration over hundreds of kilometres takes place entirely at or north of the latitudinal treeline, have in recent decades witnessed the symptoms attributed by scientists to a warming climate, such as later freeze up in autumn, earlier thaw in spring and warmer winters characterized by more frequent and intense rain-on-snow events (Forbes and Stammler, 2009; Bartsch et al., 2010). The latter can result in ice-encrusted pastures and significant losses (up to 25 per cent) of herds (Bartsch et al., 2010). Yet, so far, herders feel that this variation in weather does not represent a trend and does not endanger their survival in the foreseeable future. On the other hand, they are much more concerned about the impacts on their livelihoods from massive hydrocarbon extraction activities on their traditional territories (Rees et al., 2008; Forbes et al., 2009; Forbes and Stammler, 2009; Kumpula et al., 2012) or policies prescribing large reductions in herd size.

We believe that more circumspect use of the term 'vulnerability' with respect to indigenous peoples is required. The ability of systems to adapt to global climate change is often discussed in terms of both vulnerability and resilience. In the AR4, the IPCC defines vulnerability as 'the degree to which a system is susceptible to, and unable to cope with, adverse effects of climate change, including climate variability and extremes' (IPCC, 2007b: 89; see also Adger, 2006: 268). This definition emphasizes the importance of considering two factors: exposure to stress and an inability to cope.

Another approach differentiates among vulnerability's three constituent parts: exposure, sensitivity and adaptive capacity (Prno et al., 2011; see also Eriksen, Brown and Kelly, 2005; Parkins and MacKendrick, 2007; Tschakert, 2007; Forbes, 2008; Ford et al., 2008; Keskitalo, 2008; Young et al., 2010). According to the IPCC: '"exposure" relates to the degree of climate stress upon a particular unit; "sensitivity" is the degree to which a system will be affected by, or responsive to climate stimuli, either positive or negative; and "adaptive capacity" refers to the potential or capability of a system to adjust to climate change' (Rosenzweig and Hillel, 2008: Box 8.1). In the context of indigenous communities, exposure and sensitivity refer to the 'presence of potentially problematic conditions (exposure) and the occupancy and livelihood characteristics that make individuals and communities susceptible to these exposures (sensitivity)' (Prno et al., 2011: 7364; see also Smit and Wandel, 2006). Adaptive capacity relates to 'both local determinants – e.g. availability of human and financial capital, access to technology, local institutions – and the larger context within which the community operates – e.g. the terms of self-government and federally sponsored programs' (Prno et al., 2011: 3).

Thus, rather than describing indigenous groups as 'vulnerable' to climate change, it would be more accurate to emphasize their high degree of exposure-sensitivity, while drawing attention to their considerable adaptive capacity. Adaptive capacity contributes to resilience in that it relates to a people's ability to modify their behaviour and environment to manage and take advantage of changing climatic conditions (Ford et al., 2006).

Indigenous Knowledge in the Face of Climate Change: Imminent Demise or Source of Resilience?

While indigenous knowledge is gaining recognition in climate change decision-making, the announcement of its imminent demise is already circulating. Today it has become commonplace for participants in public forums to assert that indigenous peoples' knowledge and practices will soon become obsolete. At first glance, the logic behind their position seems sound: as global climate change will transform the environment beyond lived experience, the experience developed by indigenous peoples for dealing with environmental change will soon be outpaced. As a result, they will be, more than ever, climate change victims relying upon external aid to provide them with solutions to the new challenges they will face.

Indigenous peoples, however, do not share this view. At global climate change forums, indigenous peoples have long maintained two positions: first, that their homelands are being transformed irreversibly by climate change; and second, that they have valuable contributions to make towards climate decision-making due to their extensive experiential knowledge. Indigenous peoples furthermore state that their cultures and traditions are inherently resilient, and that heightened vulnerability is often a result of external forces, a combination of political and social pressures, along with competing land use (such as large-scale commercial agriculture, mining or forestry, urban expansion, etc.) that erode their resource base and institutions. Ford et al. (2007) argue that even though climate change may call into question specific elements of traditional expertise, the mainstays of life on the land will remain the same. Indeed, in-depth knowledge of the land, familiarity with home territories and basic skills for safety and survival will become all the more essential in the face of changing conditions.

This incessant discourse about the demise of indigenous knowledge is also rooted in a fundamental misunderstanding about the nature of this knowledge. It assumes that knowledge among indigenous peoples is a static set of information that is handed down with little change from one generation to the next. This obscures its dynamism and the important role played by each generation that assesses and adapts 'old' knowledge while creating and accumulating 'new' knowledge. It also ignores the existence of knowledge as a shared system that is collectively re-shaped, enriched and exchanged by a web of social actors. Finally, it only treats knowledge as fixed bytes of information, data or facts, whereas indigenous knowledge is largely about *process*. Rather than *what* one knows, it is more about *how* one learns and how information about one's surroundings is compiled and renewed.

Scientists characterize science, not by *scientific data* (destined to be outdated and surpassed), but by the *method* of scientific enquiry. Similarly, indigenous knowledge is a dynamic system of understanding that is continually called into question and refreshed. Through learning-by-doing, experimenting and knowledge building (Berkes, 2012), knowledge holders are able to adjust and modify their actions in response to environmental change. The adaptive capacity of the Wemindji Cree of Subarctic Canada is based both on their in-depth knowledge and continuous interaction with the land. This allows them to read critical signs from the environment and to detect the 'unusual' against the familiar backdrop of the 'normal' (Berkes, 2009).

As suggested by Takano (2004, cited in Ford et al., 2007), elements of knowledge per se may be of lesser importance than the cultural attitudes and values that shape the ways in which knowledge is acquired, transformed and deployed. Yup'ik elders in Western Alaska emphasize certain 'attitudes' are of critical importance while out on the land, such as persisting in a task, never giving up and not succumbing to fear, no matter what the circumstances (Fienup-Riordan and Rearden, 2010). The crux of indigenous knowledge is not the specific elements of information it encompasses, but rather the enduring values of patience, persistence, calmness, respect for elders and respect for the environment, which allow indigenous communities to remain resourceful and resilient in a changing world (Ford et al., 2007).

As such, we have to recognize the dynamism of indigenous knowledge and its inherent capacity to adapt in the face of change. If we consider indigenous knowledge, not as static bytes of accumulated information, but instead as a vehicle for passing on social values and attitudes that *reinforce* resilience, then the announced demise of indigenous knowledge in the face of climate change is revealed for what it is: misleading and, in fact, seriously erroneous.

It also amounts to a double standard. The principal argument for dismissing indigenous peoples' knowledge and experience is that climate change will impose conditions 'beyond lived experience'. But the 'lived experience' of climate scientists (and scientists in general) will also be surpassed by climate change. Yet no one is dismissing the role that science can play in combating and adapting to climate change. One cannot help but suspect that these skewed and oft-repeated views may be nourished by an inherent ethnocentric bias.

Linking Indigenous Knowledge and Climate Science: A Few Examples and Words of Caution

Recognizing the role played by indigenous peoples' knowledge in climate change observation, adaptation and mitigation is an important first step. However, basing climate change decision-making and action on the 'best available knowledge', both scientific and indigenous, brings with it additional challenges. Scientific observations and understandings may at times concur with those of indigenous peoples; at other times differ but be complementary; and at still other times may be totally at odds. What synergies can be developed between knowledge systems that are anchored in distinctive worldviews and ontologies? How can joint decision-making draw upon such diverse sets of knowledge, in part irreconcilable, to broaden the intellectual foundations and societal relevance of climate action?

Early efforts to understand indigenous climate knowledge reveal some of the challenges and complexities of bridging the gap between indigenous and scientific understandings. These case studies underscore the need for caution and careful reflection. Indeed, first impressions and standard procedures may favour incomprehension and derail efforts to establish a dialogue between indigenous knowledge holders and scientists.

One insight comes from the pioneering efforts of Ben Orlove, an anthropologist, working in collaboration with Mike Cane, a meteorologist, and John Chiang, a climatologist. Orlove, Chiang and Cane (2000, 2002) mobilized their collective cross-disciplinary expertise to comprehend an ancient practice that indigenous Andean farmers continue to use today to forecast the nature of the forthcoming rainy season. For centuries, Andean farmers have scrutinized the Pleiades star cluster, when it appears close to the horizon in the pre-dawn sky of late June, in order to decide to maintain or delay the planting of potatoes, their principal subsistence crop. This interpretation of the stars allows them to forecast, several months ahead, the incidence of rain (normal or late arrival; steady or erratic precipitation) during the next growing season from October to March.

By all measures of the imagination, this indigenous forecast seemed far-fetched. How could the observation of stars, light-years distant from the Earth, influence precipitation in one corner of the planet, and this several months hence? Orlove had participated in the farmers' observation of the Pleiades during the 1970s, but as an anthropologist, he focused his attention on the social and ritual aspects of this tradition. It was only two decades later, once he teamed up with Cane and Chiang, that the requisite interdisciplinary expertise was assembled to grasp the empirical and pragmatic dimensions of this age-old tradition. Knitting together ethnographic observations, physical science hypotheses and cutting-edge analyses of atmospheric data, the interdisciplinary scientific team concluded that the ritualized star-gazing of the Andean potato farmers provided a mechanism for assessing levels of humidity in the upper atmosphere. This proved in turn to be an indicator for El Niño years, when domestic crops in the Andes suffer from diminished and less reliable rainfall.

The apparent size and brightness of the Pleiades varies with the amount of thin, high cloud at the top of the troposphere, which in turn reflects the severity of El Niño conditions over the Pacific. Because rainfall in this region is generally sparse in El Niño years, this simple method provides a valuable forecast, one that is as good or better than any long-term prediction based on computer modelling of the ocean and atmosphere.

(Orlove et al., 2002: 428)

The efficacy of this traditional knowledge of Andean farmers reminds us to be circumspect about pre-conceived notions that we may harbour about indigenous knowledge. The challenge for scientists who try to understand indigenous knowledge and practice is that they are rarely equipped for the job. Disciplinary expertise, even highly developed, illuminates only one facet of a multi-dimensional ensemble. The full meaning of the Andean farmers' ritualized encounter with the Pleiades constellation could only be revealed through the collaborative work of an anthropologist, a meteorologist and a climatologist. This pooling of expertise from several science fields is one of the challenging requirements for appreciating and deciphering indigenous knowledge, including about climate.

In another case study, Inuit people from the community of Clyde River in Arctic Canada reported that weather conditions have become increasingly variable and difficult to predict during recent decades (Gearheard et al., 2010). In particular, they report stronger and more constant winds than in the past, as well as a change in the prevailing wind direction. They date the beginning of these changes with remarkable specificity to the early 1990s. Despite the certitude and unanimity among Clyde River Inuit, no similar trends were detected by scientific climate studies from anywhere in the Arctic (Weatherhead et al., 2010). From Clyde River itself, time series wind data since 1977 from the local meteorological station showed no significant change, contradicting the observations made by Inuit (Gearheard et al., 2010).

Intrigued by this lack of coherence between what climate scientists and Inuit from across the Arctic were reporting, Gearheard and Weatherhead teamed up to broaden investigation beyond the metrics that climate scientists habitually use. A *new* metric called weather 'persistence' was developed – the likelihood that an unusually warm (or cold) day will be followed by another such day. Applying this metric to meteorological data from another Inuit community of Baker Lake and focusing on the culturally important spring hunting season, Weatherhead et al. (2010) found a significant drop in weather persistence that begins in the late 1980s and early 1990s, and accentuates in the following decades – just as the Inuit report. Weatherhead et al. (2010) were able to determine, like Orlove et al. (2000), that indigenous reports may indeed be based on solid empirical data. Beyond underscoring once more the need for interdisciplinary expertise – in this case a cultural geographer (Gearhead) and a meteorologist (Weatherhead) – the Inuit case illustrates that standard scientific measurements may be inadequate to detect what indigenous peoples are experiencing. If Weatherhead et al. had not explored and designed new metrics, scientists and indigenous knowledge holders would be condemned to talk past each other with little hope of engaging in a meaningful dialogue.

Observations made by nomadic pastoralists in Mongolia about recent changes affecting their herds and livelihoods bring this point home (Marin, 2010). As in the Clyde River Inuit case, the reports of Mongolian pastoralists differ and even contradict standard meteorological records. Scientists make quantitative measures of rainfall and calculate annual means. But these measurements are of limited utility for herders. What is important to them and the survival of their herds is the *quality* of the rain: *how* and *where* it falls.

Herders attribute a major degradation in their milieu, which they have observed since 1999, to a decrease in soft rains (*shivree boroo*), which penetrate into the soil, and an increase in hard rains (*shiruun boroo*), which quickly run off. They also express concern about an increased patchiness of rainfall. Rather than raining and regenerating pastures across large areas, the rains which they refer to as 'silk embroidery' rains (*torgnii hee boroo*) only fall in limited areas, leaving extensive zones dry and devoid of pasture for their herds (Marin, 2010: 167).

While these qualitative parameters are essential for the herders, they remain invisible to the scientists. Standard meteorological measurements record rainfall as mean annual precipitation, which has hardly changed over the years. They do not record qualitative features such as the intensity of rainfall, nor the patchiness of its distribution. So even if indigenous peoples and scientists are observing the impacts of climate change over the same time period and in the same region, they are not necessarily attentive to the same phenomena nor measuring in the same ways.

This underscores another lesson to be learned: that which is meaningful to an indigenous hunter, herder or farmer about changes in weather or climate may have little to do with what scientists are knowing and recording. The point being made here is not that all indigenous observations are empirical and reliable, nor that they require scientific validation to be acknowledged by the international community. Rather, it is simply to point out that, even when there is a shared basis for indigenous and scientific understandings, bringing that common ground to light may require considerable dedication and effort. Where mutual comprehension can be attained, this often requires inputs from several scientific disciplines, both biophysical and social.

For sure, when indigenous and scientific observations and interpretations point in different directions, it is tempting to attribute shortcomings to the indigenous data set (but see Huntington et al., 2004). Detractors may point to the alleged inconsistency of information collected by different indigenous observers with little coordination apparent among them: analyses that would seem to be largely subjective, and data that are more qualitative than quantitative. In contrast, today's science seems to promise decision makers a multitude of answers based on high-resolution satellite imagery available from all corners of the globe, time series data covering several decades, and sophisticated theoretical models. But a closer assessment also reveals shortcomings that limit the applicability of science models: a reductionist approach focusing on a limited set of variables; extrapolations from narrow data sets; a restricted set of metrics; focus on mean values when the extremes may be the critical factor; and difficulties downscaling data to suit local needs.

Understanding the impacts of climate change and efforts to adapt through indigenous peoples' own systems of knowledge is an endeavour that is only just getting under way. The insights from farmers in the Andes, hunters in the Arctic and herders in Mongolia remind us that understanding weather and climate is vital for their livelihoods and well-being, and that as a result their capacities to observe, interpret and forecast are highly developed. Indigenous people's forecasting practices may seem preposterous (as in the Andes case) or their observations may contradict the findings of climate scientists (as in the Arctic and Mongolia). But first impressions can be deceiving. Considerable effort and ingenuity, interdisciplinary collaboration and the elaboration of new measurements are required to come to a comprehensive and mutually respectful understanding of indigenous peoples' knowledge of climate change impacts and adaptations.

About This Volume

This introductory chapter has traced some of the major milestones towards international recognition of the knowledge, practices and worldviews of indigenous peoples and local communities in the global climate change debate. While the emergence of this recognition is recent, it has been welcomed and relayed by individual and institutional actors at national, regional and international levels.

Yet global recognition brings with it expectations, challenges and responsibilities for indigenous peoples and local communities, natural and social scientists, and policy and decision makers at all levels. Certain questions immediately spring to mind. What is the specific nature and content of indigenous and local knowledge, and how does it enhance our collective understanding of climate change impacts and options for adaptation? What needs to be done to create conditions for a fruitful dialogue among indigenous knowledge holders, scientists and state actors? By what procedures, methods and partnership arrangements can climate change action be transformed into a collaborative effort that fully involves local communities?

The chapters in this volume try to fulfil some of the expectations raised and to bring initial answers to the many questions posed. The assembled authors offer a regionally diverse perspective with contributions concerning indigenous peoples' knowledge from across the globe. They confront some of the misconceptions and hastily formulated assertions about climate change impacts and responses with on-the-ground observations and accounts from various indigenous communities. And they demonstrate the need for interdisciplinary efforts that cross disciplinary boundaries and build linkages across knowledge systems and worldviews.

The book is organized into four major parts, which also interconnect with each other. The first part on 'Knowing our weather and climate' speaks to indigenous peoples' knowledge, conceptions and practices about weather and climate, an emerging field of study referred to by some as ethnometeorology or ethnoclimatology (Orlove et al., 2002). Such detailed knowledge of weather cycles is vital to community livelihoods, production systems and spiritual life.

In the second part entitled 'Our changing homelands', case studies are presented that reflect upon indigenous resources (both natural and cultural) and the multiple factors, including a changing climate, that impact upon them.

Part III, 'Confronting extreme events', presents case studies that focus on understanding and responding to extreme weather events and disasters. Indigenous knowledge in relation to disaster and disaster risk management is a rapidly growing area of interest with multiple applications in an increasingly disaster-prone world. Finally, in 'Sources of indigenous strength and resilience', contributors discuss different factors that contribute to enhancing resilience at the community level.

The chapters in the volume also collectively underscore some important gaps and shortcomings that need to be addressed. In this introduction, we have profiled some of the misconceptions that are increasingly encountered in climate change circles about indigenous peoples and indigenous knowledge. This intellectual 'no-man's land', where information is scarce and speculation can gain the upper hand, needs to be confronted with real experiences and rigorous data, such that policies and decisions are not perpetuated on the basis of past stereotypes and conventional but erroneous wisdom.

Acknowledgement

Special thanks to Tanara Renard–Truong Van Nga for compiling the data for Figures 1.1 and 1.2.

References

ACIA (Arctic Climate Impact Assessment). 2005. *Arctic Climate Impact Assessment*. Cambridge: Cambridge University Press.
Adger, W. N. 2006. Vulnerability. *Global Environmental Change*, 16(3): 268–81.
Bartsch, A., Kumpula, T., Forbes, B. C. and Stammler, F. 2010. Detection of snow surface thawing and refreezing using QuikSCAT: Implications for reindeer herding. *Ecological Applications*, 20: 2346–58.
Bennett, T. M. B., Maynard, N. G., Cochran, P. et al. 2014. Ch. 12: Indigenous peoples, lands, and resources. In Melillo, J. M., Richmond, T. C. and Yohe, G. W. (eds.). *Climate Change Impacts in the United States: The Third National Climate Assessment*. US Global Change Research Program, pp. 297–317.
Berkes, F. 2009. Indigenous ways of knowing and the study of environmental change. *Journal of the Royal Society of New Zealand*, 39(4): 151–6.
Berkes, F. 2012. *Sacred Ecology*, 3rd edn. New York: Routledge.
Berkes, F. and Armitage, D. 2010. Co-management institutions, knowledge and learning: Adapting to change in the Arctic. *Etudes/Inuit/Studies*, 34: 109–31.

Cobo, M. 1986. *Study of the Problem of Discrimination Against Indigenous Populations*, Preliminary Report to the UN Sub-Commission on the Prevention of Discrimination of Minorities /CN.4/Sub.2/1986/Add.4.

Cochran, P. 2008. Indigenous perspectives on snow and ice. *Mother Earth Journal*, http://mother-earth-journal.com/contributors-op-ed/2301patricia-cochran/

Crate, S. A. and Nuttall, M. (eds.) 2009. *Anthropology and Climate Change: From Encounters to Actions*. Walnut Creek, CA: Left Coast Press.

Crate, S. A. 2011. Climate and culture: Anthropology in the era of contemporary climate change. *Annual Review of Anthropology*, 40: 175–94.

Eriksen, S. H., Brown, K. and Kelly, P. M. 2005. The dynamics of vulnerability: Locating coping strategies in Kenya and Tanzania. *Geographical Journal*, 171: 287–305.

Fienup-Riordan, A. and Rearden, A. 2010. The ice is always changing: Yup'ik understandings of sea ice, past and present. In Krupnik, I. et al. (eds.) *SIKU: Knowing Our Ice: Documenting Inuit Sea Ice Knowledge and Use*. Dordrecht: Springer, pp. 303–28.

Forbes, B. C. 2008. Equity, vulnerability and resilience in social-ecological systems: A contemporary example from the Russian Arctic. *Research in Social Problems and Public Policy*, 15: 203–36.

Forbes, B. C. and Stammler, F. 2009. Arctic climate change discourse: The contrasting politics of research agendas in the West and Russia. *Polar Research*, 28: 28–42.

Forbes, B. C., Stammler, F., Kumpula, T. et al. 2009. High resilience in the Yamal-Nenets social-ecological system, West Siberian Arctic, Russia. *Proceedings of the National Academy of Sciences*, 106(22): 41–8.

Ford, J. D., Smit, B. and Wandel, J. 2006. Vulnerability to climate change in the Arctic: A case study from Arctic Bay Canada. *Global Environmental Change*, 16(2): 145–60.

Ford, J. D., Pearce, T., Smit, B. et al. 2007. Reducing vulnerability to climate change in the Arctic: The case of Nunavut, Canada. *Arctic*, 60(2): 150–66.

Ford, J. D., Smit, B., Wandel, J. et al. 2008. Climate change in the Arctic: Current and future vulnerability in two Inuit communities in Canada. *The Geographical Journal*, 174(1): 45–62.

Ford, J. D., Bolton, K., Shirley, J. et al. 2012. Mapping human dimensions of climate change research in the Canadian Arctic. *Ambio*, 41(8): 808–22.

Ford, J. D., Cameron, L., Rubis, J. et al. 2016. Including indigenous knowledge and experience in IPCC assessment reports. *Nature Climate Change*, 6: 349–53.

Gearheard, S., Pocernich, M., Stewart, R., Sanguya, J. and Huntington, H. P. 2010. Linking Inuit knowledge and meteorological station observations to understand changing wind patterns at Clyde River, Nunavut. *Climatic Change*, 100: 267–94.

Grabherr, G. 2009. Biodiversity in the high ranges of the Alps: Ethnobotanical and climate change perspectives. *Global Environmental Change*, 19: 167–72.

Hall, E. F. and Sanders, T. 2015. Accountability and the academy: Producing knowledge about the human dimensions of climate change. *Journal of the Royal Anthropological Institute*, 21(2): 438–61.

Harrison, D. K. 2007. *When Languages Die: The Extinction of the World's Languages and the Erosion of Human Knowledge*. Oxford: Oxford University Press.

Herlander, E. and Mustonen, T. (eds.) 2004. *Snowscapes, Dreamscapes. Snowchange Book on Community Voices of Change*. Study Materials 12. Tampere, Finland: Tampere Polytechnic Publications.

Huntington, H. P. (ed.) 2000. *Impact of Changes in Sea Ice and Other Environmental Parameters in the Arctic. Report of the Marine Mammal Commission Workshop*. Girdwood, AK, 15–17 February 2000. Bethesda, MD: Marine Mammal Commission.

Huntington, H. P., Callaghan, T., Fox, S. and Krupnik, I. 2004. Matching traditional and scientific observations to detect environmental change: A discussion on Arctic terrestrial ecosystems. *Ambio*, 13: 18–23.

IIPFCC (International Indigenous Peoples' Forum on Climate Change). 2014. *Executive Summary of Indigenous Peoples' Proposal to the UNFCCC COP 20 and COP 21*. Lima, Peru. 30 November 2014. www.iwgia.org/images/stories/int-processes-eng/UNFCCC/ExecutiveSummaryIPpositionFINAL.pdf

IISD (International Institute for Sustainable Development). 2000. *Sila Alangatok. Inuit Observations on Climate Change*. Video. Winnipeg, Canada.

IPCC. 1990. *Climate change 1990. Report prepared for Intergovernmental Panel on Climate Change by Working Group II*. [McG. Tegart, W. J., Sheldon, G. W. and Griffiths, D. C. (eds.)] Canberra: Australian Government Publishing Service, 294 pp.

IPCC. 1995. *Climate Change 1995 – Impacts, Adaptations and Mitigation of Climate Change: Scientific-Technical Analyses Contribution of Working Group II to the Second Assessment Report of the Intergovernmental Panel on Climate Change*. [Watson, R. T., Zinyowera, M. C. and Moss, R. H. (eds.)] Cambridge, UK and New York: Cambridge University Press, 879 pp.

IPCC. 2001. *Climate Change 2001: Impacts, Adaptation and Vulnerability. Contribution of Working Group II to the Third Assessment Report of the Intergovernmental Panel on Climate Change* [McCarthy, J. J., Canziani, O. F., Leary, N. A., Dokken, D. J. and White, K. S. (eds.)] Cambridge, UK and New York: Cambridge University Press, 1033 pp.

IPCC. 2007a. *Climate Change 2007: Impacts, Adaptation and Vulnerability. Contribution of Working Group II to the Fourth Assessment Report of the Intergovernmental Panel on Climate Change* [Parry, M. L., Canziani, O. F., Palutikof, J. P., van der Linden, P. J and Hanson, C. E. (eds.)] Cambridge, UK: Cambridge University Press, 976 pp.

IPCC. 2007b. *Climate Change 2007: Synthesis Report Contribution of Working Groups I, II and III to the Fourth Assessment Report of the Intergovernmental Panel on Climate Change. Core Writing Team* [Pachauri, R. K. and Reisinger, A. (eds.)] Geneva: IPCC, 104 pp.

IPCC. 2014. *Climate Change 2014: Impacts, Adaptation, and Vulnerability. Part A: Global and Sectoral Aspects. Contribution of Working Group II to the Fifth Assessment Report of the Intergovernmental Panel on Climate Change* [Field, C. B., Barros, V. R., Dokken, D. J. et al. (eds.)]. Cambridge, UK and New York: Cambridge University Press, 1132 pp.

Keskitalo, E. C. H. 2008. *Climate Change and Globalization in the Arctic: An integrated approach to vulnerability assessment*. Sterling, VA: Earthscan.

Krupnik, I. and Jolly, D. 2002. *The Earth is Faster Now*. Fairbanks, AK: Arctic Research Consortium of the United States.

Kumpula, T., Forbes, B. C., Stammler, F. and Meschtyb, N. 2012. Dynamics of a coupled system: Multi-resolution remote sensing in assessing social-ecological responses during 25 years of gas field development in Arctic Russia. *Remote Sensing*, 4: 1046–68.

Larsen, J.N., Anisimov, O.A., Constable, A. et al. 2014. Polar regions. In Barros, V. R., Field, C. B., Dokken, D. J. et al. (eds.) *Climate Change 2014: Impacts, Adaptation, and Vulnerability. Part B: Regional Aspects. Contribution of Working Group II to the Fifth Assessment Report of the Intergovernmental Panel on Climate*. Cambridge, UK and New York: Cambridge University Press, pp. 1567–612.

Lawrence, A. 2009. The first cuckoo in winter: Phonology, recording, credibility and meaning in Britain. *Global Environmental Change*, 19: 173–5.

Lazrus, H. 2012. Sea change: island communities and climate change. *Annual Review of Anthropology*, 41: 285–301.

Macchi, M., Oviedo, G., Gotheil, S. et al. 2008. *Indigenous and Traditional Peoples and Climate Change*. Gland, Switzerland: International Union for Conservation of Nature (IUCN).

Maldonado, J. K., Shearer, C., Bronen, R., Peterson, K. and Lazrus, H. 2013. The impact of climate change on tribal communities in the US: Displacement, relocation, and human rights. *Climatic Change*, 120(3): 601–14.

Marin, A. 2010. Riders under storms: Contributions of nomadic herders' observations to analysing climate change in Mongolia. *Global Environmental Change*, 20: 162–76.

Marino, E. and Schweitzer, P. 2009. Talking and not talking about climate change in northwestern Alaska. In Crate, S.A. and Nuttal, M. (eds.) *Anthropology and Climate Change: From Encounters to Actions*, Walnut Creek, CA: Left Coast Press, pp. 209–17.

Maslin, M. 2014. *Climate Change: A Very Short Introduction*. Oxford, UK: Oxford University Press.

McDonald, M., Arragutainag, L. and Novalinga, Z. 1997. *Voices from the Bay: Traditional Ecological Knowledge of Inuit and Cree in the James Bay Bioregion*. Ottawa: Canadian Arctic Resources Committee and Environmental Committee of the Municipality of Sanikiluaq.

McDowell, G., Stephenson, E. and Ford, J. 2014. Adaptation to climate change in glaciated mountain regions. *Climatic Change*, 126(1–2): 77–91.

Montenegro, R. and Stephens, C. 2006. Indigenous health in Latin America and the Caribbean. *Lancet*, 367: 1859–69.

Nakashima, D. and Roué, M. 2002. Indigenous knowledge, peoples and sustainable practice. In Munn, T. (ed.) *Encyclopedia of Global Environmental Change*. Chichester: Wiley and Sons, pp. 314–24.

Nakashima, D. J., McLean, K. G., Thulstrup, H. D., Castillo, A. R. and Rubis, J. T. 2012. *Weathering uncertainty: Traditional knowledge for climate change assessment and adaptation*. Paris and Darwin: UNESCO and UNU.

Nickels, S. 2005. *Unikkaaqatigiit: Putting the human face on climate change: Perspectives from Inuit in Canada*. Joint publication of Inuit Tapiriit Kanatami, Nasivvik Centre for Inuit Health and Changing Environments at the Universite Laval and the Ajunnginiq Centre at the National Aboriginal Health Organization.

Orlove, B., Lazrus, H., Hovelsrud, G. K. and Giannini, A. 2014. Recognitions and responsibilities: On the origins of the uneven attention to climate change around the world. *Current Anthropology*, 55(3): 249–75.

Orlove, B. S., Chiang, J. C. H. and Cane, M. A. 2000. Forecasting Andean rainfall and crop yield from the influence of El Niño on Pleiades visibility. *Nature*, 403: 69–71.

Orlove, B., Chiang, S., John, C. H. and Cane, M. A. 2002. Ethnoclimatology in the Andes. *American Scientist*, 90: 428–35.

Parkins, J. R. and MacKendrick, N. A. 2007. Assessing community vulnerability: A study of the mountain pine beetle outbreak in British Columbia, Canada. *Global Environmental Change*, 17(3–4): 460–71.

Prno, J., Bradshaw, B., Wandel, J. et al. 2011. Community vulnerability to climate change in the context of other exposure-sensitivities in Kugluktuk, Nunavut. *Polar Research*, 30: 1–21.

Rees, W. G., Stammler, F. M., Danks, F. S. and Vitebsky, P. 2008. Vulnerability of European reindeer husbandry to global change. *Climatic Change*, 87: 199–217.

Retter, G-B. 2009. Norwegian fisheries and adaptation to climate change. *Climate Change and Arctic Sustainable Development: Scientific, Social, Cultural and Educational Challenges*. Paris: UNESCO, pp. 88–93.

Rocheleau, D. 1991. Gender, ecology and the science of survival: Stories and lessons from Kenya. *Agricultural and Human Values*, 8(1): 156–65.

Roncoli, C., Crane, T. and Orlove, B. 2009. Fielding climate change in cultural anthropology. In Crate, S. A and Nuttall, M. (eds.) *Anthropology Climate Change from Encounters to Actions*. Left Coast Press, pp. 87–115.

Rosenzweig, C. and Hillel, D. 2008. *Climate Variability and the Global Harvest: Impacts of El Niño and Other Oscillations on Agro-Ecosystems*. Oxford: Oxford University Press.

Rudiak-Gould, P. 2013. *Climate Change and Tradition in a Small Island State: The Rising Tide*. New York: Routledge.

Salick, J. and Byg, A. 2007. *Indigenous Peoples and Climate Change*. Oxford: Tyndall Centre for Climate Change Research, http://tyndall2.webapp3.uea. ac.uk/sites/default/files/Indigenous%20 Peoples%20and%20Climate%20 Change_0.pdf

Salick, J. and Ross, N. 2009. Traditional peoples and climate change [Introduction to Special Issue], *Global Environmental Change*, 19: 137–9.

Schroeder, H. 2010. Agency in international climate negotiations: The case of indigenous peoples and avoided deforestation. *International Environmental Agreements: Politics, Law and Economics*, 10(4): 317–32.

Sillitoe, P. (ed.). 2007. *Local Science vs. Global Science: Approaches to Indigenous Knowledge in International Development*. New York: Berghahn Books.

Sillitoe, P., Bicker, A. and Pottier, J. (eds.) 2002. *Participating in Development: Approaches to Indigenous Knowledge*. London: Routledge.

Smit, B. and Wandel, J. 2006. Adaptation, adaptive capacity and vulnerability. *Global Environmental Change*, 16(3): 282–92.

Tschakert, P. 2007. Views from the vulnerable: Understanding climatic and other stressors in the Sahel. *Global Environmental Change*, 17(3–4): 381–96.

UNDP (United Nations Development Programme). 2011. *Human Development Report 2011: Sustainability and Equity – A Better Future for All*. New York: Palgrave Macmillan.

UNFCCC (United Nations Framework Convention on Climate Change). 1992. *United Nations Framework Convention on Climate Change* (FCC/INFORMAL/84/Rev.1), Bonn, Germany.

UNFCCC. 2004. *Promoting Effective Participation in the Convention Process. Note by the Secretariat*. (FCCC/SBI/2004/5), 16–25 June. Bonn, Germany.

UNFCCC. 2009. *Issues Relating to Indigenous People and Local Communities for the Development and Application of Methodologies* (FCCC/SBSTA/2009/MISC.1/Add.1), 1–10 June. Bonn, Germany.

UNFCCC. 2010. *The Cancun Agreements: Outcome of the work of the Ad Hoc Working Group on Long-term Cooperative Action under the Convention* (FCCC/CP/2010/7/Add.1), 29 November–10 December 2010. Cancun, México.

UNFCCC. 2013. *Best Practices and Available Tools for the Use of Indigenous and Traditional Knowledge and Practices for Adaptation, and the Application of Gender-Sensitive Approaches and Tools for Understanding and Assessing Impacts, Vulnerability and Adaptation to Climate Change* (FCCC/TP/2013/11), 31 October. Bonn, Germany.

UNFCCC. 2014. *Report on the Meeting on Available Tools for the use of Indigenous and Traditional Knowledge and Practices for Adaptation, Needs of Local and Indigenous Communities and the Application of Gender-Sensitive Approaches and Tools for Adaptation* (FCCC/SBSTA/2014/INF.11), 4–15 June. Bonn, Germany.

UNFCCC. 2015. *Adoption of the Paris Agreement* (FCCC/CP/2015/L.9/Rev.1), 30 November– 11 December. Paris, France.

UNPFII. n.d. *Who are Indigenous Peoples?* Factsheet – Indigenous Peoples, Indigenous Voices, www.un.org/esa/socdev/unpfii/documents/5session_factsheet1.pdf

Warren, D. M., Slikerveer, L. J. and Brokensha, D. (eds.) 1995. *The Cultural Dimension of Development: Indigenous Knowledge Systems*. London: Intermediate Technology Publication.

Weatherhead, E., Gearheard, S. and Barry, R. G. 2010. Changes in weather persistence: insight from Inuit knowledge. *Global Environmental Change*, 20: 523–8.

Weller, G., Anderson, P. and Wang, B. 1999. *Preparing for a Changing Climate: The Potential Consequences of Climate Variability and Change, Alaska*. Fairbanks, AK: Alaska Center for Global Change and Arctic System Research, University of Alaska.

Young, G., Zavala, H., Wandel, J., et al. 2010. Vulnerability and adaptation in a dryland community of the Elqui Valley, Chile. *Climatic Change*, 98(1): 245–76.

Part I

Knowing our Weather and Climate

2

Forest, Reef and Sea-Level Rise in North Vanuatu: Seasonal Environmental Practices and Climate Fluctuations in Island Melanesia

Carlos Mondragón

This chapter offers a critical perspective on the status of indigenous knowledge in relation to adaptation in Island Melanesia. The focus of my discussion centres on the Torres Islands, a remote, small island society in the Vanuatu archipelago (Figure 2.1). Through this case study I contend that the adaptive capacity of local communities is intrinsic to the form that their environmental knowledge takes, and as such it is not separable from the broader cultural frames, ethical values and existential principles that inform indigenous worldviews. One of my guiding concerns is that 'too often adaptation is imagined as a non-political, technological domain', and consequently

> there is a danger that adaptation policy and practice will be reduced to seeking the preservation of an economic core, rather than allowing it to foster the flourishing of cultural and social as well as economic development, or of improved governance that seeks to incorporate the interests of future generations, non-human entities and the marginalised.
>
> (Pelling, 2011: 3)

The basic argument in this text is that climate policy design must emerge from the 'symmetrical co-production of knowledge' between scientific and indigenous actors, with equal recognition for each party's epistemologies and practices (see Jasanoff, 2004; Lahsen, 2010). This approach necessarily challenges aspects of the scientific and technical frames that inform international adaptation initiatives, insofar as it is based on the argument that knowledge is not and cannot be kept separate from politics and cultural values (Domínguez Rubio and Baert, 2012: 2). For this reason, indigenous knowledges often expose the 'impatience on the part of policy-makers with complexity and dispute' (Leach, 2012: 79). But only by taking indigenous knowledge on its own terms, as a fully rounded, contested and entangled field of politics, social values and environmental engagements can policy designers aim at successful community-based interventions that go beyond the rhetoric of consultation, effectively assisting communities 'to help themselves' and 'come up with their own solutions to problems' (Warrick, 2011).

I begin with a summary description of the wealth of human–environmental relations that constitute the Torres Islands' lived world in order to highlight the complexity and diversity of small islands' environs. The aim is to upset widespread notions of fragility and

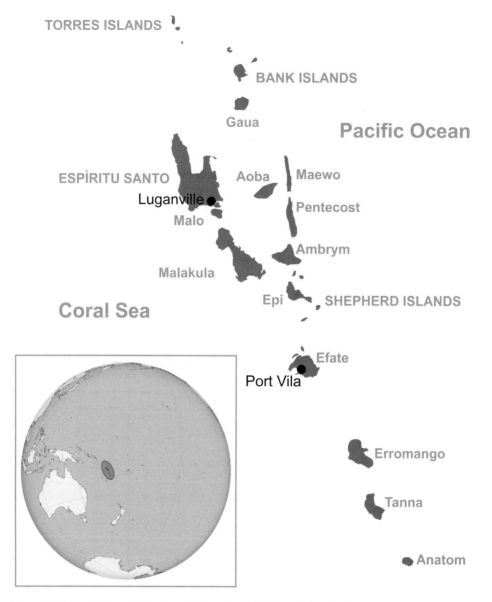

Figure 2.1 Map of Vanuatu showing the location of the Torres Islands.

simplicity by which small island societies are imagined. These stereotypes diminish local actors and their capacities, and can lead to single-issue interventions, such as the current focus on sea-level rise, to the detriment of broader chains of human–environmental complexity. I then describe a key example of culturally informed adaptation, namely an indigenous ritual cycle that has a long history of unfolding in synchronization with weather extremes, in order to lay out the multifaceted nature of local environmental engagements. In the final section I focus on the contrast between local and non-local perceptions of environmental change in the Torres Islands. I contend that existing asymmetries between knowledge regimes must be addressed if climate policies are to be successfully grounded and appropriated by local communities. Unfortunately, this is not a new argument, and it speaks to the continuing lack of dialogue between anthropologists and policy designers (for ethnographic studies that have highlighted the importance of indigenous knowledge in relation to the environment in Oceania, see Kwa'ioloa and Burt, 2001; Taylor, 2008; Jacka, 2009; Wescott, 2012; Rudiak-Gould, 2013; McCarter and Gavin, 2014).

It bears noting that I avoid talking about local knowledge as a coherent corpus of 'tradition' that must be preserved or protected from outside intrusion. Instead I emphasize the contingent, processual and relational nature of local and extra-local frames of reference. In this respect I am aware that traditional environmental knowledge 'embodies a sense of place and worldview that may be challenged by climate change, may enhance adaptive capacity to environmental changes, or, alternatively, be maladapted to contemporary rates of environmental change' (Lazrus, 2012: 290).

Even as local knowledge has increasingly come to be regarded as key to 'community-based adaptation', it continues to be approached as a culturally neutral and uncomplicated phenomenon. Consequently, in most climate-related interventions local knowledge is characterized as a grab bag of local 'skills' and practices which appear to be 'suitably "scientific"' (Castree, 2015), and therefore useful to technical solutions and policy design. In one after another example of adaptation initiatives in the Pacific, the representation of indigenous knowledge, when it is recognized, is reduced to a piecemeal selection of technically relevant 'environmental markers' and skills that can be detached from their original epistemic and social frames and unproblematically incorporated into broader objectivist models (for a range of examples where this practice is in evidence, see, for Vanuatu: Hay et al., 2003; Cronin, 2004; Muliagetele, 2007; Bartlett, 2009; and for the broader region see Nakalevu, 2006; GIZ-ACCPIR, 2009; SPREP, 2011). In the worst cases, indigenous knowledge is dismissed as 'superstition', proper to primitive ways of engaging with 'objective' natural phenomena (e.g. Ballu et al., 2011). At issue are not only competing truth claims about reality, but the fact that, unlike the natural sciences, indigenous epistemologies do not try to separate environmental knowledge from political, moral and social considerations.

The Torres Islands: A Dynamic, Anthropogenic Environment

I have chosen the theme of humanized landscapes as a point of departure for my description of the Torres Islands' environment in order to emphasize that 'understanding climate change is helped by a deeper temporal perspective' (Lazrus, 2012: 288), a necessary reminder that the environments of Oceania are anything but simply 'natural'. Instead, they are the ongoing result of productive, affective and spiritual human engagements. Such humanized landscapes give rise to forms of flexibility that are not always evident because they transcend narrow understandings of what constitutes indigenous adaptive capacities.

In fact, however, the Melanesian islands – the region in which the Torres are located, which is also home to almost 80 per cent of the population of the Pacific Islands – tend to be dominated by land masses that rise well above the maritime horizon and are graced with multifarious microclimates, soil types, forest and marine resources. This geographical spread gives rise to a plethora of socioenvironmental imbrications, each with its particular forms of vulnerability, risk and adaptive capacities (e.g. Waddell, 1975; Davies, 2002, Galipaud, 2002, Torrence, 2002). In the following paragraphs, I offer a description of one such local world.

Mid- and long-term environmental fluctuations have long been a part of ni-Vanuatu engagements with the physical world: 3,200 years before the present in the case of the Torres Islands (Galipaud, 1998). The human–environmental histories of the Torres Islands' community encompass not only ancestral practices, understood as stable, long-term forms of environmental stewardship, but also the cumulative effects of pre-European, colonial and postcolonial interventions including: (1) the widespread colonial process of forced displacement of people from scattered hilltop settlements to concentrated coastal villages that face a constant risk from storm surges and tsunamis; (2) the radical transformation of land-use patterns, from almost fully subsistence agroforestry to intermingled regimes of domestic gardening and cash cropping (copra, kava, coffee, etc.) and cattle farming; and (3) the introduction, over hundreds of years, of new species of flora and fauna – a process that was greatly accelerated since contact with Europeans.

It follows that local knowledge in the Torres Islands is neither simply 'traditional' nor purely local, because for at least 150 years it has been defined in increasing interactions with extra-local agents and interventions. Recognizing that the physical milieu of the Torres undergoes constant modification by a combination of geophysical, climatic and human factors is an indispensable first step to understanding islander communities as engaged actors, not just passive receptors, of changing environmental conditions.

The Torres group is made up of six small islands – Toga, Lo, Linua, Tegua, Metoma and Hiu – located on the northern political border of the Vanuatu archipelago (Figure 2.2). They consist of uplifted limestone terraces with a limited surface area (111.8 km^2), heights that range in elevation between 150 and 360 metres above sea level, and rocky shorelines that are interrupted here and there by small white-sand beaches. Importantly, they are the ongoing product of the convergence between the Australian and Pacific tectonic plates, hence, 'while many of the Vanuatu islands are typical subaerial arc volcanoes related to present-day subduction, others, including the Torres group, appear to have emerged due to collision-driven uplift … [and are estimated] to have uplifted by more than 100 m in the last 125,000 yrs' (Ballu et al., 2011: 1).

The Torres lie in a scarcely populated maritime region that is distant from Vanuatu's main centres of government, transport and commerce, and suffer from very poor delivery of basic services – health, education and transport. They currently sustain an approximate population of over 1,000 people that speak two distinct Austronesian languages, Lo-Toga and Hiu, and are distributed across eight coastal settlements. The actual number of people identifying themselves as Torres islanders probably exceeds 1,000 at present,

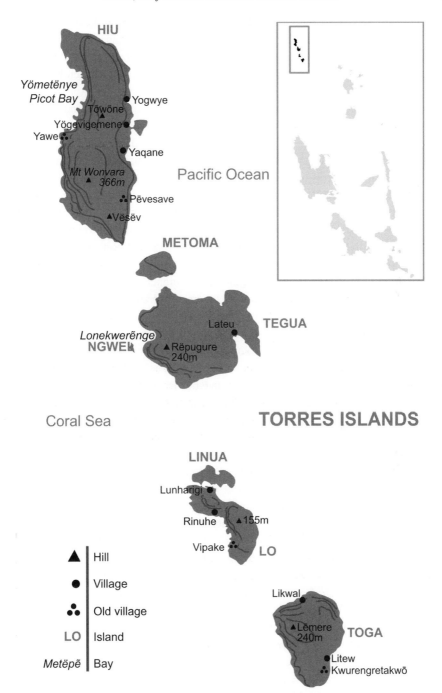

Figure 2.2 Map of the Torres Islands, Vanuatu.

but not all of them reside within their home communities. Several dozen islanders spend variable periods of time, ranging from months to years, living and working in Luganville (Santo Island), while a very small minority reside in Vanuatu's capital, Port Vila, more than 400 km away to the south. (The population numbers cited here are taken from my latest field census in combination with official Health Ministry figures from April 2012.)

According to a recent survey on local adaptive capacity, 'in comparison to many islands and communities in Vanuatu, Torres island communities have retained an abundance of their own traditional skills related to minimizing vulnerability to climate and environmental stress ... Remoteness plays an important role in this respect; many traditional skills still prevalent in the Torres have been lost from less remote nearby islands in the Banks' (Warrick, 2011: 7–8). While the gist of this observation is accurate, my own ethnographic analysis suggests that the robust state of local knowledge and cultural practices are not due specifically to remoteness, which would suggest the 'preservation' of traditional wisdom in direct proportion to the presumed scarcity of outside influence. Rather, there is an intense history of islander–outlander relations in the Torres that highlight the development of effective, long-term strategies for engaging with foreigners, including circular migration to urban and overseas places for reasons of income generation, which have always been combined with a revitalization of internal practices related to land tenure and inheritance. I discuss aspects of these practices in the final part of this chapter.

The climate of the Torres Islands follows an equatorial monsoon regime that is driven by two prevailing wind seasons common to the South Pacific. These are the predominant south-easterlies (from May to November) and the irregular north and north-westerlies which are typical of the cyclone season (from December to April). Wet and dry seasons are determined by prevailing wind patterns, and so, by extension, is the annual planting cycle of the primary food gardens (Mondragón, 2004). Wind, however, is not simply wind; each wind, known generically as *leng*, possesses a certain *mene* (Oceanic *mana*; generative power over living things) and is associated with a cardinal vector. These vectors constitute part of a sophisticated, horizon-based wind rose by which Torres people can frame and engage with certain meteorological phenomena in ways that sometimes combine human actions with physical and perceived supernatural forms of causation. For instance, while they are understood to be a recurrent fact of life, the specific ways in which hurricanes, storms and even bolts of lightning take form and make themselves manifest (where they originate, where they make landfall, what specific damage they generate) are never perceived as being entirely separate from potential human or supernatural intentionality (François, 2013; Mondragón, 2004, 2015).

Taken as a whole, this way of experiencing and engaging with the surrounding environs does not sit comfortably with a strictly naturalistic worldview. To examine how local people experience these causes and effects in relation to their environment, I turn now to a more detailed description of the environs of the Torres Islands.

Following past and present land-use patterns, the constricted surface area of the Torres can be differentiated into (1) a narrow but productive nearshore and shoreline environment which includes reefs, including a shallow but extensive tidal flat with associated areas of seagrass, and several small mangrove lagoons; (2) a relatively low-lying backshore on whose soils people plant their principal food-producing gardens (root crops) and coconut plantations for cash cropping; and (3) extensive hilltop areas on whose darker soils islanders engage in various forms of surplus horticulture and arboriculture. Each of these environs is the subject of the following subheadings.

Human–Animal Relations and Agroforestry

The Torres Islands are almost uniformly covered by a dense, humid forest whose flora and fauna are proper to the northern Vanuatu rainforest ecoregion, which extends into the neighbouring South-East Solomons (CSIRO, n.d.; Wheatley, 1992: 6–8). The qualities and transformations of the forest are not detachable from human–animal interactions, a fact that goes without saying for Torres islanders, whose production systems 'suggest a functional integration between cultivation practices (including the cultivation of tree crops) and the management of the forest itself' (Hviding and Bayliss-Smith, 2000: 17). A useful summation of the complex, interleaving history of humans and the forest can be gleaned from the human–animal relations and the dispersal patterns of tree species, to which I now turn.

In line with the gradual reduction in the range of species as one moves from the Western to the Central and Eastern Pacific, the trees of the Torres do not host high rates of biodiversity; their canopy is small and the variety of flora sparse. The fauna of the Torres are largely represented by: (1) birds, among which the most prominent are the *nawimba* (*Ducula pacifica*), the kingfisher (*Halcyon chloris*), the ubiquitous feral fowl (*Gallus gallus*), a certain number of parrot species, a local type of barn owl (*Tyto alba*), the critically endangered *namalao* (*Megapodius freycinet*), and several migratory types such as the *tuwia* (*Tringa incana*), the *nasivi* (*Limosa lapponica*), and at least two varieties of reef heron (*Ardea sacra*); and (2) large terrestrial fauna, which are largely circumscribed to domestic dogs and cats, domestic boars (there are almost never feral pigs running wild in the Torres), chickens and groups of semi-feral cats. Overall, the greatest variety of species are to be found in the nearshore, mangrove, seagrass and coastal reef systems, which host a wide range of echinoderms (starfish, sea urchins, sea cucumbers), molluscs (gastropods and cephalopods), numerous types of reef fish, the occasional dugong, eels, stingrays, turtles and arthropods – arachnids but also crabs, of which islanders distinguish at least a score of local varieties, with the most notorious being the coconut crab (*Birgus latro*). Culturally relevant species in and beyond the reef include sharks, manta rays, barracuda and cetaceans – the occasional whale and numerous pods of dolphins.

Humans engage in everyday relations with almost all of these animals. While many are often ignored, others are regularly harvested for food (eggs, in the case of fowl and megapods, and meat, in other cases) or cash (coconut crab, sea cucumber, trocha). Some, like the *nawimba* or parrots, are kept occasionally as pets. Six species stand out as being culturally salient. They include the large flying mammal known as *nikwerot*, a local variety of flying fox (*Pteropus fundatus*); the coconut crab, with whom humans had key predatory and kin relations in the mythic past; and spiders, whose attributes – sting, web-building, speed and stealth – are associated with the entity known as *Merawehih* (*Meravtit*, in the language of Hiu), the greatest culture hero of the Torres group (see François, 2013: 222, for the key relevance of spiders (*marawa*) and 'Stories of Spider' across north Vanuatu). Eels, dolphins and sharks also feature prominently in the origin and territorial narratives of various kin groups, and are known to be able to sustain salient forms of inter-species relationships with humans. The presence of these creatures as animals, supernatural entities and, in some contexts, even socially prominent persons remind us of the unexpected processes by which Torres islanders engage with their surrounding world, in ways that are not separable from ethical, moral and historical considerations.

Torres islanders have long been aware of the presence of different tree and animal species on different sides of their islands. The windward sides are more humid and tend to be dominated by taller, broad leaved and evergreen varieties that include: *takë* (*Kleinhovia hospita*; *Bislama namatal*; wood used for fuel, house rafters and leaves for traditional medicine), *var* (*Hibiscus tiliaceus*; *B. burao*; house posts, fences, firewood), *nov* (*Gyrocarpus americanus*; *B. kenu tri*; predominant in canoe building), *nor* (*Casuarina equistefolia*; *B. oktri*; hard wood used for carving ceremonial staves), and *kwoh* (*Cordia dichotoma*; *B. glutri*; leaves used for medicinal/magical purposes). Conversely, the leeward – north and west – sides of the islands tend to be drier and are characterized by semi-deciduous trees, bamboo groves, scattered shrubs and coarse grasses. Islanders also possess a considerable body of botanical knowledge which arises from the medicinal and ritual use of plants, but for reasons of space I will not go into this ethnobotanical corpus here.

The dispersal pattern of tree species summarized above is also influenced by a number of human actions, the most prominent being preferential planting, in the areas adjacent to garden plots, of trees that provide edible fruits, nuts and other resources, such as bark and sap. Other related factors include the itinerant settlement patterns characteristic of ni-Vanuatu villages, with their associated planting, clearing and management of flora, including domestic plants and herbs (e.g. Rodman, 1987). Finally, there is periodical damage by hurricanes, which provoke subsequent regrowth that has favoured sturdier colonizing species.

Patterns of Land Use and the Dispersal of Horticultural Risk

Horticulture and arboriculture are the main spheres of productive activity which, following Hviding and Bayliss-Smith's definition, I refer to as 'agroforestry', in order to emphasize 'agricultural practices which achieve an integration of trees with food plants, either through intercropping (crops grown in close juxtaposition to trees) or through shifting cultivation (crops grown after a tree fallow)' (Hviding and Bayliss-Smith, 2000: 17). Agroforestry in the Torres is dominated by root crops, namely the yam cultivar, *Dioscorea alata*, whose qualities, not being tied to the seasonal year, give rise to multiple, staggered harvests, which are supplemented mainly with the cultivation of giant taro (*Alocasia macrorrhiza*) and the opportunistic use of 'wild yam' (principally, *D. nummularia*). The resulting abundance of

Table 2.1 *Torres Islands garden and soil types*

Name	Quality	Subtype	Kind	Soil type
Lete li venie ('earth garden')	'hard'/hilltop soils	*Tenë wetagë*	Most common	Black topsoil, red underneath
		Ten lave	Large garden	Grey soil, volcanic (very fertile)
		Ten mevë	'Heavy' garden	Black, rich topsoil
Lete Lo ('soft garden')	'soft'/coastal soils	They are planted first and constitute the more accessible food-producing gardens		

garden produce has been referred to as 'subsistence plus', given that 'a fair proportion of production' is directed to the maintenance of localized food exchange systems and ceremonial feasting, on which local cultural forms are predicated (Weightman, 1989: 29; see also Rio, 2007: 105–31).

Torres islanders cultivate at least two major types of garden, one hilltop and one coastal, per annum. Lowland garden soils are more productive and represent the main sources of daily food production. Hilltop gardens are the mainstay for surplus produce which provides the alimentation that is directed at ceremonial exchange and long-term storage. Categories for different garden types are based on the kind of soil and approximate height above sea level on which they are situated. But while Torres islanders recognize two major soils, 'hard' (hilltop) and 'soft' (coastal), they also commonly distinguish at least four relevant subtypes in relation to location and specific plantings (Table 2.1; see also Weightman, 1989: 9–10, for a general listing of soil types across Vanuatu).

It has been observed that local livelihoods and cultural values are integral to the adaptive capacity of Torres communities; these values, often enshrined as *kastom* (the most common Vanuatu Bislama pidgin term for 'culture', 'tradition'), provide 'a sociocultural framework that enables good local disaster management and food-production systems that are resilient to environmental variability and uncertainty' (Warrick, 2011: 29–30). The details of this 'sociocultural framework' have largely remained absent from the literature on Torres Islands territorial practices. The following summary offers an initial outline of the social dimensions of risk dispersal in relation to land use.

Most nucleated families (*metaviv* in the Lo-Toga language) plant and tend anywhere from three to seven different gardens per year. Fallowing times in traditional horticultural practice average between seven and twenty years per plot, although at present these are sometimes reduced to between three and five years. This intensification follows more than four decades of interventions by agricultural experts associated with the central government, who have pushed for greater productivity by advising a reduction in fallowing times, soil fertilization and of planting and harvesting cycles.

The garden plots proper to any given *metaviv* are not contiguous, but are distributed irregularly across coastal and hilltop areas. This discontinuous distribution arises from the staggered patterns of yam cultivation described above and from patterns of matrilineal inheritance, in which use rights to specific plots of land shift from one lineage to another every second generation. Because no *metaviv* ever simultaneously cultivates the full range of plots over which they have use rights, the overall result is a patchwork of horticultural soils whose condition – and by extension that of the surrounding forest – is constantly changing in time (following the activation of specific use rights) and space (in terms of being in a state of fallowing or active cultivation).

Because this changing patchwork of land use implies a distribution of environmental risk, the scattered pattern of food production to which it gives rise has long been critical to enhancing resilience in relation to unexpected climate fluctuations, such as drought and excessive rainfall, as well as mitigating against soil impoverishment.

Reefs, Tides and Hydrodynamics

The undersea mounds upon which the Torres Islands rest are steep and drop sharply into the depths. Consequently, a majority of the reef systems that surround the islands are squeezed into a very narrow ecological niche that lies between the shoreline and the colder seawater below. This means that the nearshore environment of the Torres is particularly vulnerable to extreme changes in sea level, average sea surface temperatures and ocean acidification. However, local people have not yet begun to perceive this kind of damage to their reefs. This might be because most of the Vanuatu reefs have historically been subject to regular damage from hurricanes and storm surges, and because local communities are aware that their reefs may be in diverse states of health at any given time.

The reefs of the Torres Islands provide an important share of people's quotidian protein consumption. This is obtained primarily through spear-fishing; a low-impact activity whose patterns, like those proper to agroforestry, are carried out by both men and women and are closely tied to the seasonal patterns of the predominant winds. The windward-facing reefs are exploited during the austral summer (hurricane) season, November to March, when the south-easterly winds and swell are at a minimum and the predominant winds are in the north and north-west. Conversely, the leeward-facing reefs are reserved for the April to October period. Torres islanders are well aware that these seasonal shifts allow fish stocks on different reefs several months a year to recover from human exploitation. Deep-water fishing is not practised regularly, and is generally reserved for the November to March period of sustained calm, which is also when interisland voyaging is at its most intense.

There are at least two major coastal sites, on Lo and Hiu, which are dominated by mangrove lagoons and seagrass shallows which, as previously stated, concentrate much of the high-productivity coastal and marine tenure of the Torres islanders. The displacement of land and sea animals within these mangrove ecosystems are intimately tied to tidal levels, which in turn follow the changing position of the moon in the sky. This may help to explain the existence of a highly developed set of indigenous categories for tides (Table 2.2).

Table 2.2 *Torres Islands tides. Carlos Mondragón 2011*

Category	Name	Type
Liave	*Liave li vu liave*	Beginning of any high tide
High tide	*Liave me tave*	Morning tide
	Liav revrev	Evening tide
	Liave li lowate	Midday tide
	Liavi ihar	Highest tide of the year
	Liave mëren	High tide during daylight
Met	*Nu metmet*	Beginning of any low tide
Low tide	*Met metave*	Morning tide
	Met mëren	Midday tide
	Met revrev	Evening tide
	Met melige	New moon
	Met mëgagë	Full moon
	Met mëgagë metavu toten	Low tide during the day and during waning quarter moon

In addition to determining the various times of the day, the week, the month and indeed the solar year that are dedicated to harvesting the sea, the tides also affect the freshwater lens under the islands, and thence the relative moisture of coastal soils. It is no surprise that Torres people are well aware of the links between tidal levels and the health of their lowland garden soils. I have elsewhere organized some of the principal ritual and social dimensions of their food-producing cycles in the form of a lunar horticultural 'calendar' (Mondragón, 2004).

Socioclimatic Synchronizations: ENSO and Ritual Cycles

It is important to emphasize that Torres islanders' knowledge forms are not organized according to any fixed system. They are not taught wholesale in formal settings but are the result of emerging engagements with the panoply of entities, forces and moral values that define the world. Some of these knowledge types are transmitted orally by different specialists in different contexts; some are public, some private, some uniquely ritual. They are living, shifting and contested bodies of experience and moral value, which remain extant as common domains of shared praxis insofar as they emerge from the most relevant of locally grounded activities – namely food production and ritualized exchange.

The results of extended ethnographic observations in the Torres Islands throughout the past fifteen years have revealed that there may be a unique form of synchronization between ritual activity and El Niño Southern Oscillation (ENSO)-related (seven to eight year) periods of drought and extreme rainfall. Specifically, La Niña inversions, which often follow an ENSO year, tend to present above average rainfall. This favours the production of bumper crops of ceremonial yams and kava, both of which are critical components for the most important life cycle ceremonies of the islands, known as *lehtemet* (Mondragón, 2009). In seeking to corroborate this possibility, I carried out a reconstruction of El Niño and La Niña events over the past half century, and contrasted it with the oral narratives of various Torres elders in relation to their memory of *lehtemet* ceremonies held during the past century. The results suggest that ENSO fluctuations are indeed closely related to surplus production of kava and yams, which in turn facilitate the organization of *lehtemet*.

The social importance of *lehtemet* is directly linked to the concept of *mena* (better known as *mana* in other Oceanic contexts), a moral and practical quality that is associated, among other things, with people who are skilful at evoking 'living growth' (*vavelete*) from key spheres of the lived world (in other contexts, it is also associated with Christian exegesis, ritual rank and political power). In the context at issue, *mena* is a 'power' which people acquire through a lifetime of efficacious horticultural production and of successful social alliances and exchange relations. As I have discussed elsewhere (Mondragón, 2004), *mena* is a moral and practical quality attributed to efficacious nurturers (of people) and producers (of food, which is the life stuff that allows the daily reproduction of society), and as such it constitutes an essential cultural value. *Lehtemet* is performed only twice during a person's lifetime, and tends to coincide with periods of bumper crops brought on by ENSO climate extremes. Its importance clearly cannot be measured only in empirical terms, as 'adaptation' understood as a technical response or skill, because it is entangled in moral, political and complex socioenvironmental moments of opportunity and cultural creativity.

Finally, a notable and recent trend in local perceptions of hydroclimatic variability has been the apparent increase of sudden, extreme rainstorms that do not seem to follow any previously known seasonal pattern. This is a phenomenon that Torres islanders only began to take note of since 2006, which is interesting because unseasonal rainfall was not a part of climate change discourse at that time. It was only after 2007 that extreme and damaging rainfall began to appear in the printed and oral dissemination of climate awareness programmes reaching the Torres Islands. This would suggest an exceptional example of local perceptions preceding exogenous information about climate change.

The likeliest explanation for extreme rainfall appears to be the growing frequency of northward shifts of the South Pacific Convergence Zone (SPCZ) towards equatorial waters. The SPCZ is the southern hemisphere's most persistent rain band, which usually lies at subequatorial latitudes. But according to recent compilations of historical trends and various climate models, there has been a growing pattern of northerly displacements of the SPCZ that seem to be precipitated by conditions linked to climate change (Cai et al., 2012; see also IPCC, 2013: 1219).

This suggests that extreme rainfall, rather than sea-level rise, constitutes the single most relevant risk factor for Torres islanders' subsistence systems in terms of climate change.

Afloat on a Bottomless Sea: Environmental Perceptions, Politics and Frames

As previously stated, the seismic uplift that created and continues to shape the Torres group is a product of the convergence dynamics of the plate interface zone that lies within the deep-sea Torres Trench. Located a mere four kilometres offshore, at 6,000 metres below sea level, this area marks the active boundary between the Australasian and Pacific Plates. Recently, two violent earthquakes took place in the Torres group, giving rise to two very different sets of environmental perceptions, local and extra-local, the first of which offer insights into the very different principles by which islanders produce environmental knowledge.

In 1997, a magnitude Mw 7.8 earthquake provoked a process of inter-seismic subsidence of 117 ± 30 mm which affected a part of Lo island and possibly a part of the shoreline of Tegua island, on which lay a small village known as Lateu. This subsidence took place gradually, at a rate of about 20 mm per annum, between 1997 and early 2008, when a second earthquake abruptly inverted its effects (see Ballu et al., 2011). During those years, as the shorelines of north Lo and around Lateu became increasingly waterlogged, many of the coconut palm trees and mangrove plants in the affected areas declined, lost their foliage and died. Importantly, since before the 1997 earthquake the inhabitants of Lateu had already begun to formulate plans to relocate further inland owing to: (1) the exposure of their coastal settlement to the high waves produced by the deep bay in which their village was situated; and (2) ongoing internal disputes over political and territorial influence in relation to chiefly authority. The debate over relocation had simmered for years for reasons of political expediency, and gradual, rather than urgent, environmental necessity.

Between the years 1999 and 2003, prior to the arrival of climate change discourses, most of my Torres Islands' interlocutors perceived the inundation of the Lo-Linua shoreline to be entirely consistent with previous alterations to other parts of the coast over the past half century. In relation to Lateu, it had always been clear that the site on which the village stood, at the end of a narrow bay, was subject to strong wave damage and swampy conditions. Sure enough, as I compiled a detailed register of place names across the Torres group, several toponyms stood out as marking sudden coastal changes in previous decades. The reasons offered by islanders for these changes were almost invariably interpreted as having been the consequence of wave action, erosion and the accumulation of sand, soil and plants – in other words, this was the result of direct observation, consistent with broader ni-Vanuatu ways of relating truth claims, at least partially, to first-person experience (see Rubinstein, 1981). When pressed to explain the ultimate causes of coastal change, however, many islanders also mentioned the possible intervention, in some instances, of saltwater spirits or man-made weather magic. In other words, their empiricism was sometimes combined with ideas about the concatenation of supernatural and human intentionality.

By far the most striking response that I have documented in relation to environmental change and instability is the perception that the Torres Islands – and indeed all islands in the known world – are not anchored to the sea floor, but are floating pieces of land upon an oceanic cosmos that has no solid bottom. Importantly, this idea was proffered by middle-aged and elderly islanders who never went through primary schooling and lacked direct experience or previous received knowledge of the Earth being a round planet with an ocean bottom; and consequently of the relation between sea level and shorelines based on absolute, fixed territorial foundations.

In mentioning this last perception, I do not aim to portray local worldviews as radically exotic. The broader point is that Torres islanders' environmental knowledges are multifarious – in the above case, partly dependent on age and forms of education – and are grounded on principles of truth and experience in which 'people's interactions with the environment [are] processual rather than classificatory' (Hviding, 1996: 180). In other words, knowledge is relational, empirical and contingent, not absolute; as such, it is not linked to an overarching, objective view of the 'natural' world as a socially neutral medium, but subject to contestation and modification (see Rubinstein, 1981).

In 2001 a small team from the Secretariat of the Pacific Regional Environment Programme (SPREP) visited the Torres. They quickly came to the mistaken conclusion – later rectified in official documents and subsequent assessments – that Lo and Tegua were experiencing accelerated sea-level rise as a result of global warming, and rushed to inform the global media about the fact that Lateu villagers had become the world's first 'climate refugees' (UNEP, 2005; Nakalevu, 2006). As a result of this visit, Torres islanders first began to hear about climate change. Subsequently, some islanders began to associate the 'sinking' of their islands with sea-level rise, climate change and earthquakes.

Just what different Torres people actually understood climate change to be, in light of the worldview described above, was and remains open to debate. Over the course of several conversations during different visits throughout the past decade my interlocutors have variously associated climate change with rough seas, with earthquake activity and, more vaguely, with the idea that the ice that melts in one part of the world then 'runs down' and floods their part of the ocean (see Warrick, 2011: 45 for a similar example of local interpretations regarding climate change among Torres people). Importantly, despite the intense 'awareness-raising' that has taken place since SPREP's original visit, perceptions of climate change remain open to contestation and competing truth claims.

On 2 October 2009, a new 'seismic crisis consisting of three major earthquakes (magnitudes Mw 7.6, 7.8 and 7.4) lifted the islands by approximately 200 mm, bringing the relative sea level back down to about its 1998 level' (Ballu et al., 2011: 3). Soon, the waterlogged coastal area on Lo began to dry up, and within a few years new vegetation – mostly mangrove shoots – began to return to the area, partly as a result of Lo islanders' active efforts to restore the affected shallows given that conditions were again adequate for mangrove plants. In the meantime, the people of Tegua island had successfully removed themselves from Lateu to two new settlements. The smaller of these settlements, known as Tenia, had been decided on by only one extended family, as a result of the long-running political and territorial dispute mentioned above, while the other, known as Lirak, was formally arrived at after a consultation process in which SPREP assisted the villagers. The specifics of relocation and the building of new homes were left to Tegua islanders, but in the case of Lirak the most significant assistance that they received came in the form of water tanks, which made it easier to obtain and store fresh water.

An independent assessment of the Tegua relocation concluded that 'from the perspective of donors, adaptation to climate change needs to maintain and increase the (well-developed) ability of the community to deal with their own environmental uncertainties. This requires initiatives that assist the community to sustain and improve adaptive capacity, on their own terms and in their own way' (Warrick, 2011: 52). The object of this chapter has been to begin to explain some of the 'terms' and the 'ways' by which Torres islanders generate knowledge and consequently respond to environmental risk – including what outside agencies define as 'adaptation to climate change'.

Conclusion

The data presented in this text point to the fact that the most important impacts resulting from environmental change over the coming years in the Torres Islands and more generally across Island Melanesia will arise from patterns of drought and excessive rainfall, rather than sea-level rise. In this regard the most important impact on ni-Vanuatu rural communities from climate change over the coming years is likely to take the form of diminishing horticultural produce, rather than changes to the local shoreline. Beyond the physical specifics of climate and environment, however, social and historical factors remain the key to understanding the adaptive capacities of local people and their territories.

My principal contention in this regard has been that climatic fluctuations have never been explained by Torres islanders in relation to a detached – because morally and socially neutral – 'environmental' frame, but are associated with broader chains of relations between humans and other entities, as well as with a diverse set of ideas regarding the shape and nature of the world. More important, islander frames are not premised on a single, coherent model of knowledge, but are processual and relational, insofar as Melanesian epistemologies operate not only 'from an absence of a nature-culture dichotomy … but in processual chains of observations of causal linkages which tend to be deduced and postulated with little or no separation of the "magic" from the "real"' (Hviding, 1996: 178; see also Rubinstein, 1981).

The continuing underestimation of indigenous epistemologies and responses is not an issue of adequate representation of the local – as though indigenous knowledge were simply resolved through a politically correct representational quota system. Instead, it points to a critical limitation of the international development and natural sciences' communities to take seriously the diversity of ways in which environmental knowledge is entangled with local productive, affective, moral and spiritual worlds.

Over their 3,000 year history, Torres islanders have learned to adapt to abrupt transformations to their land and seascapes. The case study that I put forward in my contributions to this book (as that of Damon, Chapter 10 in this volume) offers a powerful example of the resilience and adaptability of humans and environments even in cases that seem extreme, i.e. small island environments under pressure from both climate change and destructive resource extraction. When adequately accounted for, existing forms of environmental flexibility give rise to a more positive outlook for local communities' capacity to adapt to sudden climatic transformations. Indeed, we are tempted to proffer the argument that these Melanesian communities appear to be far more likely to successfully adapt to abrupt climate change than the encompassing, globalized societies and institutions that are seeking to help them in this process of environmental crisis and transition.

References

Ballu, V., Bouin, M.-N., Siméoni, P. et al. 2011. Comparing the role of absolute sea-level rise and vertical tectonic motions in coastal flooding, Torres Islands (Vanuatu). *Proceedings of the National Academy of Sciences*, 108(32): 13019–22.

Barnett, J. and Campbell, J. 2010. *Climate Change and Small Island States: Power, Knowledge and the South Pacific*. London: Earthscan.

Bartlett, C. 2009. *Emergce, Evolution and Outcomes of Maine Protected Areas in Vanuatu: Implications for social-ecological governance*, unpublished PhD Thesis, James Cook University, Queensland, Australia.

Cai, W., Lengaigne, M., Borlace, S. et al. 2012. More extreme swings of the South Pacific convergence zone due to greenhouse warming. *Nature*, 488(7411): 365–9.

Castree, N. 2015. The 'three cultures' problem in global change research. *EnviroSociety*. 9 March, www.envirosociety.org/2015/03/the-three-cultures-problem-in-global-change-research

Cronin, S. J., Gaylord, D. R., Charley, D. et al. 2004. Participatory methods of incorporating scientific with traditional knowledge for volcanic hazard management on Ambrym Island, Vanuatu. *Bulletin of Vulcanology*, 66(7): 652–68.

CSIRO (n.d.) *The Environments of Vanuatu. A Classification And Atlas of the Natural Resources of Vanuatu and their Current Use as Determined from VANRIS*. Brisbane: Commonwealth Scientific and Industrial Research Organisation (CSIRO Brisbane) and the Australian Department of Primary Industries Forest Service.

Davies, H. L. 2002. Tsunamis and the coastal communities of Papua New Guinea. In Torrence, R. and Grattan, J. (eds.) *Natural Disasters and Culture Change*. London: Routledge, pp. 28–42.

Dominguez Rubio, F. and Baert, P. (eds.). 2012. *The Politics of Knowledge*. New York: Routledge

François, A. 2013. Shadows of bygone lives. The histories of spiritual words in northern Vanuatu. In Mailhammer, R. (ed.) *Lexical and Structural Etymology: Beyond Word Histories, Studies in Language Change*. Berlin: De Gruyter Mouton, pp. 185–244.

Galipaud, J-C. 1998. Recherches archéologiques aux îles Torres. *Journal de la Société des Océanistes*, 107(2): 159–68.

Galipaud, J.-C. 2002. Under the volcano: Ni-Vanuatu and their environment. In Torrence, R. and Grattan, J. (eds.) *Natural Disasters and Culture Change*. London: Routledge, pp. 162–71.

GIZ-ACCPIR. 2009. *GTZ Adaptation to Climate Change in the Pacific Island Region, Summary*. Deutsche Gesselschaft für Internationale Zusammenarbeit (GIZ). www.spc.int/lrd/index.php?option=com_docman&task=cat_view&gid=210&Itemid=48

Hay, J. E., Mimura, N., Campbell, J. et al. 2003. *Climate Variability and Change and Sea-level rise in the Pacific Islands Region: A Resource Book for Policy and Decision Makers, Educators and other Stakeholders*. Apia, Samoa: SPREP.

Hviding, E. 1996. Nature, culture, magic, science: On meta-languages for comparison in cultural ecology. In Descola, P. and Pálsson, G. (eds.) *Nature and Society: Anthropological Perspectives*. London: Routledge, pp. 165–85.

Hviding, E. 2005. *Reef and Rainforest: An Environmental Encyclopedia of Marovo Lagoon, Solomon Islands/Kiladi oro vivineidi ria tingitonga pa idere oro pa goana pa Marovo*. Paris: UNESCO.

Hviding, E. and Bayliss-Smith, T. 2000. *Islands of Rainforest: Agroforestry, logging and eco-tourism in Solomon Islands*. Farnham, UK: Ashgate.

IPCC (Intergovernmental Panel on Climate Change). 2013. *Climate Change 2013: The Physical Science Basis. Contribution of Working Group I to the Fifth Assessment Report of the Intergovernmental Panel on Climate Change* [Stocker, T. F., Qin, D., Plattner, G.-K. et al. (eds.)] Cambridge: Cambridge University Press.

Jacka, J. 2009. Global averages, local extremes: The subtleties and complexities of climate change in Papua New Guinea. In Crate, S. and Nuttall, M. (eds.) *Anthropology and Climate Change: From Encounters to Actions*. Walnut Creek, CA: Left Coast Press, pp. 197–208.

Jasanoff, S. 2004. The idiom of co-production. In Jasanoff, S. (ed.) *States of Knowledge: The Co-production of Science and the Social Order*. New York: Routledge, pp. 1–12.

Kwa'ioloa, M. and Burt, B. 2001. *Our Forest of Kwara'ae. Our Life in the Solomon Islands and the Things Growing in our Home*. London: The British Museum Press.

Lahsen, M. 2010. The social status of climate change knowledge: An editorial essay. *Wiley Interdisciplinary Reviews: Climate Change*, 1(2): 162–71.

Lazrus, H. 2012. Sea change: Island communities and climate change. *Annual Review of Anthropology*, 41: 285–301.

Leach, J. 2012. 'Step inside: Knowledge freely available'. The politics of (making) knowledge objects. In Dominguez Rubio, F. and Baert, P. (eds.) *The Politics of Knowledge*. New York: Routledge, pp. 79–95.

McCarter, J. and Gavin, M. C. 2014. Situ maintenance of traditional ecological knowledge on Malekula Island, Vanuatu. *Society & Natural Resources: An International Journal*, 27(11): 1115–29.

Mondragón, C. 2004. Of winds, worms and mana: The traditional calendar of the Torres Islands, Vanuatu. *Oceania*, 74(4): 289–308.

Mondragón, C. 2009. Encarnando a los espíritus en la Melanesia: La innovación como continuidad en el norte de Vanuatu. In Fournier, P., Mondragón, C. and Wiesheu, W. (eds.) *Ritos de Paso: Antropología y Arqueología de las Religiones*, vol. 3. México: Escuela Nacional de Antropología e Historia, pp. 121–49.

Mondragón, C. 2012. *The Meanings of Mana in North Vanuatu: Cosmological dualism, Christianity and Shifting Forms of Efficacious Potency in the Torres Islands*. Unpublished paper presented at the 111th Annual Meeting of the American Anthropological Association, San Francisco.

Mondragón, C. 2014. *Un Entramado de Islas: Persona, Territorio y Cambio Climático en el Pacífico Occidental*. México: El Colegio de México.

Mondragón, C. 2015. Concealment, revelation and cosmological dualism. Visibility, materiality and the spiritscape of the Torres Islands, Vanuatu. *Cahiers d'Anthropologie Sociale: Montrer/ occulter. Visibilité et Contextes Rituels*. Paris : Éditions L'Herne, pp. 38–50.

Muliagetele, J. R. 2007. *An Assessment of the Impact of Climate Change on Agriculture and Food Security in the Pacific. A Case Study in Vanuatu Apia, Samoa*: Pacific Environment Consultants Ltd., prepared for the Food and Agriculture Organization of the United Nations, Subregional Office for the Pacific Islands (FAO SAPA).

Nakalevu, T. 2006. *Capacity Building for the Development of Adaptation Measures in Pacific Island Countries Project, Final Report*, Apia, Samoa: SPREP.

Pelling, M. 2011. *Adaptation to Climate Change. From Resilience to Transformation*. London: Routledge.

Rio, K. 2007. *The Power of Perspective. Societal Ontology and Agency on Ambryn Island, Vanuatu*. Oxford: Berghahn.

Rodman, M. 1987. Moving houses. Residential mobility and the mobility of residences in Longana, Vanuatu. *American Anthropologist*, 87: 56–72.

Rubinstein, R. L. 1981. Knowledge and political process on Malo. In Allen, M. (ed.) *Vanuatu. Politics, Economics and Ritual in Island Melanesia*. Sydney: Academic Press Australia, pp. 135–72.

Rudiak-Gould, P. 2013. Climate change and tradition in a small island state. *The Rising Tide*. London: Routledge.

SPREP. 2011. *Pacific Islands Framework for Action on Climate Change 2006–2015*, 2nd edn. APIA, Secretariat of the Pacific Regional Environment Programme (SPREP). www.sprep.org/attachments/Publications/PIFACC-ref.pdf

Taylor, J. 2008. *The Other Side. Ways of Being and Place in Vanuatu*. Honolulu: Hawai'i University Press.

Torrence, R. 2002. What makes a disaster? A long-term view of volcanic eruptions and human responses in Papua New Guinea. In Torrence, R. and Grattan, J. (eds.) *Natural Disasters and Culture Change*. London: Routledge, pp. 292–312.

UNEP (United Nations Environment Programme). 2005. Pacific Island villagers first climate change refugees. *UN Environment Builds Bridges Between Vulnerable Peoples in the Arctic and Small Islands*. UNEP Press Release. www.unep.org/Documents.Multilingual/Default.Print.asp?DocumentID=459&ArticleID=5066&I=en

Waddell, E. 1975. How the Enga coped with frost: Responses to climatic perturbations in the Central Highlands of New Guinea. *Human Ecology*, 4(4): 249–73.

Warrick, O. 2011. The adaptive capacity of the Tegua island community, Torres Islands, Vanuatu. *Report for the Department of the Environment*, Australian Government. www.environment.gov.au/climate-change/adaptation/publications/adaptive-capacity-tegua-island-community-torres-islands-torba-province-vanuatu

Weightman, B. 1989. *Agriculture in Vanuatu*. Cheam, UK: The British Friends of Vanuatu.

Wescott, J. 2012. *The Good Lake, the Possible Sea: Ethics and Environment in Northern Vanuatu*. Unpublished PhD dissertation, University of California, San Diego.

Wheatley, J. I. 1992. *A Guide to the Common Trees of Vanuatu. With lists of their Traditional Uses and ni-Vanuatu names*. Port Vila: Vanuatu Department of Forestry.

3

Annual Cycles in Indigenous North-Western Amazon: A Collaborative Research Towards Climate Change Monitoring

Aloisio Cabalzar

Since 2005, a collaborative research has been developed on the ecological and socio-economic cycles in the north-western Amazon. This cross-cultural and interdisciplinary research involves a team of indigenous and non-indigenous researchers, and methodologies aimed at an effective communication and collaboration between indigenous knowledge and Western science. This initiative aims to: (1) describe the economic-ecological and sociocultural calendar of indigenous peoples of this region, from observations and recordings done by the indigenous researchers over the years, identifying and analysing their patterns and variations through the annual cycles; (2) investigate possible regional effects of more extensive climate change, bridging indigenous and scientific knowledge; (3) point out how indigenous ecological-economic practices are adapted to these annual variations and how they would cope with more drastic changes; (4) delineate methodologies to monitor climate change based on networks of indigenous and scientific knowledge initiatives within the Amazon basin; and (5) as a future perspective, propose environmental governance policies for indigenous territories in the light of climate change scenarios.

This collaborative research aims at understanding how indigenous peoples are engaged with the environment through their current management practices. The notion of life cycle, integrating both sociocultural and natural aspects, is privileged as a cross-cultural research methodology. Indigenous peoples have developed detailed and sophisticated knowledge on ecological processes, which could provide a rich and territorialized comprehension of ecosystem transformations and climate change. The research departs from a simple methodology based on daily written diaries kept by indigenous dwellers (trained as researchers) from various communities along the same river (about a 400 km stretch) and some of its tributaries. These indigenous researchers, usually young adults, also search for additional information and interpretations with elder knowledge holders, about the phenomena they are observing.

Briefly, for the indigenous peoples of the Northwest Amazon, the annual cycle, the sequence of seasons, the environmental management suitable to each one and the changes that are being perceived over a shorter or longer timescale are associated with certain economic-ecological practices and the shamanic management of the world performed by ritual specialists (healers of the world, *kumua*).

In light of the recent history of these peoples (see below) and the current social and cultural situation, this research project advances two main strategies: (1) interregional exchanges within the Northwest Amazon with the view to strengthening the networks of indigenous knowledge; and (2) joint research programmes, exploring relations between indigenous knowledge and Western scientific knowledge in interdisciplinary research platforms. The focus on life cycles and on the annual calendar allows for an integrative and understandable perspective. The main challenge is the strengthening of indigenous knowledge within the context of increasing communications and interrelations with the global society.

This collaborative research emerged in the context of long-term partnership relations between regional indigenous organizations, FOIRN (Federation of Indigenous Organizations of the Rio Negro) and the Instituto Socioambiental (ISA), a large Brazilian NGO committed to environment and collective rights issues, linking together advocacy, research and technical support. FOIRN, ISA and some of the subregional indigenous associations affiliated with FOIRN have had a long-term partnership, jointly developing projects in areas such as sustainable community development, capacity building, territorial and environmental governance and environmental management. The author is an anthropologist working for ISA and has been involved in research in this region for the last twenty-five years, mainly supporting indigenous research teams.

Indigenous Populations and Environment in the Northwest Amazon

The indigenous peoples of the north-western Amazon (Figure 3.1) have lived there for centuries, probably millennia – as their narratives of origin and some archaeological research suggest (see the collection of narratives edited by FOIRN and ISA in the bibliographical references; Neves, 1998). This area is subject to serious ecological limitations: acid soils and waters that are nutrient-poor and low in productivity (Goulding et al., 1988) and extensive areas covered by heath forest (known locally as *caatinga*) which is very restrictive for agricultural practices (Cabalzar and Ricardo, 2006: 17). These two factors – antiquity of occupation and serious ecological limitations – have led the indigenous peoples to a long process of adaptation (Reichel-Dolmatoff, 1979; Ribeiro and Kenhíri, 1987; Moran, 1991; Chernela, 1993), finding effective and sophisticated forms of management of the land including forests, agriculture, fish and game. Some travellers who visited the region in the eighteenth and nineteenth centuries described the vitality and dynamics of these populations, demonstrated by the size and population of their longhouses (*malocas*), their grand intercommunal ceremonies and the sophistication of their material culture.

Historically, the Rio Negro basin is a region of early colonization, with Iberian incursions and a sparse presence beginning in the seventeenth century. The overall violence of this process led to the abandonment and drastic depopulation of the lower and middle Rio Negro. In more recent times, compulsory work for the extractive industry has been succeeded by Salesian missionaries in the first decades of the twentieth century, as the main external agent among the Amerindians of the region (Hugh-Jones, 1981; Meira, 1997). After the

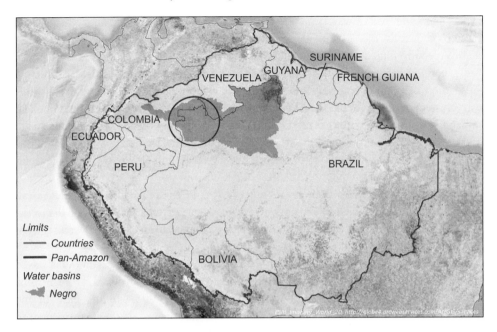

Figure 3.1 The Rio Negro basin and the Northwest Amazon.

peak of the Salesian educational and catechetic project in the 1950s and 1960s, there followed a period of decadency which culminated in the closure of the boarding schools. By the 1980s and 1990s, the struggle for indigenous rights guaranteed by the Brazilian Constitution of 1988, together with the emergence of numerous local and regional indigenous organizations, eventually reoriented local policies, enabling the demarcation of indigenous lands (about 150,000 square kilometres) and the reform of school education in the region towards autonomy in its management and curriculum.

Nowadays, even though distant from the main Amazon deforestation fronts (but under threat from mining interests), indigenous peoples of the Northwest Amazon region are also challenged by different development models and associated public policies. The federal government, to a large extent, tends to apply universalist policies such as income distribution to indigenous peoples, considering them as poor and not recognizing their specificities. These policies have negative effects such as generating new cycles of migration to cities, language and cultural losses and the abandonment of indigenous economic and ecological practices. Indigenous associations and communities, as well as their partners, pursue sustainable development projects that also strengthen their sociocultural heritage, but they find it hard to obtain support. A set of collaborative research efforts involving indigenous and non-indigenous researchers led to a proposal for an Indigenous Knowledge and Research Institute that the federal government is prone to support, but with its habitual slowness.

The Northwest Amazon region is known as a comprehensive and multi-ethnic social system comprising about thirty linguistic groups. The Tukanoan (situated in the centre of this area) and the Arawak (to the north and south) are riparian sedentary fishers and farmers, whereas the Maku, more mobile, exploiting more dispersed resources, mainly occupy inter-fluvial areas. The long-standing proximity among these peoples has generated sociocultural continuities. Their main subsistence activities are shifting cultivation of bitter manioc by the women and fishing by men. A complex agroforestry system was developed, making this region one of the centres of bitter manioc and pineapple domestication. As a result of its poisonous composition, on the one hand, and its energetically high contribution to the local diet, on the other, a well-developed culinary preparation of bitter manioc was developed (Hugh-Jones, C., 1979; Dufour, 1988; 1990).

One of these sociocultural continuities relates to the extensive network of exchange and commerce that existed and still exists to a certain extent between the various parts of the Rio Negro basin, reinforced by specializations in handicraft and different eco-regional management practices (Ribeiro, 1995).

The ideology of descent has strong territorial connections with genealogies based on 'houses', expressed in the transformation trajectories of narratives, from origins to contemporary longhouses and villages (Hugh-Jones, 1995). In the case of the Tukanoan peoples, the central narrative is the ancestral history of the Transformation Canoe, usually embodied as an Anaconda Canoe, going from Lago do Leite (Milk Lake) until the Uaupés River waterfalls and its tributaries, where it emerges and gives rise to the People of Transformation, as the eastern Tukanoan peoples regard themselves (Andrello, 2012).

The model of the longhouse is multi-scalar and conceptually projected on the territory. Many important places, usually associated with waterfalls, rapids and mountains, are conceived as pillars of the longhouse-world (Ortiz et al., 2012). The shamanic management of the world should take into account these references, each descent group protecting and managing their territory. The concept of management of the world is a translation of indigenous conceptions concerning the ways shamanic specialists observe, protect and heal the environment, according to each season of the annual cycle.

Cosmology, Ritual Cycle and Annual Calendar

The expression 'manejo do (del) mundo', largely used in Spanish and Portuguese by indigenous researchers and leaders in the Northwest Amazon, translates to 'management of the world'. This is a problematic translation, but it is difficult to find alternatives. This concept, which refers to traditional management of life cycles and of the annual calendar will be explained below.

The narratives of the origin of the world and their interpretation by indigenous knowledge holders provide evidence for the idea of a certain symmetry or (initial) identity among human beings and other peoples. Within this concept, the differentiation among beings and of humanity in relation to other beings – for example, humanity in relation to fishes – is not permanent and experiences frequent transformation and instability: fish that change into land animals, especially snakes, to find food on land; fish into monkeys to collect fruits from riverside trees; and individuals that have their souls stolen by fish-people. The very origin of humanity is narrated as occurring in two phases, one of transformation of birds into fish in the Lago do Leite and the other of fish into humans within the holes of transformation that are situated in various waterfalls in the Vaupés River and its tributaries (Azevedo, and Azevedo, 2003; Cabalzar, 2005; Andrello, 2006).

This aspect is relevant here, since the life cycles of plants and animals are largely understood in terms of analogy with the indigenous culture and its social calendar – longhouses, ritual objects, family relations, festivities, ceremonies, work and reproduction. For example, reproductive phenomena, such as pairing and spawning of fish, are considered as festivities, ceremonial dances of the fish-people, when ritual substances are consumed and social relations celebrated. There are places in the rivers such as rapids or isolated stones (generally those where fish gather for reproduction) that are considered the *caapi* pots of the fish. *Caapi* is a hallucinogenic drink, prepared from a vine *Banisteriopsis caapi*, that is used during ceremonial dances.

Another example is the festivity period of the Wooga-people (a toad species), during which the knowledge holders must protect the people of the community. It is thought that the drinks they use during these occasions are poisonous for people, causing illnesses. The *kumua* has (shamanic) prescriptions that avoid contact with these substances. It is important to note that these other beings and their masters are potentially sources of illness (Buchillet, 1988; Cabalzar, 2005; AEITY and ACIMET, 2008; AIETU, 2009).

This sociocosmological symmetry between human society and other species, which also interfere in the course of the seasons, is fundamental to understanding the shamanic management of the annual cycles. It is conceived that substances, performances and relationships are the same across interspecific rituals, and can be exchanged in order to ensure the normal temporal process, free from misfortune and disease. These exchanges mean mediation and the possibility to avoid the negative unfolding of these relations. These concepts are shared among many indigenous groups in the Northwest Amazon; not just by the eastern Tukanoan peoples but also among the Arawak, Maku and others. In this sense, Ignacio Valencia, Makuna knowledge holder from this region, states:

We do treatments and prevention to be in spiritual alignment with the animals [and their masters]. We humans are also animals, because we eat certain forest fruits, just as they do. In order to do this, we first prevent illnesses. This is how the ecological calendar works and nobody can reorganize or destroy that which has been given since the beginning. And this is how it is, this is what happens, because since the creation, the same animals have been protagonists in the origin stories, accompanied by the *Ayawa* [creators], in different stories.

In this way, we know that we indigenous peoples are connected to the animal and vegetable world, including the very space where we live. This is our own method for managing our territory. Knowing and practicing our ancestral knowledge, *kubua baseri keti oka*. For now we say that while the *kubua* [shamans] are alive, our management will continue to be alive. So that this knowledge will continue to be living and current, we are reinforcing and building upon our management with this research. It is extremely important for future generations.

(Valencia, 2010)

The indigenous peoples that live in the Northwest Amazon organize their social, economic and ritual life using astronomical and ecological cycles as time references (Cardoso, 2007; Azevedo et al., 2010; Cabalzar et al., 2010). These cycles are described in narratives of indigenous knowledge holders on the annual calendar, particularly in the talks of the elders. They express indigenous concepts of the seasonal cycles, the rotation of the world (*bureko watotire* in the Tuyuka language), the sequence of seasons with their corresponding beings and occurrences, leaving it to the specialists (spiritual healers and healers of the world) to protect their relatives and their surroundings in order for them to pass without setbacks, misfortunes or diseases (Ribeiro and Kenhíri, 1987; Pãrõkumu and Kehirí, 1995; Azevedo, M. and Azevedo, A. N., 2003; Garcia Rodriguez, 2010; Valencia, 2010; Valle, 2010). This cycle was established when the world originated and is conceptualized as its proper way of working.

Traditionally, the Tukanoan peoples go through an annual cycle of ceremonies for protection of the longhouse or of the local community. These are festivities with singing and dancing by the *baya* (the master of ceremony) and drinking of *caxiri* (manioc beer), *caapi*, *ipadu* (coca leaves powder), and various other ritual substances, in which the *kumua* (ritual healers) undertake the blessings specific to each season. They burn resin (called *breu* in the regional Nheengatú) and beeswax by way of supporting the healing. The ceremony brings up the essence of the ritual substances, recalling and reviving the knowledge of the ancestors. The knowledge holders follow the movement of the constellations (Hugh-Jones, 1982; Epps and Oliveira, 2013), associated with the cycles of forest fruits, of fluctuations of the rivers, of the lives and reproduction of fish and animals, of the plots and plant seedlings. They undertake the ceremonies of protection with these cycles in mind, which can be thought of as rituals for environmental management, in that they strive to maintain this movement of the world and its cycles (Hugh-Jones, C., 1979; Hugh-Jones S., 1979).

This annual ceremonial cycle of festivities has been unevenly carried out across the region. While along the Pirá-Paraná River in Colombia these practices have not been abandoned, as the area is more isolated and the interventions of missionaries occurred later and were less intense than in other parts, in the Brazilian section of Vaupés Basin sociocultural changes were deeper. It means that the ceremonial cycle was simplified or partially or even totally abandoned in a considerable number of communities, limiting it to more discrete individual protection, for example, at birth or at the time of first menstruation. Knowledge and awareness of the cycles or calendar, however, continue to be present and there is interest on the part of the youth to fully understand them. These ritual transformations have come with broader sociocultural and political changes which also impact on ecological and economic practices. In addition, a growing proportion of the indigenous population has been more engaged with non-traditional activities as part of their routine, like commuting to the urban centre for commerce or paid work. This is the context in which this collaborative research has been developing, apprehending the life cycles and social relations through a broader perspective. To think 'climate change' locally also means taking into account these important dimensions of current socioenvironmental transformation.

Research on the Calendar and Life Cycles

The concepts of life cycles and the annual calendar enable the description and translation of cosmological ideas and indigenous management knowledge, and provide a guiding principle in research platforms that involve young indigenous and non-indigenous researchers. The research on the indigenous calendar in the Tiquié River area (Tukano, Desana and Tuyuka peoples) and in other regions of the Northwest Amazon, such as the Pirá-paraná and Içana Rivers, emerged within the context of the increasing interest of indigenous organizations in establishing, through intercommunity agreements, environmental and territorial management plans – taking charge of the governance of their territories within the current political context. Likewise, the indigenous environmental management agents (AIMAs) emerged who, being residents of the communities and mostly young adults, were in charge of promoting this theme. Currently, there are seventeen AIMAs in the Tiquié who work with three shamans who are also directly involved in the research. They participate in a knowledge-exchange program through workshops that were set up by these associations and schools, in partnership with the Instituto Socioambiental. On the Colombian side, in the Pirá-Paraná region, similar alliances and initiatives occur between the Association of Chiefs and Traditional Authorities of the Pirá-Paraná River (ACAIPI) and the Gaia-Amazonas Foundation (FGA). The indigenous associations of the Tiquié and Pirá-paraná Rivers and their partners cooperate within the context of CANOA – Cooperation and Alliance in the Northwest Amazon – including with regard to the research on calendars.

Methodologies

Research instruments and methodologies are employed in the observation and registration of the phenomena of the annual cycle. This initiative situates the indigenous researchers in an intermediate position between their own knowledge traditions and those of the Western sciences, benefiting from this interface. Since 2005, these cycles have been observed, registered and systematized in joint research practices undertaken by knowledge holders, and both indigenous and non-indigenous researchers. From the point of view of the younger indigenous researchers, it is a way to give meaning to the management practices, a space for learning with the elders and sources of knowledge. For the non-indigenous researchers, the focus on the calendar allows an understanding, in a more systematic manner, of the relations between indigenous peoples and their environment, practices and related knowledge. Moreover, it can be a way to explore the interfaces between indigenous and Western knowledge, converging to study life cycles, both 'natural' as well as social.

The basic instrument in this research is an observation diary in which each AIMA records daily observations on: (1) rain; (2) level of the river, extent of the low and high water levels, navigating conditions; (3) name of the season in indigenous language; (4) phenology of important plants (cultivated as well as wild): their flowering and fructification, when they are ripe and when they are being consumed by people, fish and animals; (5) cycles of fish (and land animals): migration, spawning and mating seasons, maturity stages, food they eat and fishing techniques; (6) reproduction of insects, amphibians, mammals

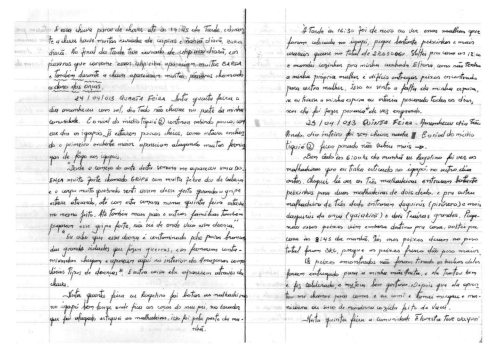

Figure 3.2 Excerpt from an AIMA diary.

(e.g. the swarming of ants, spawning of frogs, appearance of edible caterpillars). Also daily events of the communities are recorded: (1) community meals; (2) the results of hunting and fishing; (3) the agricultural calendar: felling of old-growth or secondary forest, burning, planting, harvesting; (4) community festivities and rituals; and (5) the most common diseases (Cardoso, 2007; AEITY and ACIMET, 2008; AEITU, 2009; Cabalzar et al., 2010).

Figure 3.2 shows an example of the diary notes in which some of the above-mentioned themes are recorded, such as the name of the season, level of the river, weather conditions, activities and illness. The reference to illness is associated with far-away 'white people' activities leading to pollution and environmental degradation inside the indigenous territory – it is remarkable that new global threats have been perceived as the causes of local health conditions.

The regular, daily note-taking enables an understanding of management practices and their variations during the annual cycle, the seasonal availability of natural resources, the interdependency between socioeconomic organization, life cycles and climate. Since the AIMAs are present in approximately twenty communities along a stretch of almost 400 km of the river in different environmental conditions (Cabalzar and Lima, 2005), this research permits analysis at local (community) and regional scales.

The information gathered in the diaries is collectively systematized, providing a description of each annual cycle in accordance with the astronomical calendar and classification of the seasons by the Tukano, Desana and Tuyuka (Figure 3.3). Certain procedures are

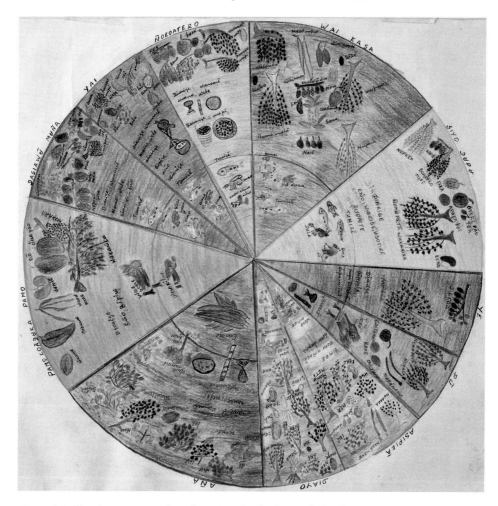

Figure 3.3 Circular representation of an annual calendar made by the AIMA. (A black and white version of this figure will appear in some formats. For the colour version, please refer to the plate section.)

adopted: writing of texts in indigenous language, summarizing the cycles of the seasons and their description (AEITY and ACIMET, 2008; AEITU, 2009), and development of a timeline that contains a graph of the level of water in the river and the name of each season associated with the various phenomena of the life and phenological cycles.

Figure 3.4 is an extract that visualizes this last way of summarizing the information from the diaries. To the left, '*dia pairo*', '*dia keoro*' and '*dia perogã*' mean, respectively, 'big river', 'midwater' and 'small river' (or dry); the line follows the level of water in the river day by day. Along the line, the names of the seasons are noted, such as *Aña Pue* (flood of the *jararaca* viper snake), *Mene K¡ma* (summer of the *inga* tree), *Pamo Pue* (flood of the armadillo). In the upper part, the boxes below the month and day are

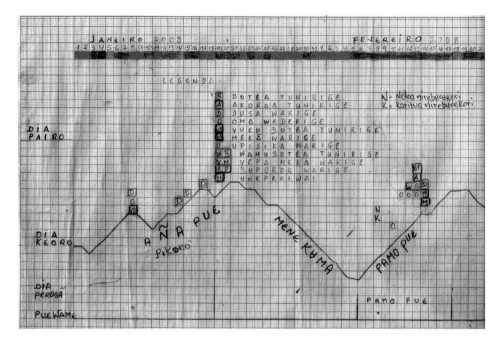

Figure 3.4 Excerpt of the timeline for 2008, showing season names along the water level line, as well as the timing of some phenological phenomena. (A black and white version of this figure will appear in some formats. For the colour version, please refer to the plate section.)

colour-coded to indicate whether the day was sunny, rainy, cloudy or drizzly. The legend, using abbreviations and colours, refers to phenomena such as the mating of *botea tunirige*, the aracu fish (*Leporinus* genus) and of *akoroa tunirige*, the characin (*Cyphocharax* genus), the swarming of *dusa warige*, an indigenous bee, and so on. Following this legend, the events are situated along the curve.

Added to this information are measurements of rainfall and water level in this river, collected during the same three seasons by the Geological Service of Brazil (CPRM).

Even with a simple methodology the volume of information is enormous and requires a lot of effort to systematize and analyse. Five years of data (2006–2010) are currently being compared and analysed. The intention is to maintain a long-term register of the diaries.

Preliminary Results

As a result of being on the equator and due to the high rainfall levels, the north-western Amazon region has a complex climatic regime, with seasons that are not sharply delimited, characterized by a regime of quite uneven rains and droughts with substantial variations from year to year. The indigenous calendar takes these fluctuations into account, applying a flexible terminology. *Repiquetes* (rapid rises in river water levels) interspersed with 'summer' days can be distinguished by that part of the constellation that is waning (*poero dihase*) on the horizon in the early evening.

Table 3.1 presents the main sequence of the yearly seasons, together with some of their characteristics as they are known to the indigenous peoples (Tukano, Desana, Tuyuka, Miriti-tapuya) of the Tiquié River. Floods are always associated with the waning, in the early evenings, of the constellation with the same name. The names of the 'summers' are related to phenological phenomena.

This research is making possible a more accurate description of the annual cycle and its variations. To this end, it seeks a convergence between indigenous knowledge and the more scientific methodology of registering and organizing data. This combined research has an additional inquiry, to test such relations and their effects.

In the view of the knowledge holders, longer-term changes have been observed, above all, the migration of birds that happens in certain seasons. In recent years, more humid periods have been experienced, which makes burning for opening new agricultural fields difficult (various cleared areas failed to burn), as well as significant changes in the phenology of fruit species that are important for food. The second half of 2010 was atypically dry, enabling the earlier burning of old growth forest for new plots. These phenomena have not yet generated significant or permanent problems for the communities. To the extent that climatic changes modify these natural cycles, they influence not only the economic activities of the indigenous populations but also the whole ritual life and sociocultural experience of these peoples, as these are two sides of the same coin.

The food security of these populations depends on the cultivation of cassava and on fishing, the staple food of the indigenous populations. The production and reproduction conditions of these resources need to be observed within the context of climatic changes. This research and monitoring will make it possible to evaluate such situations and formulate proposals for future actions.

Indigenous agriculture is based on shifting cultivation with slash and burn practices. Small areas (less than 0.5 ha on average) of old growth or secondary forest (*capoeira*) are cleared, left to dry, and during the 'summers' (consecutive days without rain) burned. Depending on the type of cleared forest and the intensity of the sun, three to ten days of summer are necessary. Once burned, they start to plant upon the first rains. After the cassava has matured, the plot can be used for two to three years, being replanted regularly. After this period of production, the forest is left to grow back (fallow) and there is a waiting period before using the same area another time.

In this region, characterized by acid and rather infertile soils, the burning of old growth forest, makes it possible to cultivate more demanding plants such as peach palm (*pupunha*), Amazon grapes (*cucura*), banana, pepper, sugar cane, tubers etc. The absence of sufficient 'summer periods' can compromise this burning and subsequently the results of indigenous agriculture, a situation that has already been recorded for a number of years. If this occurs for only one year, it is possible to compensate the losses with fields that are already in production. Difficulties can emerge when these extreme phenomena become more frequent, which seems will be the case (Lewis et al., 2010). In this region with high

Table 3.1 *Calendar of the indigenous peoples of the Tiquié River*

Main season	Characteristic
Jararaca [snake] Flood *Aña poera*	Beginning of the yearly cycle (corresponding to the beginning of the month of November)
	First reproductions of fish, amphibians and insects, in relation to heavy rains and consistent river level rise (heavy rainfall starting at 3 a.m. until 1 p.m. in the afternoon, on this day the swamps will already start to flood; the *daracubis* earthworms, spiders, insects and snakes will climb into the swamp trees). The first spawn of the scratched-aracu and the three-band-aracu (genus *Leporinus*) and various other species appear; these first spawns (*ñeekoese*) are considered the cleaning of the spawning places (the longhouses where the fish are having their festivities). In this period, only some of the fish are mature enough to spawn. Flocks of birds, called 'those of the jararaca', appear and descend into the pastures, thereafter they fall in the river and become fish. Fishing is difficult (hardly productive); the elders say that 'the fish go into the anus of the jararaca' – as if it were a structure, a longhouse, a house where they will live; the anus is the gate. Also many jararaca snakes appear. The bite of the jararaca in this period cannot be cured; as they are transformed through the lightning of the thunder, they are from the house of heaven. Bites from those that are born on Earth can be cured.
	The constellation of the jararaca is the biggest of the astronomy of the Tukanoan peoples. It is divided in parts of which the main ones are: lumen (*aña siõkha*), head (*aña dʉpoá*), eggs (*aña diepá*), tail (*aña pihkorõ*). These parts can indicate small floods (*repiquetes*), separated by a few days of drought (*kʉmataro*)
Inga [tree] Summer *Mere kʉma*	The cultivation fields as well as virgin forest are burned, resulting in strong sun and soft wind
Tatu [armadillo] Flood *Pamo poero*	There are no differences compared to the floods of the jararaca constellation, it follows the reproductive period of the fish (spawning) insects (swarming) etc. The only distinct characteristic is the landslides of the river banks. Trees fall from the shores, the water remains dirty, muddy and with scum
	It is the second constellation of the annual cycle, divided into two parts: a piece of the armadillo bone (*pamõ oãdʉka*) and the body of the armadillo (*pamo opʉ*)
Pupunha [peach palm] Summer *Ʉre kʉma*	This is a long summer with a very dry period, impeding navigation on the rivers that have stones and rapids
Jacundá fish Constellation *Mʉha poero*	The time of hot summers is in decline and the time of strong floods in the annual cycle is approaching; during this flood there is much spawning and reproduction of animals

Table 3.1 (*continued*)

Main season	Characteristic
Umari [tree] Summer *Wamʉ kʉma*	Short summer with duration of only a few days, at the most a week, with low water levels
Shrimp Flood *Dahsiʉ poero*	In the narrative of origin, the Jacundá fish will swallow the shrimp when he precipitates
Onça [jaguar] Flood *Yai poero*	This is the first big flood in the annual cycle of the constellations, all the swamp forests will be flooded with the exception of only the highest lands, which is where the animals will survive, the common agouti, the red agouti, the spotted paca, armadillos and others. The big flood is stable and will not get much lower or higher during a week or more. In the phase of the Body of the Jaguar all fish can spawn. Thereafter comes the Abiu (*Sapotaceae*) summer. When that is over, the rain of the Jaguar Tail will come. It is customary to then have *dabucuri* (offering festivity) of miriti-palm and swamp-forest-açaí fruits
	The jaguar flood is divided into: jaguar beard (*yai ehseka poari*); jaguar head (*yai dʉhpoa*), jaguar body (*yai ohpʉ*), jaguar tail (*yai pihkorõ*)
Plêiades Flood *Ñokoatero poero*	Last spawning of the season, the most intensive of the year. Hereafter the first schools of fish start to come, such as the scratched-aracu, the three-band-aracu and others. Thunder and lightning occur during this period. Sometimes it rains a lot but other times not so much. The river floods a lot. During the rains of this constellation the *tamiria* birds appear
Fish rack Flood *Wai kahsa poero*	In this period occurs the winter of laziness, cold weather with frequent showers for a week or less. Windstorms, lightning and thunderstorms occur
Adze Flood *Sioyahpu poero*	Migrating schools of all fish species: scratched-aracu, three-band-aracu, minnows, *pirandira*
Otter Flood (*Diayoa*), Bodó fish (*Yahka*), Bihpia, Ñamia, Folhas (*Puri*), Thoto, Sihpia, Jabuti [turtle] (*Uú*), Kaisariro	These floods occur without major differences. Some of these constellations are little known
Largatas-do-cunurizeiro [cunuri-tree-caterpillars] summer *Iña kʉma*	This summer can last up to two weeks before the Egret Flood
Garça [egret] Flood *Yhe poero*	This is the end of the annual cycle

humidity and low seasonality (short and not so discernible seasons), ten days without rain occur only few times a year. Between 2008 and 2010, for example, summers of more than nine days occurred only in 2010. The indigenous farmers get round this problem by burning only partially and compensating with the practice of heap burning (*coivara*, clearing and collecting of twigs and small burnable materials) after the initial burning. This is labour-intensive, however, and increases work pressure on a family. This situation is very different from that encountered in the southern Amazon region, with long dry seasons, more disturbed environments and forest fires that are difficult to control.

The production of fish is much more sensitive, as it depends on the climatic variations and the level of the river associated with the annual cycle, as well as on the history of fisheries management (the threat of overfishing). Contrary to agriculture, where an excess of rainfall (or the absence of consecutive sunny days) is the most vulnerable aspect, for fishes, it is the absence of rain, in sufficient amounts during certain periods of the year. The most consumed species (genus *Leporinus*, aracu) are ready to reproduce from November, which corresponds to the occurrence of the *Aña* (*jararaca* viper snake) constellation, but they only spawn (locally called *piracema*) when the level of the river is consistently and uninterruptedly rising during the entire day, until the end of the day. The aracu only reproduce in the afternoon or early evening; but if the level of the river does not continue to rise until this hour, reproduction does not occur. When the period of rains is delayed or if they are of low intensity, the reproduction of these species can be compromised. Another sensitive aspect on the biology of migrating species, which are among the most important economically, is the period of migration up river. Between June and July, the river has to reach its highest levels to provide adequate conditions for these movements. Otherwise, the abundance and subsequent reproductive season would be affected.

Discussion

In the Northwest Amazonian region, on the Brazil–Colombia border, there is a comprehensive social system of about thirty indigenous peoples that have been related to each other for many generations through intensive exchanges and interrelations (of kinship and trade, knowledge and ritual alliances).

The indigenous concepts of management of the world contain two fundamental aspects that are interrelated. First, a strong territorial sense, in that they consider this region (basins of the Negro and Apapóris Rivers) as their place of origin, from where their names, capacities, health and well-being originate and emerged. There exists a vital relation between people and places which makes each one indispensable to the other. In the words of a Piraparaná leader:

This great territory of Yuruparí, where the land, the air, the animals, the vegetation species, the visible and invisible beings, the humans, the crops, contain the *Hee* [ancestral] spirit, the power or essence of life that makes the World exist. The territory of the Jaguars of Yuruparí, *Hee Yaia Godo*, has been structured since the Origin, since the same creators, the Ayawa, established the planet, the cosmos.

(ACAIPI and FGA, 2010)

There is a need to manage the world through the annual cycle, to keep the balance of each season with its people, food, substances, diseases. This management of the world is carried out by those who know how, those who are prepared for it and have this prerogative: knowledge holders, *kumua*, shamans. They know the rituals that seek to heal, protect, mediate, repair, communicate, share. They are undertaken in each community or longhouse, and also in connection, in networks, within the outreach of the broader territory.

This network of knowledge and knowledge holders has been unevenly affected by external interference over the last century. In general, these populations are getting increasingly involved with national society, and are confronted with conflicting understandings and attitudes. In relation to this, Maximiliano Garcia, a *makuna* of the Pirá-Paraná River said that:

the world sees natural resources as a source of money; this is what we suffer from these days. It is not just climate change; it is changes in the way of thinking. The sacred places are a part of us, whereas the world sees these sites as a source of monetary resources to extract gold, wood. That is why climate change is abrupt, if we do not have our sacred places there is no life, for it is there that there is air, food, cures. Today it starts to rain when it is not supposed to rain, it is hot out of season, this is what causes poverty. Because there is lack of food when it is very dry and very full, there are no crops, there are no places to hunt or fish. Nature regulates itself, there will be no fertility in the land if it rains a lot.

Indigenous knowledge and practices will always contribute to the development and maintenance of the great socioenvironmental diversity of the Amazon. Today, in circles of indigenous organizations, new forms of governance of their territories are being sought, in line with their knowledge and practices of environmental management, and based on new alliances. The cooperation between indigenous organizations of the Tiquié and Pirá-Paraná Rivers has generated renewed flows of knowledge for management, strengthening indigenous knowledge and favouring its circulation within new generations. The main challenge is the strengthening of indigenous knowledge within a context of an increasing communication and connection with the global society.

References

ACAIPI (Asociación de Capitanes y Autoridades Tradicionales Indígenas del Pirá-Paraná) and FGA (Fundación Gaia Amazonias). 2010. *Postulación del Conocimiento Tradicional para el Manejo del Mundo de los grupos indigenas del Rio Pirá-Paraná – Hee Yaia ~Kubua Baseri Keti Oka*. Unedited.

AEITU (Associação Escola Indígena Tuyuka Utapinopona). 2009. *Bureko Watotire*. São Paulo/São Gabriel da Cachoeira, Brasil: Instituto Socioambiental (ISA)/Federação das Organizações Indígenas do Rio Negro (FOIRN).

AEITY (Associação Escola Indígena Tukano Yupuri) and ACIMET (Associação das Comunidades Indígenas do Médio Tiquié). 2008. *Marĩ kahtiri pati kahse ukuri turi*. São Paulo/São Gabriel da Cachoeira, Brasil: ISA/FOIRN.

Andrello, G. 2006. *Cidade do Indio: Transformações e Cotidiano em Lauaretê*. São Paulo/Rio de Janero, Brasil: UNESP/ISA/NuTI.

Andrello, G. (ed.). 2012. *Rotas de Criação e Transformação. Narrativas de Origem dos Povos Indígenas do Rio Negro*. São Paulo/São Gabriel da Cachoeira, Brasil: ISA/FOIRN.

Azevedo, M. [Ñahuri] and Azevedo, A. N. [Kumarõ] (2003). *Dahsea Hausirõ Porá Ukushe Wiophesase Merã Bueri Turi. Mitologia Sagrada dos Hausirõ Porã. Coleção Narradores Indígenas do Rio Negro, vol. 5*. São José I/São Gabriel da Cachoeira, Brasil: UNIRT/FOIRN.

Azevedo, V. V. B., Oliveira, M., Azevedo, D. A. et al., 2010. Calendário astronômico do médio Rio Tiquié. Conhecimento para educação e manejo. In Cabalzar, A. (ed.) *Manejo do Mundo. Conhecimentos e Práticas dos Povos Indígenas do Rio Negro, Noroeste Amazônico*. São Paulo/São Gabriel da Cachoeira, Brasil: Instituto Socioambiental (ISA)/Federação das Organizações Indígenas do Rio Negro (FOIRN), pp. 56–66.

Buchillet, D. 1988. Interpretação da doença e simbolismo ecológico entre os índios Desana. *Boletim do Museu Paraense Emilio Goeldi*, 4(1): 7–32.

Cabalzar, A. (ed.) 2005. *Peixe e Gente no Alto Rio Tiquié. Conhecimentos Tukano e Tuyuka, Ictiologia, Etnologia*. São Paulo: Instituto Socioambiental (ISA).

Cabalzar, A. (ed.) 2010. Manejo ambiental e pesquisa do calendário anual no Rio Tiquié. In *Manejo do Mundo. Conhecimentos e práticas dos Povos Indígenas do Rio Negro, Noroeste Amazônico*. São Paulo/São Gabriel da Cachoeira, Brasil: ISA/FOIRN.

Cabalzar, A. and Lima F. C. T. 2005. Do Rio Negro ao Alto Tiquié. Contexto socioambiental. In Cabalzar, A. (ed.) *Peixe e Gente no Alto Rio Tiquié. Conhecimentos Tukano e Tuyuka, Ictiologia, Etnologia*. São Paulo, Brasil: ISA, pp. 23–42.

Cabalzar, A. and Ricardo, C. A. 2006. *Mapa-Livro. Povos Indígenas do Alto e Médio Rio Negro, uma introdução à diversidade cultural e ambiental do noroeste da Amazônia brasileira*. São Gabriel da Cachoeira/São Paulo, Brasil: FOIRN/ISA.

Cardoso, W. T. 2007. *O céu dos Tukano na Escola Yupuri. Construindo um Calendário Dinâmico*. Ph.D. Thesis. São Paulo, Pontifícia Universidade Catolica.

Chernela, J. M. 1993. *The Wanano Indians of the Brazilian Amazon: A Sense of Space*. Austin: University of Texas Press.

Dufour, D. L. 1988. Cyanide content of cassava (Manihot esculenta, Euphorbiaceae) cultivars used by Tukanoan Indians in Northwest Amazonia. *Economic Botany*, 42(2): 255–66.

Dufour, D. L. 1990. Use of tropical rainforests by native Amazonians. *BioScience*, 40: 652–59.

Epps, P. and Oliveira, M. 2013. The Serpent, the Pleiades, and the one-legged hunter: Astronomical themes in the upper Rio Negro. In Epps, P. and Stenzel, K. (eds.) *Upper Rio Negro: Cultural and Linguistic Interaction in Northwestern Amazonia*. Rio de Janeiro, Brasil: Museu do Índio – FUNAI, Museu Nacional.

Garcia Rodriguez, M. 2010. Que ha significado la selva para nosotros? In Cabalzar, A. (ed.) *Manejo do Mundo. Conhecimentos e Práticas dos Povos Indígenas do Rio Negro, Noroeste Amazônico*. São Paulo/São Gabriel da Cachoeira, Brasil: ISA/FOIRN.

Goulding, M. A., Carvalho, M. L. and Ferreira, E. G. 1988. *Rio Negro, Rich Life in Poor Water*. The Hague, Netherlands, SPB Academic Publishing.

Hugh-Jones, C. 1979. *From the Milk River. Spatial and Temporal Processes in Northwest Amazonia*. New York: Cambridge University Press.

Hugh-Jones, S. 1979. *Amazonian Indians (The Civilization library)*. New York: Gloucester Press.

Hugh-Jones, S. 1981. Historia del Vaupés. Maguare. *Revista del Departamento de Antropoloría, Universidad Nacional de Colombia*, 1: 29–51.

Hugh-Jones, S. 1982. The Pleiades and Scorpius in Barasana Cosmology. In Aveni, A. F. and Urton, G. *Ethnoastronomy and Archaeoastronomy in the American Tropics*. New York: New York Academy of Sciences, pp. 183–202.

Hugh-Jones, S. 1995. Inside-out and back-to-front: The androgynous house in Northwest Amazonia. In Carsten, J. and Hugh-Jones, S. (eds.) *About the House: Lévi-Strauss and Beyond*. Cambridge: Cambridge University Press, pp. 226–52.

Lewis, S. L., Brando, M., Nepstad, D. P., Phillips, O. L. and Van der Heijden G. M. 2011. The 2010 Amazon drought. *Science*, 331(6017): 554.

Meira, M. 1997. *Índios e Brancos nas Águas Pretas. Histórias do Rio Negro. Seminário Povos Indígenas do Rio Negro*. São Gabriel da Cachoeira: Terra e Cultura/FOIRN.

Moran, E. 1991. Human adaptive strategies in Amazonian blackwater ecosystems. *American Anthropologist*, 93: 361–82.

Neves, E. 1998. *Paths in the Dark Waters: Archeology as Indigenous History in the Upper Rio Negro Basin, Northwest Amazon*. Ph.D. Thesis, Bloomington, Indiana University.

Ortiz, N., Avila, E., Marín, R. and Rodriguez, G. (2012). Tras las Huellas de Nuestro Territorio. In Andrello, G. (ed.) *Rotas de Criação e Transformação. Narrativas de Origem dos Povos Indígenas do Rio Negro*. São Paulo/São Gabriel da Cachoeira: ISA/FOIRN.

Pãrõkumu, U and Kehirí, T., 1995. *Antes o Mundo não Existia. Mitologia dos Antigos Desana-Kêhíripõrã*. São Gabriel da Cachoeira: UNIRT/FOIRN.

Reichel-Dolmatoff, G. 1979. Cosmology as ecological analysis. A view from the rainforest. *Man*, 2(3): 207–318.

Ribeiro, B. G. 1995. *Os índios das águas pretas: modo de produção e equipamento produtivo*. São Paulo, Brasil: Edusp.

Ribeiro, B. G. and Kenhíri, T. 1987. Chuvas e constelações. *Ciência Hoje*, 6(36): 26–35.

Valencia, I. 2010. Calendario Ecológico. La selva, los animales, los peces, el hombre y el río, en cada época del año. In Cabalzar, A. (ed.) *Manejo do Mundo. Conhecimentos e Práticas dos Povos Indígenas do Rio Negro, Noroeste Amazônico*. São Paulo/São Gabriel da Cachoeira: ISA/FOIRN.

Valle, D. 2010. Proteção das Maloca (casas cerimoniais). Basawiseri wanoare makañe wederige. In Cabalzar, A. (ed.) *Manejo do Mundo. Conhecimentos e Práticas dos Povos Indígenas do Rio Negro, Noroeste Amazônico*. São Paulo/São Gabriel da Cachoeira: ISA/FOIRN.

4

Indigenous Knowledge in the Time of Climate Change (with Reference to Chuuk, Federated States of Micronesia)

Rosita Henry and Christine Pam

In order to understand how social resilience might be achieved in the face of climate change, it is crucial to consider how people employ everyday 'local' and 'indigenous knowledge' to deal *in practice* with uncertainty and risk in their lives. Focusing on responses to climate change discourse in the Pacific, with particular attention to Chuuk, Federated States of Micronesia (FSM), we call for more fine-grained ethnographic studies on how the global discourse of climate change transforms knowledge and practice at the local level.

According to the 2010 FSM Census, the state of Chuuk has a population of 48,651, or approximately 47 per cent of the total population of the Federated States of Micronesia, distributed unevenly over about forty islands. Chuuk Lagoon, which lies at the heart of the state, consists of a barrier reef encircling nineteen small 'high' volcanic islands, including the capital, Weno. Around this heart are numerous coral atolls and 'low' coralline islands.

This diversity of high and low islands poses complex socioenvironmental challenges for the FSM in relation to the country's capacity to respond to extreme climatic events. The state of Chuuk, in particular, provides instructive case material on how the discourse of global warming is being taken up and interpreted at the local level and on the conjuncture between contemporary scientific predictions of climate change and indigenous theories of causation and understandings of climate variability and extreme climatic events. What kinds of knowledge and what sorts of practices might island-based communities employ at the grass roots to deal with the impacts of climate change? We consider this question on the basis of research conducted in Chuuk; in the capital of Weno and on the small low-lying coral island of Moch, in the Mortlocks Region to the south-east of Chuuk Lagoon.

Indigenous Knowledge

There have been numerous scholarly deliberations about the definition of indigenous knowledge, in tandem with debates about 'the validity of the concept of indigeneity' (Dove, 2006: 191). It is beyond the scope of this chapter to review these debates yet again (Agrawal 1995; Ellen et al., 2000; Merlan, 2009). Nevertheless, it is important to consider the nature and substance of indigenous knowledge, or any category of knowledge for that matter, if one is to understand how such knowledge might be employed in the face of change. In other words, behind this chapter lie deep philosophical questions about the nature of knowledge per se and how or where it might find tangible expression.

'Indigenous knowledge' is a category produced in the context of contemporary representations of identity and indigenous rights politics, but it also has substance. That substance is more than particular traditional technologies and taxonomies, or even particular practices, associated with the natural environment. That substance is knowledge of the relatedness and interdependency among things (human and non-human). As such, indigenous knowledge includes the cosmological understandings, values and principles that *inform* the use of technologies, that *generate* taxonomies and that *find expression* in practices of social obligation and responsibility, including in practices of negative reciprocity.

Thus our interest is not only in knowledge as it manifests itself in social practice, but also in how knowledge categories themselves (local, indigenous, scientific) come to be strategically employed in a global politics of climate change. In order to understand how people might respond to climate change predictions, as well as deal with actual climatic events, it is necessary to explore how global scientific discourse is taken up at the local level, and the transformative effect it has on local knowledge. For example, as Rudiak-Gould notes, among Marshall Islanders 'scientific knowledge of climate change has powerfully influenced local perceptions of environmental change: precisely that which seems to best exemplify "pure" traditional ecological knowledge' (2010: 12).

Systems and Practices

The predominant approach among scientists (including many social scientists) to understanding responses to climate change has been a systems-based one, where social and ecological variables are not analysed in isolation but in terms of their interrelation as part of a system. Thus, emphasis has been placed on concepts of the 'vulnerability', 'adaptation' and 'resilience' of socioecological systems. Such concepts 'do not exist in isolation from the wider political economy' (Adger et al., 2006: 5). It is important to recognize that complex historical, political and economic factors constitute the social dimension of any socioecological system. In the context of the Pacific one must keep in mind the shattering effects of the Second World War and catastrophic anthropogenic factors such as atomic testing in the Marshall Islands, as well as the more recent impacts of globalization, population movement, rapid population growth and other threats.

Adger et al. (2005: 1036) define resilience as

the capacity of linked social-ecological systems to absorb recurrent disturbances such as hurricanes or floods so as to retain essential structures, processes and feedbacks. Resilience reflects the degree to which a complex adaptive system is capable of self-organization (versus lack of organization or organization forced by external factors) and the degree to which the system can build capacity for learning and adaptation.

They argue that the concept of resilience marks 'a profound shift in traditional perspectives, which attempt to control changes in systems that are assumed to be stable, to a more realistic viewpoint aimed at sustaining and enhancing capacity of social-ecological systems to adapt to uncertainty and surprise' (Adger et al., 2005: 1036).

Yet, what are the human qualities, or characteristics, which define this capacity to adapt? What is the relationship between such adaptive capacity and knowledge systems (whether scientific or indigenous)? In considering vulnerability and resilience it is necessary to take into account the social practices that not only underpin and generate knowledge, but also constitute knowledge systems in and of themselves (knowledge of how to live in the world as social beings).

Although systems-based approaches have generated useful models, they are not able to effectively capture the complexities of people's real-life experiences and responses. Models must be informed by studies of actual practices of contemporary everyday life. While climate change is a global phenomenon, there is great regional and local variability in terms of impact. As noted in the Third Assessment Report of the Intergovernmental Panel on Climate Change (IPCC):

Climate models are simplifications of very complex natural systems; they are severely limited in their ability to project changes at small spatial scales, although they are becoming increasingly reliable for identifying general trends. In the face of these concerns, therefore, it would seem that the needs of small island states can best be accommodated by a balanced approach that combines the outputs of downscaled models with analyses from empirical research and observation undertaken in these countries.

(McCarthy et al, 2001: 870)

However, in the Fifth Assessment Report, Nurse et al. (2014: 1626) note that while there has been 'a problem in generating formal climate scenarios at the scale of small islands because they are generally much smaller than the resolution of the global climate models', this has improved with the use of 'the new Representative Concentration Pathway (RCP) scenario General Circulation Models (GCMs) with grid boxes generally between 100 and 200 km^2 in size'.

Nevertheless, because local conditions vary widely on different island types it is difficult to predict climate change impacts. The issue is not just geographical and meteorological variability. The way people deal with any kind of change is dependent on pre-existing local knowledge, as well as on their adoption and interpretation of new knowledge, including science-based theories of climate change. It is also dependent on locally *experienced* historical, social, cultural and political contexts and administrative environments (Torry, 1979). For example, Marshall (1979) provides a case study on how the distribution of aid in the wake of a natural disaster can prove problematic when 'local administrative politics' are not taken into account. He documents the distribution of relief in the Mortlock Islands (22 May–7 July 1976) by the United States government and the Red Cross following cyclone Pamela and shows that more politically powerful island communities received the most aid even though they were the least impacted by the cyclone. Although Marshall conducted his study over forty years ago, his general point still holds – political and administrative contexts play a significant role in how effectively people are able to respond to climate change impacts. Thus, systems-based research on climate change must be supported by practice-based research at the community level that will provide an understanding of the impact of local political realities on response strategies (see also Barnett and Busse, 2001). As Nurse et al. (2014: 1635) argue 'Recent moves toward participatory approaches that link scientific knowledge with local visions of vulnerability offer an important way forward to understanding island vulnerability in the absence of certainty in model-based scenarios.'

Pacific States and the Discourse of Global Warming

Climate models predict that low-lying atoll states in the Pacific, and elsewhere, are highly vulnerable (Barnett and Adger, 2003; Pittock, 2005; Adger et al., 2014; Nurse et al., 2014). States such as Tokelau, Tuvalu, the Marshall Islands and Kiribati, which are composed entirely of low-lying coral atolls that rarely exceed 3–4 m above present mean sea level, are considered most at risk (Barnett and Adger, 2003). However, states that include high islands, such as the Federated States of Micronesia, also face disastrous consequences because of climate change, including sea-level rise, rise in sea surface temperature and increased intensity and frequency of extreme weather events. On the high islands, human populations are concentrated in the coastal zone, making them vulnerable to sea-level rise which 'will exacerbate storm surge, erosion and other coastal hazards, threaten vital infrastructure, settlements and facilities, and thus compromise the socio-economic well-being of communities and states' (Mimura et al., 2007: 689). One of the key matters of concern is the impact of climate change on water security (Nurse et al., 2014: 1622–3).

Climate science predictions are well accepted in the Pacific, with Pacific Island states responding by developing a regional framework for action on climate change, climate variability and sea-level rise (2000–2004), superseded by the *Pacific Islands Framework for Action on Climate Change 2006–2015*. There has also been wide acceptance of popular media representations among Pacific Islanders of their plight as victims of global warming. For example, Rudiak-Gould (2010) discusses how climate change doomsday prophesies have been taken up in the Marshall Islands. Some states have appropriated a discourse of vulnerability and enlisted the metaphor of 'smallness' in challenges to global debates on climate change (Kempf, 2009). Some states have also strategically employed the media as a means of seeking international recognition and assistance. Connell (2003) expresses concern about this, arguing that the acceptance of 'doomsday' predictions blinds people to immediate, local (including anthropogenic) causes of environmental degradation. Barnett and Adger (2003: 329) suggest that acceptance of the inevitability of imminent disaster and 'lost confidence in atoll-futures' may be more of a problem than the physical impact of climate change itself, in that it may lead to 'changes in domestic resource use and decreased assistance from abroad'.

Figure 4.1 Moch Island, Mortlock Islands, Chuuk, FSM. (A black and white version of this figure will appear in some formats. For the colour version, please refer to the plate section.)

Chuukese Concerns About Climate Change

Because Chuuk and other small island states have long faced more immediate socio-economic problems (including poverty, high unemployment, housing, education and health care), one might expect that climate change issues would be accorded a low priority. Yet, our research on Chuuk reveals that government officers and other Chuukese are deeply concerned about the potential effects of climate change. During our pilot study and the longer-term ethnographic fieldwork that followed, we interviewed people living in the high island capital of Weno as well as members of the Mochese community living predominantly on their home island in the Mortlocks (Figure 4.1). The pilot study was conducted in 2008 and involved in-depth interviews with government officers in Weno, and a field trip to Moch at the invitation of the Mochese research officer at the Chuuk Historic Preservation Office (Henry et al., 2008). Pam's doctoral research was an extension of this study, conducted primarily with the Mochese community living on their home island, but also among the extended community living in Weno. Her research was based on participant observation over a period of eight months between 2009 and 2011, and was especially informed by her involvement with a climate change project initiated by the Moch municipal government.

Because of urbanization processes, many of those we interviewed who were living in Weno originally hailed from outer atoll islands in Chuuk. They conveyed great fears for the future of their home islands, emphasizing that these were in the process of gradually disappearing:

It's a low flat island ... When I go back it looks like it's getting worse because some of the beach I saw before is gone and the island is getting *smaller and smaller* because the waves wash away the sand. It's not like before ... Our main food is taro and the waves come into the taro patch and kill off the crops.

Little by little our shore is disappearing and coconut trees are standing in water. It's happening more and more over the years. We heard that pretty soon we'll be all underwater, the whole of the Mortlocks. They say this is the time of evacuation. They say this is the time to evacuate people to higher ground. Right now we are suffering erosion and we are suffering the high tide. In our language we call it *setupul*.

Research participants described their experiences of storm surges and unusually high tides as well as their memories of particularly extreme climatic events such as cyclones (typhoons). In particular, cyclone Pamela, which hit the Mortlock Islands in May 1976, was well remembered. It washed away a quarter of the island of Kuttu, a small coral island (0.2 square miles with a population of about 350), which forms part of the Satawan Atoll. According to Marshall (2004: 65), all the islands in the Mortlocks were badly affected by this cyclone. Namoluk, where he conducted his field research, was completely submerged by waves that swept across the island for fifteen to eighteen hours.

The saltwater entered the taro swamps and killed all the taro plants, and a high percentage of breadfruit, coconut and banana trees perished outright from the combination of salt, wind and waves. Equally devastating was the substantial damage to the atoll's reefs and reef organisms many of which serve as a regular food source.

(Marshall, 2004: 65)

The cyclone resulted in the formation of a new islet, which was immediately named 'Pamela' by 'local wags' (ibid.).

For people living on Moch, typhoons loom large in the social memory of the community. While discussing the effects of a tidal surge on the island in late 2008, one woman said she was scared when the waves came close to her house because it reminded her of Typhoon Pamela. Others remembered listening to the old people tell stories about the typhoon that hit the island in 1907 (Spennemann, 2007); how they heard the screams of people from the islands of Satawan and Ta as they were swept out through the channel and into the ocean, and how afterwards they found people dead on the reef.

Clearly, storm surges, high tides and cyclones are not new phenomena in this environment, and people in times past developed various cultural responses and strategies to cope with their impact. The dynamic potential for transformation of the seascape is recognized within Micronesian cosmology, as reflected in narrative knowledge of islands disappearing into the sea, or being deposited within the seascape by ancestral beings (Lessa, 1961; Mitchell, 1973; Lessa, 1980; Goodenough, 1986). According to Mitchell (1973: 23) there is a recurrent motif in these narratives of islands being 'fished up' from under the sea with a hook or a net. Other narratives tell of islands being 'kicked' out of their original place by a culture hero or of being formed from a rock or sand being thrown onto the water or dropped into the ocean (Lessa, 1961: 275).

Yet, the concern today among Chuukese people is that climate change will result in this dynamic world becoming so drastically transformed that it will no longer be habitable. Certainly there was a general consensus among people living on Moch that the high tide and waves were not as they were before. Many men noticed that high tide was getting higher each year and the low tides that exposed the reef happened less often. One man remembered that when he was younger the high tide did not 'go up inland' but that now it was very easy; 'the high tides just wash up inland'. Women also noticed the changes to the tides – that a few years ago the tide was low and you could walk out on the reef, but that now the tide stayed high – and there was agreement that the changes were happening rapidly, that every year the tide was higher and every year it was getting worse.

Additionally, there was agreement within the Mochese community that the land was getting smaller. As one man said, not only was there increasingly more people and more houses on the island, but the land itself was being eroded by the waves. People said that each year they watched the waves take away land, and they were concerned for the loss of significant food and medicinal plants that grow on the shore.

People were worried about the prospect of having to abandon the atoll islands altogether and of having to evacuate to the high islands. While their chief concern was the gradual disappearance of the islands, they were also troubled about the immediate effect of sea-level rise on water resources:

There is only one deep well on the island that is not brackish, but all the deep wells around the shore have been affected by saltwater. The water in the ground itself is salty and this has killed the taro. All the deep wells around the shore are all half seawater and half fresh. They say, 'Don't drink; you drink fresh water catchment.'

Concerns about saltwater inundation extend to the immediate impact this has on food security on atoll islands, especially given the importance of the freshwater lens for the cultivation of swamp taro, the mainstay of subsistence agriculture for island communities. This was recognized by some as the primary concern for the Mochese community:

First we have to think about our survival – food. That's what we have to think about first. Ten years from now breadfruit and taro patch and whatever will be dying out, and after that the island is sinking ... But first thing is the plants, food.

Chuukese government officers we interviewed in 2008 (including the Chuuk State Historic Preservation Officer, the Disaster Coordinator, Federal Emergency Management Authority and the Executive Director, Environmental Protection Authority) conveyed concern about the condition of the freshwater lenses on the more vulnerable outer islands and their capacity to maintain their current populations. They also expressed apprehension about the limited capacity of Chuuk State to handle the socioeconomic and political consequences of resettlement within the state.

Figure 4.2 Land reclamation on Weno, Chuuk Lagoon, FSM. (A black and white version of this figure will appear in some formats. For the colour version, please refer to the plate section.)

The money is not enough [from the Compact of Free Association with the United States of America]. So when we heard about this climate change, we are worried about these outer islands, because we do not have enough [money] … We need ships, we need planes to evacuate the people.

While people on Moch made it clear they do not want to leave their island, there was talk of evacuation that expressed similar concerns. In particular, one man who identified himself as a teacher of climate change criticized the government for not having a plan for the island communities:

It's good for our mayor and our senators to … find the money to start to move the people to somewhere else, to immigrate to the US, to immigrate to Hawaii, Guam, wherever we accepted to immigrate. I think that's the only way. [But] there is no such thing because as I told you, there is no future plan for thirty years on this island. There is no plan.

The Executive Director of the Environmental Protection Authority noted that the high islands were already densely populated and that the migration of increasing numbers of outer islanders to the capital of Weno was causing competition for land to become acute. Thus, it is not surprising to find land reclamation activity along much of the foreshore of Weno (Figure 4.2), a practice that has caused concern because it is thought to place 'communities and infrastructure in positions of increased risk' (Nurse et al., 2014: 1623).

The social situation of the landless in Chuuk is unenviable. As Hezel writes: 'If land was life for traditional Micronesians, then loss of land was a form of death; in Chuuk, according to one author (Parker, 1985] it was lamented with the cry "I am no longer alive"' (Hezel, 2001: 34–5).

In its 1997 Climate Change National Communication, the FSM National Government acknowledges that 'land ownership is the most valued right in Micronesia; the landless person has much lower status than the landowner' (FSM, 1997). Thus, land is valued not only for its economic use-value but also for the social status and political power that derives from its ownership. Moreover, the concept of personhood, what it means to be *fully human*, is tied to one's association with land, as are the very qualities of a person. For example, the lagoon island of Tonoas is said to be a reposing man and people living in the district around the highest peak of the island, which represents the head of the man, are said to have large heads through scheming and 'planning evil deeds' (Young et al., 1997: 16).

Traditionally, a complex system of social security based on ceremonial exchange networks, kinship practices and complex land tenure systems assured rights to and association with land for most people in this region of small islands. Land is, in principle, inherited matrilineally but patrilineal succession is recognized, with men being able to give land to their sons while they are still alive. Claims can arise from living on land, cultivating it, having ancestors born or buried on it and so on. People can have interest in many plots or trees in different locations. Cultivable land is divided into named tracts owned by different lineages, as are the offshore waters and reefs (Murdock and Goodenough, 1947).

However, such cultural practices may not be able to keep up with and/or sustain the large population shifts that have occurred in Micronesia over the past forty years as, for example, ethnographically documented by Marshall (2004) on the island of Namoluk in the Mortlock Islands. While Nurse et al. (2014: 1625) note that there is currently 'no unequivocal evidence that reveals migration from islands is being driven by anthropogenic climate change', leaders we interviewed on Moch Island considered that migration may well be inevitable and that eventually the whole island may have to be evacuated. Chuukese government officers in Weno expressed concern over the great pressure that migration already exerts on state resources and particularly on infrastructure in the high islands of the lagoon. Outer islanders in the capital were worried about their ability to support their kin should it be necessary to evacuate them from their home islands: 'See, it's a really sad situation. If they come in, the money that we have is not enough to share around. So they have to stay out there and we send them what we can.'

Outer island respondents living in Weno expressed a strong connection to their home island. While they feared that evacuation might eventually be necessary, they stated that it was important to maintain their families on their home islands in order that these places, so significant to their identities, would continue to be cared for and nurtured. As Marshall notes in relation to the people of Namoluk, fewer than 40 per cent of whom live on the island itself, individuals continue to derive their personal identities 'from a collective community identity that is rooted in place: their island home. They carry Namoluk with them when they move – wherever they move' (2004: 134).

This is the case also for the Mochese community, which prioritizes sharing as 'a way of living' and as a means through which to activate a network of relationships that extend well beyond the home island. For example, our research evidences the profusion of energy invested by people within the community, both on Moch and Weno, to sustain their connections through the ongoing sharing of resources. Certainly the boat that travelled between Moch and Weno always had people on-board taking local foods to relatives living in Weno, and likewise the boat from Weno to Moch was loaded with processed foods, and with building materials, fuel, letters from loved ones and parcels from relatives living in the USA. Indeed, while much is written about the 'dependence' of small island communities on support from off-island relatives (Connell, 2010: 122–3; Birk, 2012: 89), it is the intentional activities of sharing – *to and from* the home/island – that establishes identity and a way for belonging.

Providing support for kin on the home islands is not only important in terms of identity but also for more practical reasons. If coral islands are not inhabited, if people do not dwell there, then these places may become even more vulnerable. For example, Rainbird (2004: 171–2) attributes recent coastal transgression in some areas in the Chuuk Lagoon to the fact that, following adoption of a western cash economy, 'coastal lowlands are not being maintained and consequently the sea is reclaiming the space it had enjoyed prior to human intervention' (Rainbird, 2004: 171).

Similarly, a preliminary report to the Conference of the Parties for the Convention on Biological Diversity by the Federated States of Micronesia (FSM, 2001: 17) notes that 'urbanisation removes people from day to day contact with natural resources so that there is no longer a feedback system between people and their environment that could alert people to problems'.

Pam's research reveals an intense relationship between people and a coral landscape, a relationship that is realized through practices that maintain the home island as a successfully and intensely inhabited place – as a place worthy of attachment and belonging. These practices of engagement with coral included attending to taro pits and seawalls, weeding the main path, spreading clean white coral around buildings and keeping socialized areas open and free from overgrowth and debris (see also Besnier, 2009).

Knowledge for Orienting to Climate Change

In 2001, a Chuukese task force appointed to report on the impact of 'unusual sea-level rise and its adverse effects' recommended that in order to adapt to global climate change it is important to:

Re-orient ourselves with our own traditions and cultural values to be the driving force on how we counter this onslaught of events. Our main problem stems from the loss of our values and the old ways whereas we take great pride in the things that we planted and reap from the ground as opposed to buying and relying entirely on cash (money) economy.

(Billimont et al., 2001)

The IPCC has considered the significance and relevance of indigenous knowledge for dealing with climate change. For example, in its contribution to the Fourth Assessment Report of the IPCC, Working Group II provides case studies on 'Indigenous knowledge for adaptation to climate change' in the report on small islands (Parry et al., 2007). However, Mimura et al. (2007: 712) in the same report note that while the use of traditional knowledge as a means of adaptation to climate change has been advocated in relation to small island states, further research on whether such knowledge can realistically enhance 'adaptive capacity and resilience' is required, given the dire future scenarios predicted:

> With respect to technical measures, countries may wish to pay closer attention to the traditional technologies and skills that have allowed island communities to cope successfully with climate variability in the past. However, as it is uncertain whether the traditional technologies and skills are sufficient to reduce the adverse consequences of climate change, these may need to be combined with modern knowledge and technologies, where appropriate.
>
> (Mimura et al., 2007: 712)

The relevance of indigenous knowledge has been further discussed in the IPCC Fifth Assessment Report, where it is concluded that:

> As in previous IPCC assessments, there is continuing strong support for the incorporation of indigenous knowledge into adaptation planning. However, this is moderated by the recognition that current practices alone may not be adequate to cope with future climate extremes or trend changes. The ability of a small island population to deal with current climate risks may be positively correlated with the ability to adapt to future climate change, but evidence confirming this remains limited.
>
> (Nurse et al., 2014: 1636)

Certainly, climate change scenarios paint a dismal picture for small Pacific Island states. However, one of the problems with such representations is that they serve to erase recognition of 'the agency, resourcefulness and resilience' of Pacific Islanders (Farbotko, 2005). As Barnett and Adger (2003: 333) note, 'the challenge is to understand the adaptation strategies that have been adopted in the past and which may be relevant for the future in these societies'.

In relation to this, Rainbird (2004: 94, 171) presents archaeological evidence of past human transformations of the seascape and of practices of protection and maintenance in Micronesia. Rainbird argues that the first human settlers

> altered the very nature of the landscape, by manipulating the vegetation so as to cause erosion and thereby lay the foundations for the subsistence systems … This approach to the landscape by the initial settlers would be responsible for creating conditions of high sediment transport and the progradation of the shoreline onto the reef flats.
>
> (Rainbird, 2004: 95).

Atoll islands were purposefully cultivated into rich dwelling places. This is evidenced by the central taro patch typical of atoll islands (Figure 4.3), developed, mulched and cared for over centuries (Rainbird, 2004: 163). A workshop in the Mortlock Islands in 2006 documented practical knowledge and cultivation of thirty-two varieties of swamp taro.

Figure 4.3 Healthy taro patch. Moch Island, Chuuk, FSM. (A black and white version of this figure will appear in some formats. For the colour version, please refer to the plate section.)

While different islands are known for having different cultivars, one man on the island of Ta, where there has been damage to taro due to salinity, like on the island of Moch (Figure 4.4), grew up to twenty cultivars. At the workshop, salt-resistant cultivars were proposed as a means of countering the effects of sea-level rise (Levendusky et al., 2006).

In their contribution to the Fifth IPCC report, Adger et al. (2014: 766) review recent anthropological studies. While many studies argue that 'mutual integration and co-production of local and traditional and scientific knowledge increase adaptive capacity and reduce vulnerability' and that local and traditional knowledge 'contributes to mitigating the impact of natural disasters', 'maintaining domestic biodiversity' and 'developing sustainable adaptation and mitigation strategies', other studies point out the limitations of indigenous knowledge for dealing with climate change.

Our research on Moch and in Weno reveals that Mochese people strive to orient themselves to climate change in practical ways by drawing on their own empirical observations, lived experiences and deep connection with their island home. Knowledge is not something fixed, but dynamic and changing. Moreover, it is enacted in terms of place-based practices of engagement with other people and things of the world. Thus we argue for a concept of indigenous knowledge that goes beyond the realm of ideas, to practices of relatedness

Figure 4.4 Saltwater where there was once a taro patch, Moch Island, Chuuk, FSM. (A black and white version of this figure will appear in some formats. For the colour version, please refer to the plate section.)

that foster social security (see also Henry and Jeffery, 2008). In other words, knowledge does not exist in a vacuum, but thrives or withers as part of a wider political economy of principles and practices.

For example, one of the strategies that people often used in the past to deal with extreme climatic events was to invoke the principle of reciprocity by turning to their kinship and exchange networks across the region. According to Rubenstein (2001: 75): 'Micronesian island communities accommodated to climate extremes and natural disasters through the development of social and political linkages between the more vulnerable coral atolls and the neighbouring high islands.' In some parts these linkages became institutionalized in terms of formal ceremonial exchange systems, such as the *sawai* system in the western Carolinian islands (Darcy, 2006). While travel and/or migration were common practices among Micronesians in the past, emigration from the Federated States of Micronesia became a significant social phenomenon after the implementation of the Compact of Free Association in 1986. Thousands of FSM citizens left for Guam, the Commonwealth of the Northern Mariana Islands, Hawaii and the mainland of the United States. The explanation for this was primarily employment and education (Hezel and Lewin, 1996; Marshall, 2004). They did not conceive of themselves as leaving their home islands permanently.

They maintained close social bonds through exchange with their kin so that they could return home if and whenever they wanted:

> Goods flow back and forth between the home islands and the new communities as freely as people. A few years ago, Chuukese would send fish and pounded breadfruit to their relatives on Guam in ice chests that would be returned a few days later, filled with frozen chicken and other treats that could be bought cheaply on Guam. Chuukese on Guam would also send back cartons of secondhand clothes ... Goods are exchanged between migrants and their relatives back home just as they would be if all were living on the same island.
>
> (Hezel, 2001: 153).

Thus, keeping exchange paths active might be considered an adaptive strategy in the face of climate change (see also Barnett, 2001). Because kinship relations and exchange networks themselves provide a means of mitigating impacts of climate events, it is important to understand the way these networks of connection operate and the political and economic contexts under which they will be able to continue to flourish. There has been much recent anthropological research conducted on migration, transnationals, diasporas and global flows of people and goods, which is helpful for understanding the kinds of *social capital* that people might employ to cope with climate change. Such social flows are linked to a cultural seascape constructed in terms of movement, so that rather than being perceived as a barrier, the sea is conceptualized as a 'way' or 'pathway' of connection (Hau'ofa, 1994; Rainbird, 2004; Darcy, 2006).

Conclusion

Much socioeconomic modelling has been undertaken to try and estimate and quantify the damage from climate change, and 'climate anthropologists appear to be making strides at relating global warming models to everyday lives' (Brown, 1999), yet little research has been conducted to date on the creative human capacity to respond to and mitigate its effects.

In Chuuk, as elsewhere in Micronesia, strong ties to land, understood to include sea, are a vital aspect of local cosmologies. In the past, people dealt with environmental change by responding resourcefully through their land tenure regimes, economic and political institutions, and exchange networks. Our research with Mochese people reveals that they continue to do so today.

Indigenous knowledge research is crucial in order to understand how people perceive risk and orient themselves towards change (Pam and Henry, 2012). However, such research commands an expanded definition of what constitutes knowledge, a definition that includes cosmologies, or worldviews, and practical modes of social organization, governance and management.

Understanding human resourcefulness and resilience in the face of climate change requires fine-grained ethnographies on the political economy of knowledge. It requires a holistic approach that involves consideration of the dynamic social, political and economic relations that constitute knowledge *in practice*.

Acknowledgements

The pilot study in Chuuk was conducted in association with an Earthwatch project run by James Cook University maritime archaeologist, William Jeffery. We acknowledge the assistance and support of Bill Jeffery and the staff at the Chuuk Historic Preservation Office (HPO) for this initial study. We especially thank Mr Doropio Marar, research officer (HPO), who introduced us to his home island of Moch. We are deeply grateful to the people of Moch for their hospitality and their generous invitation to return for the longer-term ethnographic research project conducted by Christine Pam. We acknowledge the financial support of James Cook University and the Australian Research Council postgraduate scholarship that enabled Christine Pam's doctoral project.

References

Adger, W. N., Hughes, T. P., Folke, C., Carpenter, S. R. and Rockström, J. 2005. Social-ecological resilience to coastal disasters. *Science*, 309: 1036–9.

Adger, W. N., Paavola, J. and Huq, S. 2006. Toward justice in adaptation to climate change. In Adger, W. N., Paavola, J., Huq, S. and Mace, M. J. (eds.) *Fairness in Adaptation to Climate Change*. Cambridge, MA: MIT Press, pp. 1–19.

Adger, W. N., Pulhin, J. M., Barnett, J., et al. 2014. Human security. In Field, C. B., Barros, V. R., Dokken, D. J. et al. (eds.) *Climate Change 2014: Impacts, Adaptation, and Vulnerability. Part A: Global and Sectoral Aspects. Contribution of Working Group II to the Fifth Assessment Report of the Intergovernmental Panel on Climate Change*. Cambridge, UK and New York: Cambridge University Press, pp. 755–91.

Agrawal, A. 1995. Dismantling the divide between indigenous and scientific knowledge. *Development and Change*, 26(3): 413–39.

Barnett, J. 2001. Adapting to climate change in Pacific island countries: The problem of uncertainty. *World Development*, 29(6): 977–93.

Barnett, J. and Adger, W. N. 2003. Climate dangers and atoll countries. *Climatic Change*, 61: 321–37.

Barnett, J. and Busse, M. 2001. *Ethnographic Perspectives on Resilience to Climate Variability in Pacific Island Countries. (APN Project Ref: 2001–11). Final activity report*, www.apngcr.org/resources/files/original/be5630282fa16cb2ab1d33822f6dd185.pdf

Besnier, N. 2009. *Gossip and the Everyday Production of Politics*. Honolulu, Hawai: University of Hawai'i Press.

Billimont, B., Penno, I. and Osiena, R. 2001. *'La Nina' Task Force Special Report to Office of the Governor, State of Chuuk, Federated States of Micronesia*. Unpublished manuscript.

Birk, T. 2012. Relocation of reef and atoll island communities as an adaptation to climate change: learning from experience in Solomon Islands. In Hastrup, K. and Olwig, K. F. (eds.) *Climate Change and Human Mobility: Global Challenges to the Social Sciences*. Cambridge, UK: Cambridge University Press.

Brown, K. S. 1999. Taking global warming to the people. *Science*, 283(5407):1440–1.

Connell, J. 2003. Losing ground? Tuvalu, the greenhouse effect and the garbage can. *Asia Pacific Viewpoint*, 44(2): 89–107.

Connell, J. 2010. Pacific Islands in the global economy: Paradoxes of migration and culture. *Singapore Journal of Tropical Geography*, 31: 115–29.

Darcy, P. 2006. *The People of the Sea: Environment, Identity, and History in Oceania*. Honolulu, Hawai: University of Hawai'i Press.

Dove, M. R. 2006. Indigenous people and environmental politics. *Annual Review of Anthropology*, 35: 191–208.

Ellen, R., Parkes, P. and Bicker, A. 2000. *Indigenous Environmental Knowledge and its Transformations: Critical Anthropological Perspectives*. Amsterdam: Harwood Academic Publishers.

Farbotko, C. 2005. Tuvalu and climate change: Constructions of environmental displacement in the Sydney Morning Herald. *Geografiska Annaler*, 87B(4): 279–93.

FSM (Federated States of Micronesia). 1997. *Climate Change National Communication: Federated States of Micronesia*. Pohnpei, FSM, Climate Change Program, http://unfccc.int/resource/docs/natc/micnc1.pdf

FSM. 2001. *Preliminary Report to the Conference of the Parties of the Convention on Biological Diversity, May 2001*. Ponhpei, FSM, Biological Diversity.

Goodenough, W. H. 1986. Sky world and this world: The place of Kachaw in Micronesian cosmology. *American Anthropologist*, 88(3): 551–68.

Hau'ofa, E. 1994. Our Sea of Islands. *The Contemporary Pacific* 6(1): 148–61.

Henry, R. and Jeffery, W. 2008. Waterworld: The heritage dimensions of climate change in the Pacific. *Historic Environment*, 21(1): 12–18.

Henry, R., Jeffery, W. and Pam, C. 2008. *Heritage and Climate Change in Micronesia: A Report on a Pilot Study Conducted on Moch Island, Mortlock Islands, Chuuk, Federated States of Micronesia. January, 2008*. Townsville, AU: James Cook University, www.islandvulnerability.org/docs/henryetal.2008.pdf

Hezel, F. X. and Levin, M. J. 1996. New trends in Micronesian migration: FSM migration to Guam and the Marianas. *Micronesian Counselor*, 19, www.micsem.org/pubs/counselor/frames/micmigfr.htm

Hezel, F. X. 2001. *The New Shape of Old Island Cultures: A Half Century of Social Change in Micronesia*. Honolulu, Hawai, University of Hawai'i Press.

Kempf, W. 2009. A sea of environmental refugees? Oceania in an age of climate change. In Hermann, E., Klenke, K. and Dickhardt, M. (eds.) *Form, Macht Differenz: Motive und Felder ethnologischen Forschens*, Göttingen: Universitätsverlag Göttingen.

Lessa, W. A. 1961. *Tales from Ulithi Atoll: A Comparative Study in Oceanic Folklore*. Berkeley, CA: University of California Press.

Lessa, W. A. 1980. *More Tales from Ulithi Atoll: A Content Analysis*. Berkeley, CA: University of California Press.

Levendusky, A., Englberger, L., Teelander, R. et al. 2006. *Documentation of Mortlockese Giant Swamp Taro Cultivars and Other Local Foods on Ta, Moch, and Satowan, May 2006*. Kolonia, Pohnpei: Island Food Community of Pohnpei.

Marshall, M. 1979. Natural and unnatural disaster in the Mortlock Islands of Micronesia. *Human Organization*, 38(3): 265–72.

Marshall, M. 2004. *Namoluk Beyond the Reef: The Transformation of a Micronesian Community*. Boulder, CO: Westview Press.

McCarthy, J. J., Canziani, O. F., Leary, N. A., Dokken, D. J. and White, K. S. (eds.) 2001. *Climate Change 2001: Impacts, Adaptation & Vulnerability, Contribution of Working Group II to the Third Assessment Report of the Intergovernmental Panel on Climate Change*. Cambridge: Cambridge University Press.

Merlan, F. 2009. Indigeneity: Global and local. *Current Anthropology*, 50(3): 303–33.

Mimura, N., Nurse, L., McClean, R. F. et al. 2007. Small islands. In Parry, M. L., Canziani, O. F., Palutikof, J. P., van der Linden, P. J. and Hanson, C. E. (eds.) *Climate Change 2007: Impacts, Adaptation and Vulnerability. Contribution of Working Group II to the Fourth Assessment Report of the Intergovernmental Panel on Climate Change*. Cambridge: Cambridge University Press, pp. 687–716.

Mitchell, R. E. 1973. The folktales of Micronesia. *Asian Folklore Studies*, 32: 1–276.

Murdock, G. P. and Goodenough, W. H. 1947. Social organization of Truk. *Southwestern Journal of Anthropology*, 3(4): 331–43.

Nurse, L. A., McLean, R. F., Agard, J. et al. 2014. Small islands. In Barros, V. R., Field, C. B., Dokken, D. J. et al. (eds.) *Climate Change 2014: Impacts, Adaptation, and Vulnerability. Part*

B: *Regional Aspects. Contribution of Working Group II to the Fifth Assessment Report of the Intergovernmental Panel on Climate Change,* Cambridge, UK and New York: Cambridge University Press, pp. 1613–1654.

Pam, C. and Henry, R. 2012. Risky places: Climate change discourse and the transformation of place on Moch (Federated States of Micronesia). *Shima*, 6(1): 30–47.

Parry, M. L., Canzani, O. F., Palutikof, J. P., van der Linden, P. J. and Hanson, C. E. 2007. Indigenous knowledge for adaptation to climate change. In Parry, M. L., Canziani, O. F., Palutikof, J. P., van der Linden. P. J., and Hanson, C. E. (eds.) *Climate Change 2007: Impacts, Adaptation and Vulnerability. Contribution of Working Group II to the Fourth Assessment Report of the Intergovernmental Panel on Climate Change.* Cambridge: Cambridge University Press, pp. 843–68.

Pittock, B. A. 2005. *Climate Change: Turning up the Heat.* Collingwood, Victoria: CSIRO.

Rainbird, P. 2004. *The Archaeology of Micronesia.* Cambridge: Cambridge University Press.

Rubenstein, D. 2001. Climate change and relations between local communities and larger political structures in the Federated States of Micronesia. In Barnett, J. and Busse, M. (eds.) *Ethnographic Perspectives on Resilience to Climate Variability in Pacific Island Countries. (APN Project Ref: 2001–11) Final activity report*, pp. 75–6.

Rudiak-Gould, P. 2010. *The Fallen Palm: Climate Change and Culture Change in the Marshall Islands.* Saarbrücken: VDM Verlag Dr. Müller.

Rudiak-Gould, P. 2011. Climate change and anthropology: The importance of reception studies. *Anthropology Today*, 27(2): 9–12.

Spennemann, D. 2007. *Melimel: The Good Friday Typhoon of 1907 and its Aftermath in the Mortlocks, Caroline Islands.* Albury, NSW: HeritageFutures International.

Torry, W. I. 1979. Anthropological studies in hazardous environments: past trends and new horizons. *Current Anthropology*, 20(3): 517–40.

Young, J. A., Rosenberger, N. R. and Harding, J. R. 1997. *Truk Ethnography. Micronesian Resources Study.* San Francisco, CA: National Park Service.

5

Local Responses to Variability and Climate Change by Zoque Indigenous Communities in Chiapas, Mexico

María Silvia Sánchez-Cortés and Elena Lazos Chavero

The anthropological study of climate in indigenous communities initially focused on ethnoclimatology and ethnometeorology. This included understanding the cultural relationship that exists with climate, and its symbolic processes in daily experiences and practices (Katz and Lammel, 2008; Ulloa, 2011). Another approach seeks to understand knowledge of climate and climate variability in relation to the management of ecosystems and agrosystems (Toledo and Barrera-Bassols, 2008). In recent years, and in the context of anthropogenic climate change, there is concern and interest in research on the loss of traditional knowledge due to the relatively rapid climate changes occurring in ecosystems located in indigenous territories (Kronik and Verner, 2010). There is also interest in understanding the vulnerability and risk faced by indigenous groups due to the impacts of climate change, and their perceptions and responses (Katz et al., 2008; Ulloa, 2011). In relation to this last point, it is argued that climate models do not provide specific information for adapting to climate change at the local level (Oreskes et al., 2010), therefore a microlocal understanding of global climate phenomena and their local effects is extremely important (Magistro and Roncoli, 2001; Ulloa, 2013) in order to account for possible adaptation responses (Pelling, 2011).

Local indigenous knowledge allows for the documentation of experiences, practices, symbols and cultural meanings of climate in relation to the observed climate change. From this perspective, and with regard to studies carried out on indigenous groups and climate change, different research perspectives have been developed in recent years. They focus on (1) local adaptation responses; (2) climate change, gender and territories (Ulloa, 2011); and (3) studies on climate predictions and observation of climate change as elements of discussion about local and scientific knowledge.

In Mexico, systematic studies on indigenous peoples' perception of climate change are still scarce, especially those that seek to understand the responses and adaptations to perceived changes in local climate. The work presented in this chapter is based on data obtained between June 2004 and September 2009 through research on the perception of environmental change in two indigenous Zoque communities in the Chiapas state of Mexico. Located in San Pablo Huacanó and Copoya, the two communities differ in their local and environmental histories, geographical location and the language used on a daily basis. In one of them, both Zoque and Spanish are used while in the other community, the Zoque language is not spoken.

When we began our fieldwork, we asked the Zoque people about perceived environmental changes. Respondents frequently mentioned changes in local climate. They pointed out the increasing temperature and decreasing rainfall as well as their unpredictability. These responses led us to consider the perception of change in local climate as part of our research on environmental change. Here we provide a brief description of the knowledge and responses about weather and climate from farmers in two Zoque communities in Chiapas. This information complements those reported in Sánchez-Cortés and Lazos Chavero (2011, 2013) which are based on 120 semi-structured interviews conducted to understand the perception of climate change in both communities.

This chapter proposes that it is important to consider local climate calendars in studies on the perceptions of climate change (Vedwan and Rhoades, 2001), weather expectations (Rebetez, 1996; Orlove et al., 2002; Roncoli, 2006), and documenting experience with climate and its variability (Katz et al., 2008; Orlove et al., 2011; Ulloa, 2011). This is in addition to situating perception within the historical, economic, cultural and environmental contexts of each community, since climate change is not perceived in isolation and the response to it is gradual and in tune with the contexts of lived experience.

Area of Research

Our research was carried out in two Zoque villages in Chiapas: San Pablo Huacanó in the municipality of Ocotepec, and Copoya in the municipality of Tuxtla Gutiérrez (Figure 5.1). Although the location and natural environments of these communities are different, in both villages agriculture is exclusively dependent on rainwater, and therefore on climate.

San Pablo Huacanó is located in the northern mountains of Chiapas (17° 11' N and 93°12' W, 1,630 m). The topography is mountainous, the climate is semi-warm, semi-humid, with summer rains (A(C)w 1(w)) and the predominant vegetation type is montane cloud forest (Rzedowski, 1978; INEGI, 2006). The village and suburbs (*ejido*) of Ocotepec (59.6 km^2) include thirty-four rural localities characterized by a high degree of marginalization and poverty. In 2010, the population of San Pablo Huacanó was 1,427: 728 men and 699 women (INEGI, 2010), the majority being bilingual in Zoque and Spanish. The members of the *ejido* belong to the city of Ocotepec, which has an area of common use of 7,786 acres. The seventh ejidal census (INEGI, 2001) registered 566 members with individual land plots of which sixty are from San Pablo Huacanó. Access to land and extension of the plots in the *ejido* is uneven. People who do not have enough land or who do not own land must rent it. Land plots are near the town and some are located in different parts of the *ejido* at different altitudes. The Zoque people from San Pablo Huacanó produce maize (*Zea mays L.*) and bean (*Phaseolus vulgaris L.*) in their mountainous territory. Some people grow coffee (*Coffea arabica L.*). Most of the population is Catholic and a few families are Protestant.

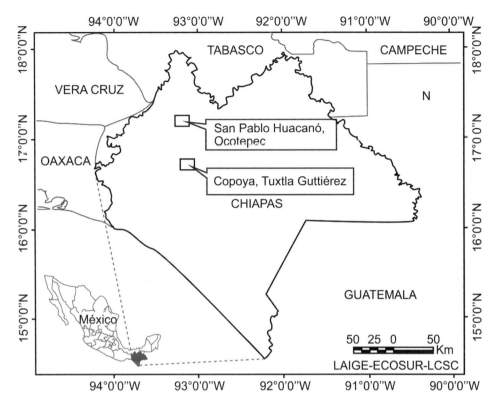

Figure 5.1 Location of study areas: San Pablo Huacanó in the municipality of Ocotepec and Copoya in the municipality of Tuxtla Gutiérrez.

While Copoya is a Zoque community, the Zoque language is no longer spoken there. It is located 5 km away from the state capital (16° 43' N and 93° 07' W, 860 msnm), in the Central Depression of Chiapas (Figure 5.1). The climate is warm and subhumid with rainfall in the summer (A (w1)) and average humidity (INEGI, 2003). The predominant vegetation type is tropical deciduous forest with forest oaks in the higher parts (Rzedowski, 1978). Of the 8,160 inhabitants, 4,021 are men and 4,139 are women (INEGI, 2010). The poverty rate is low (CONEVAL, 2009). In the Copoya *ejido* there are two areas of human settlement: the community of Copoya and Jobo. The *ejido* measures 2,278 acres, and the seventh ejidal census registered 430 members owning individual plots. Copoya is home to 270 of them but they do not all depend on subsistence agriculture. There are currently no farmers under thirty and the dominant religion is Catholicism.

Both communities are of Zoque origin and they share a Mesoamerican heritage that includes *milpa* agriculture that produces maize, bean and squash. In the narratives of both communities, some meteorological elements are present such as lightning. Snakes also feature in their narratives and these are linked to water and the lunar influence on crops. While Zoque communities share a common linguistic heritage, there are still more differences than similarities between them. This is attributed in part to their different histories. In the next section, we describe the results we obtained in both communities on knowledge of local climate and responses to perceived changes.

Cultural Knowledge of Climate

In San Pablo Huacanó, there are still elements of a climatic ethnocalendar similar to the one described for the Zoque people of Ocotepec (Báez-Jorge, 1983). Thus Sánchez-Cortés and Lazos (2011) identified several of these ethnoclimatic elements in San Pablo Huacanó related to the names of the seasons and climatic events. This calendar is linked to peoples' expectations about the nature and sequence on climatic events that include the types of topography, soils, microclimates, altitudes and crop varieties that develop in places. Of the sixty people interviewed in San Pablo Huacanó, few of them made clear reference to the ethnoclimatic calendar but some people did mention the names of the seasons (*Ningo poya* – warm period, *Tucsawa poya* – rainy weather, *Pacak poya* – cold period). In the Zoque community of Copoya there was no mention of the ethnocalendar. However, their knowledge of climate is related to their agricultural experiences, the territory, the location of their plots, the kinds of soils and predominant winds. Farmers distinguish between the dry season and the rainy season, characterized by rain showers and storms (continuous but not very heavy rainfall).

Climate Expectations

According to their knowledge of the characteristics of climate in the territory and their own forecasts, farmers calculate where, when, how much and what to plant. In both Zoque communities, we found an important relationship between the anticipated climate and the local climate. For example, if the climate is considered to be rainy in the summer, expectations are oriented towards expecting torrential rains in that period. This aspect is important because of its relationship with the perception of change in local climate. Thus different respondents first situated their referents or expected weather in this sense, and then pointed out perceived changes such as the unpredictability of rainfall or its decline. The same occurs with temperature, which is said to have increased in both warm and cold months.

Weather Forecasting

In the communities examined, forecasts are linked to the anticipated weather. The Zoque people rely on indicators in the sky, on plants or on the behaviour of wild and domestic animals. Forecasts can be long term, which corresponds to the months in which rainfall is expected, as well as short term, when forecasts refer to the same day, coming days or seasonal changes (with or without rain, storms or *nortes* corresponding to several days of rain).

Among the indicators mentioned for long-term weather forecasts in San Pablo Huacanó, there are those that relate to the sky such as the presence of lightning in the month of March[1] and its location on the horizon. In turn, other farmers use *cabañuelas* (long-term forecasts) to guide their choice of dates and places for planting. For short-term weather forecasting, the moon is observed as well as its phases. During full moon, 'rainfall stops' and when there is a new moon, it 'brings rain' (Báez-Jorge et al., 1985). Other indicators are the presence and types of clouds, the direction and intensity of the wind and whether or not it is cold. On land, animal behaviour is observed such as that of ants or the singing of birds, announcing the next rainfall or a change of season. This is the case of the *nortes*, which in San Pablo Huacanó refer to rainfall with low intensity that may persist for several days and usually occur from December to January. Thus, it is noted,

elderly people used to speak of the weather cycle (climate ethnocalendar). For their planting during the month of March, they used to notice whether a lot of birds nested in the same place. This would help them determine where the wind was going to pass and therefore they sought more protected places to plant.

In Copoya, long-term weather forecasting is mainly done through *cabañuelas* that reveal during the first days of January the characteristics of expected weather conditions throughout the different months of the coming year. Then this prediction is linked to the rest of the months. 'In the new year, elderly people paid attention to the cabañuelas on the first day in January; then on the following day for February; on the day after that for March, up until December. They paid attention to the twelve days.' Another, less widespread belief is to plant maize in a large wooden container (a canoe) from first to eight December, in order to interpret what kind of year will follow.

A division was made on the canoe to (represent) the slopes of the terrain and to see the results that would be expected when the milpa emerges. Sometimes the planting (in the canoe) shows that when it will rain a lot, it becomes yellow, but if it bends it means there will be a lot of wind.

The predictions of annual climate conditions can also be guided by the phenological behaviour of particular plants such as the totoposte (*Licania arborea Seem*), lantá (*Pseudobombax ellipticum (H.B.K.) Dugand*), candox (*Tecoma stans (L.) H.B.K.*) and the reed. When they bloom more than they normally would, it means that the following year will be good for planting as far as rainfall is concerned (Sánchez-Cortés, 2011).

In the short term the behaviour of wild and domestic animals is also very important for immediate weather predictions or in those that indicate the start of a rainy season. The presence of crowns or halos around the moon or the sun signifies the presence of rain during the day or that it will stop raining. The singing of specific birds such as the chachalaca (*Ortalis vetula*), mockingbird, quichi or quail and the cincocó announce that rain is coming.

---

[1] In Nahua communities in Veracruz such as Petlacala, Villela (2010) reports a similar event when people point out that 'the apparition of a lightning bolt on the East and the North ... are signs of good omen: if the sky is not painted with lightning there will be bad rainy weather. Among the Nahua of Atzacoaloya, if the sky is cloudy or if it rains, or if there is radiance on the Northwest, it will be a good year. If the sunlight appears on a different cardinal point, it will be a bad year: there will be grim rain, plagues will cover the land and maize will not be harvested.'

The behaviour of cows, donkeys and pigs is also significant. 'It's as if they become happy because they know it will rain … in the morning my little pig was carrying branches and was running around, that's a sign that it is going to rain.' The ant locally known as *ronda* and birds such as the *northern lama* are related to the start and the end of the rainy season, respectively (Sánchez-Cortés, 2011). As in San Pablo Huacanó, this knowledge about weather predictions is expressed differently by part of the population. In both communities the heat wave refers to a hot month without rain, which occurs from late July through August.

Requests for Rain

In San Pablo Huacanó, there was no mention of the existence of experts in requesting rain. 'There is no one in particular who prays. Each farmer does it in his plot when he is going to plant. They ask God and the Holy Earth for a good harvest.' In Copoya, it was said that in the past, the elderly used to ask for rain in a specific cave in Tuxtla Gutiérrez. They would give the owner of the hill flowers, candles, trees and drums (Sánchez-Cortés, 2011). Among the Zoque, caves are associated with the underworld just as in other Mesoamerican cultures (Katz and Lammel, 2008). At present, farmers walk outdoors through Virgencitas de Copoya to request rain by way of a pilgrimage through the town.

Agricultural Experience and Changes in Local Climate

The scenarios obtained through climatic modelling demonstrate an increase in rainfall in Ocotepec (Landa et al., 2008). In contrast, no significant increases in temperature are expected for Tuxtla Gutiérrez (Landa et al., 2011). However, local farmers and in particular the Zoque of San Pablo Huacanó in the municipality of Ocotepec describe changes in the planting season due to the loss of soil humidity and to an increase in temperature during warm and wintry months. The farming responses given have been to modify the planting season, especially for long-cycle varieties of cold environments. They have also suggested to introduce some crops like squash and coffee for local consumption in response to the warmer climate. Given the decline and lack of plots and also to obtain two harvests of maize in a year, farmers seek to rent the land as much as they can, to continue altitudinal use of the land (Sánchez-Cortés and Lazos Chavero, 2011).

In Copoya, in the municipality of Tuxtla Gutiérrez, the perception indicates an increase in temperature as well as the unpredictability and reduction of rainfall. In Copoya, no specific responses to local climate change were observed. Traditionally, farmers try to manage short- and long-cycle varieties of maize to overcome the heat wave and face droughts and days with no rain. Planting time is adjusted to the agricultural calendar and to farmers' expectations about the climate according to this calendar.

Perceptions of change in climate indicators point to the absence of birds in both communities. In San Pablo Huacanó, this is seen in the decreasing amount of clouds in the mountains or the presence of plants from warmer places (herbs) in addition to the increase in weeds and insects in different parts where maize and beans are grown. The situation of clouds is attributed mainly to the loss of vegetation. Biological indicators also correspond to their crops, thus the introduction of coffee is significant for perception. They explain that now it can be planted in the town, whereas it was not planted there before. The explanation of this change is cultural. The Earth 'became warm' after the eruption of the volcano El Chichón in the 1980s (Sánchez-Cortés and Lazos Chavero, 2011). Meanwhile, some members of Copoya believe that the use of 'liquids' to spray maize and beans explains the loss of biological indicators. These have an impact, which kills grasses and will end up finishing with birds. In turn, some people believe that animals have gone elsewhere, far away or that this obeys the will of God.

Dynamics in Land Tenure and Its Relation to Climate Change

In both communities, different interviewees mentioned that in the beginning of the twentieth century and up until the 1970s, the use of farmland was determined according to family needs. Thus, the farmer decided what to plant according to how much he could work, tend to and harvest. Later, with an increase in population they could no longer work where they wanted in terms of the location and surface used for farming. In this regard, Warman (2002) notes that starting in 1940 and internally, leasing and the sale of land among members was tolerated, and heritage was fragmented, which gave place to small-scale farming (and in the communities studied, there was a dispersed distribution of plots). Landowners sought to secure their land by inheritance or internal buyout. With the amendment to article 27 in 1992, landowners had the opportunity to capitalize on land by purchase or sale, which facilitated further changes in ownership and options for use of the land, having different impacts on natural and agricultural spaces.

In the case of Copoya, the process that followed land tenure headed towards its fragmentation into small plots as well as establishing boundaries with physical barriers, and especially plots were sold due to the proximity of the village to the capital of the State of Chiapas. Currently there are fewer persons engaged in agriculture, and those who are still engaged say that their production is meant for subsistence in the absence of external subsidies.

In San Pablo Huacanó, plots are increasingly fragmented for growing corn and members who mostly don't live in the village have also used them for livestock. Plots located in a warm climate and with access to water are in high demand. This situation led to a break in the traditional use of the Zoque territory, which allowed farmers to obtain two harvests each year. The response has been to seek plots in different microclimates through borrowing or leasing. In this case changes in land tenure, environmental changes (loss of vegetation and soil owing to erosion and fertility) and climate change are linked. While immediate climate response has been to change the planting season, this is articulated by the scarcity of land and the conditions of vulnerability of the Zoque in San Pablo Huacanó. Considering the limited access to capital, or direct inheritance of the land, members of the community have deployed mobility strategies in their territory to have access to plots. This includes an altitudinal use of the land by renting and lending land. In this sense it is necessary to acknowledge and further study historical and political factors as well as power relations and institutional aspects that influence adaptation responses and resilience in face of climate change. The answers go beyond the cause and effect diagram; they are intertwined in a socioenvironmental scenario where vulnerability is not entirely climatic since it relates with the political, economic and cultural contexts. In addition, the impacts have different effects if one has access to land or not (possession, inheritance, rent), if one is young or an adult, if one is a single woman or widow, or if one has ejidal (community) rights to the land or not, allowing access to institutional support.

Conclusion

In studies on the perception of climate change, it is suggested to consider local climate calendars and knowledge related to weather to facilitate the demarcation of referents that help locate possible changes, as well as the responses that could be immediate (reactive) or that could have to do with climate adaptation.

The perception of climate change as well as the responses and adaptation are found in cultural, historical, political and environmental contexts. Each place has its own cultural, historical, economic and environmental particularities that give place to specific responses.

References

Báez-Jorge, F. 1983. La cosmovisión de los zoques de Chiapas. Reflexiones sobre su pasado y su presente. In Ochoa, L. and Lee, T. A. (eds.) *Antropología e Historia de los Mixe-Zoques y Mayas: un homenaje a Frans Blom*. México: Universidad Nacional Autónoma de México Press, pp. 383–412.

Báez-Jorge, F., Rivera, A. and Arrieta, P. 1985. *Cuando Ardió el Cielo y se Quemó la tierra*. México: Instituto Nacional Indigenista.

CONEVAL (Consejo Nacional de Evaluación de la Política de Desarrollo Social). 2009. *Población total, indicadores, índice y grado de rezago social por localidad*. México: CONEVAL.

INEGI (Instituto Nacional de Estadística Geografía e Informática). 2001. *Resultados del VII Censo Ejidal y Agropecuario*. México: INEGI.

INEGI (Instituto Nacional de Estadística Geografía e Informática). 2003. *Cuaderno Estadístico Municipal de Tuxtla Gutiérrez*. México: INEGI.

INEGI (Instituto Nacional de Estadística Geografía e Informática). 2006. *Anuario Estadístico Chiapas. Tomo I y II. INEGI-Gobierno del Estado de Chiapas*. México: INEGI.

INEGI (Instituto Nacional de Estadística Geografía e Informática). 2010. *Censo General de Población y Vivienda. Principales Resultados por Localidad. Integración territorial (ITER)* México: INEGI, www3.inegi.org.mx/sistemas/iter/doc/fd_2010.pdf

Katz, E. and Lammel, A. 2008. Introducción. Elementos para una antropología del clima. In: Lammel, A., Goloubinoff, M. and Katz, E. (eds.) *Aires y Lluvias. Antropología del Clima en México*. México: CIESAS/CEMCA/IRD, pp. 27–50.

Katz, E., Lammel, A. and Goloubinoff, M. 2008. Clima, meteorología y cultura en México. *Ciencias*, 90: 60–7.

Kronik, J. and Verner, D. 2010. *Indigenous Peoples and Climate Change in Latin America and the Caribbean*. Washington, DC: The World Bank.

Landa, R., Magaña, V. and Neri, C. 2008. *Agua y Clima: Elementos para la Adaptación al Cambio Climático*. México: Secretaría de Medio Ambiente y Recursos Naturales and Centro de Ciencias de la Atmósfera, Universidad Nacional Autónoma de México, www.atmosfera.unam.mx/editorial/libros/agua_y_clima/agua_y_clima.pdf

Landa R., Siller, D., Gómez, R. and Magaña, V. 2011. *Bases para la Gobernanza Hídrica en Condiciones de Cambio Climático. Experiencia en Ciudades del Sureste de México*. México: ONU-Habitat.

Magistro, J. and Roncoli, C. 2001. Anthropological perspectives and policy implications of climatic change research. *Climate Research*, 19: 91–6.

Oreskes, N., Stainforth, D. A. and Smith, L. A. 2010. Adaptation to global warming: Do climate models tell us what we need to know? *Philosophy of Science,* 77: 1012–28.

Orlove, B., Chiang, J. H. and Cane, M. A. 2002. Ethnoclimatology in the Andes: A cross-disciplinary study uncovers the scientific basis for the scheme Andean potato farmers traditionally use to predict the coming rains. *American Scientist*, 90: 428–35.

Orlove, B., Roncoli, C., Kabugo, M. and Majugu, A. 2011. Conocimiento climático indígena en el sur de Uganda: Múltiples componentes de un sistema dinámico regional. In Ulloa, A. (ed.). *Perspectivas Culturales del Clima*. Bogota: Universidad Nacional de Colombia/ILSA, pp. 183–221.

Pelling, M. 2011. *Adaptation to Climate Change. From Resilience to Transformation*. Abingdon, UK/ New York: Routledge/Taylor & Francis Group.

Rebetez, M. 1996. Public expectation as an element of human perception of climate change. *Climatic Change*, 32: 495–509.

Roncoli, C. 2006. Ethnographic and participatory approaches to research on farmers' responses to climate predictions. *Climate Research*, 33: 81–99.

Rzedowski, J. 1978. *Vegetación de México*. México: Limusa.

Sánchez-Cortés, M. S. 2011. Percepciones de los Cambios Ambientales en dos Comunidades Zoques de Chiapas. Ph.D. Thesis, Universidad Nacional Autónoma de México, ME.

Sánchez-Cortés, M. S. and Lazos Chavero, E. 2011. Indigenous perception of changes in climate variability and its relationship with agriculture in a Zoque community of Chiapas, Mexico. *Climatic Change*, 107: 363–89.

Sánchez-Cortés, M. S. 2013. Percepciones del cambio en la variabilidad climática en dos comunidades Zoques de Chiapas, México. In Gay, C., Rueda, J. C., Aguirre, L. et al. (eds.) *Memorias del Segundo Congreso Nacional de Investigación en Cambio Climático*. México: Universidad Nacional Autónoma de México, pp. 921–30.

Toledo, V. M. and Barrera-Bassols, N. 2008. *La Memoria Biocultural. La Importancia Ecológica de las Sabidurías Tradicionales*. Barcelona: Icaria Editorial.

Ulloa, A. 2011. Construcciones culturales sobre el clima. In Ulloa, A. (ed.), *Perspectivas culturales del clima*. Bogota: Universidad Nacional de Colombia/ILSA, pp. 33–56.

Ulloa, A. 2013. Estrategias culturales y políticas de manejo de las transformaciones ambientales y climáticas. In Ulloa, A. and Prieto-Rozo, A. I. (eds.) *Culturas, Conocimientos, Políticas y Ciudadanías en Torno al Cambio Climático*. Bogota: Universidad Nacional de Colombia Press, pp. 71–105.

Vedwan, N. and Rhoades, R. E. 2001. Climate change in the Western Himalayas of India: A study of local perception and response. *Climate Research*, 19(2): 109–17.

Villela, S. L. 2010. Vientos, nubes, lluvias y arcoiris: simbolización de los elementos naturales en el ritual agrícola de la Montaña de Guerrero. In Lammel, A., Goloubinoff, M. and Katz, E. (eds.) *Aires y lluvias. Antropología del Clima en México,* México: CIESAS/CEMCA/IRD, pp. 121–32.

Warman, A. 2002. *El Campo Mexicano en el Siglo XX*. México: Fondo de Cultura Económica.

6

Climate Knowledge of Ch'ol Farmers in Chiapas, Mexico

Fernando Briones

In all societies, knowledge of the climate is of strategic importance: it ensures the viability of systems of production and defines strategies to deal with environmental challenges. Sillitoe et al. (2002: 113) define local knowledge as 'a unique formulation of knowledge coming from a range of sources rooted in local cultures, a dynamic and ever changing pastiche of past "tradition" and present invention with a view to the future'. In the case of the indigenous Ch'ol people (the Maya of the northern part of the state of Chiapas in Mexico), climate knowledge is directly linked to agricultural practices, and also integrated into rituals and religious festivals. Acquired through the cumulative experience of their daily agricultural labour, this traditional and local knowledge brings together local observations with historically transmitted learnings. It also demonstrates the relationships that societies establish with their milieu and their capacity to incorporate new knowledge and experiences. In addition to its role with respect to community organization and knowledge transmission, it serves a subsistence purpose: to reduce environmental risks in agricultural activities.

Climate knowledge is expressed through rituals and practices that ensure its transmission, while also facilitating local-level observations of environmental transformations. Farmers constantly observe phenological indicators that they primarily use for the cultivation of maize, and they are attentive to changes in rainfall patterns. Thus, climate knowledge has a significant potential for adaptation in the context of climate change. This chapter seeks to briefly reflect on its opportunities and limitations.

The Climate Challenge

There is overwhelming scientific evidence that the increase in global temperature is a product of human activities beginning in the industrial era (IPCC, 2013). In addition to analyses of hard data such as climate trends and inventories of greenhouse gases, social variables have been increasingly taken into account in analyses of vulnerability, mitigation and adaptation. One such variable is traditional climate knowledge – of fundamental importance for food security. This knowledge is an essential part of local livelihoods and shapes response capabilities to climate change threats such as extreme hydro-meteorological events.

Traditional knowledge is more than just social representations or perceptions. It is also an empirical mechanism for assessing risk and mobilizing capabilities for adaptation and resilience. Moreover, it also allows a greater understanding of some forms of vulnerability. Some traditional agricultural practices may fail to provide the harvests expected because of changes in rainfall patterns, droughts or temperatures due to the impact of climate change. In this regard, it is important to consider that traditional knowledge may be effective in specific contexts (on a local scale) but may not be widely applicable.

Agriculture: A History of Adaptation

Many cultures have incorporated methods of atmospheric-climate observation and prediction in their systems of agricultural production. Knowledge of the environment has given rise to ritual, religious and political systems in societies around the world (Katz et al., 1997). The observation of atmospheric conditions and identification of climate patterns have a practical purpose: to generate information for decision-making in relation to subsistence activities. This knowledge is an expression of ways of understanding nature, classifying it and integrating it into one's institutions and daily activities. But what are the limits of its effectiveness in the context of a changing climate?

There are numerous papers that show the capabilities of adaptation to the environment by different social groups. For example, in 'Impact of and response to drought among Turkana pastoralists: implications for anthropological theory and hazards research', Terrence J. MacCabe (2005) studies the way in which an ethnic group in Africa copes with the shortage of pasture for its livestock during the dry season. In 'Aires y lluvias: antropología del clima en Mexico', Lammel, Goloubinoff and Katz (2008) provide a compilation of case studies to show the significance of climate for ritual systems. With regard to climate change, in the paper 'Indigenous climate knowledge in southern Uganda: the multiple components of a dynamic regional system', Orlove et al. (2010) show the need for a knowledge dialogue that allows farmers to access scientific information to predict rainfall before the start of the rainy season.

Traditional agricultural practices are based on empirical knowledge, the optimization of available resources and the flexibility to convert these resources into cash income in contexts of precarity. However, in scenarios of a changing climate, some of these practices may no longer be useful or may require adjustments to arrive at acceptable results. For example, the use of seasonal rainfall forecasts may help farmers choose more precisely the time to carry out slash and burn, a common practice in the region that engenders a high forest fire risk in the event of drought. In addition, decisions about planting times, in cases where the rainy season arrives early or is delayed, could be accompanied with timely meteorological information.

Case Study: Ch'ol Farmers

The northern part of Chiapas characteristically receives significant levels of rainfall. However, climate change scenarios for the coming decades reveal a risk that rainfall will be reduced. Peasants with the greatest flexibility and self-sufficiency will better adapt to gradual changes in rainfall patterns than to extreme hydro-meteorological events. Moreover, adaptation strategies developed by policies and programmes related to climate change will have better results if they are based upon the previous experience and organizational skills of farmers who are used to dealing with seasonal climate variability. In this sense, it is useful to keep in mind Agrawal (2002: 295), who writes, 'It is important, if investigations of indigenous knowledges are to serve the interests of the poor and the marginalised, to bring to the fore the institutions and practices sustained by different forms of knowledge.' In other words, the focus is not only on the significance of the knowledge, but also on how community institutions put it into practice.

The case study originates from the recognition that subsistence farmers are particularly vulnerable to climate change (Lobell and Burke, 2010), and yet, they have a wealth of experience managing atmospheric-climate uncertainty (Orlove et al., 2004). The research sought opportunities for *dialogue* with local stakeholders, mainly in the community of Nueva Esperanza, in the municipality of Tila, as well as through discussions with government officials and scientific experts on climate change. Seminars and conferences represented another arena of research, which focused on the contrasting positions with respect to traditional knowledge, idealizing knowledge on the one hand and invalidating it on the other. This chapter will not address this debate, but it is necessary to point out that this dichotomy is part of the issue. We also observe that opportunities for interdisciplinary discussion are rare, leading to debates which remain epistemologically fragmented.

For the Ch'ol people, observing the environment is an integral part of their agricultural practice. This is particularly evident from rituals that combine Mayan culture and Catholic religion. The centre of the Ch'ol territory is the locality of Tila, a highly symbolic place that is also a meeting point for the indigenous and *mestizo* populations of the entire region of the northern part of Chiapas. The sanctuary of the *señor de Tila*, which is found here, houses the statue of a black Christ, venerated since the colonial era. The articulation of religious context and agricultural activity favours the observation of climate indicators and a ritualization of farming. The presence of *tlatuches* or *rezadores* (lit. people who pray), shamans who oversee the rituals related to farming activities, contributes to the persistence and transformation of knowledge related to the climate. For the Ch'ol, the farming of maize is at the heart of their identity. In fact, the world *ch'ol* means maize, that is to say 'the people of maize'.

In the constellation of ritual activities connected to the agricultural calendar there are certain activities that determine to a certain extent the success or failure of harvests. Through participatory observation, we noted that some farmers are observing that rainfall patterns are changing. As a result, they are conducting small but significant experiments, such as planting their seeds at a number of different times, using phenological indicators such as soil moisture levels and the presence of clouds to determine when to plant. In this way, if rainfall occurs earlier or is delayed, or if the rainfall is below the amount usually expected, at least part of their crop, generally sufficient for the family's consumption, is likely to succeed, thus ensuring food security.

Another practice identified was planting at different altitudes. Some farmers with lands on an incline experiment by planting seeds at higher elevations to trial their tolerance for different temperatures. The farmers interviewed also concluded that the local varieties of maize tolerate both excess and lack of moisture better than the hybrid maize. Ch'ol traditional agricultural practices are based on diversity and the utilization of the land. This is true for beans, which are planted as a cover crop on the same plot of land as maize. This increases the vegetal biomass, fertilizes the soil and improves productivity. Additionally, the high price of coffee, a crop that provides financial support, maintains an income flow for Ch'ol family agriculture. Coffee is vulnerable to excess rainfall, while, according to Ch'ol farmers, corn is more vulnerable to drought.

Can the expansion of experimentation in subsistence farming be considered a form of adaptation? According to IPCC, adaptation to climate change refers to: 'Adjustment in natural or human systems in response to actual or expected climatic stimuli or their effects, which moderates harm or exploits beneficial opportunities. Various types of adaptation can be distinguished, including anticipatory, autonomous and planned adaptation.' In this light, the Ch'ol peasants' actions are autonomous in that they respond to seasonal variations more than to climatic ones. Nonetheless, Ch'ol responses to an environment that they observe to be changing reveal a capacity for *flexible adaptation* that is an integral part of their way of life. These experiments and adjustments are anchored in the short term (the current sowing season) and not in the long term of future climate projections. Nevertheless, experiments that turn out to be successful, such as opting for the local blue and red, rather than yellow, maize because of their resistance to droughts, tend to be passed on and replicated.

The phenological markers distinguished by farmers allow them to calculate the probabilities of rainfall or drought. We already mentioned clouds and moisture, but the arrival of the rainy season is also monitored, as it is often associated with the emergence of plants and insects, and the migration of birds. The presence of ants is associated with the beginning of the rainy season. The flight of birds at lower than usual heights is associated with an imminent storm, as birds serve as a sort of barometer to indicate a decrease in atmospheric pressure. But beyond these specific indicators, it is an entire system that is known in subsistence agriculture as the *tornamilpa* (planting corn in the winter, the dry season in the rest of the country), which demonstrates the capacity to utilize resources and an optimal management that allows up to two harvests per year.

In their agricultural practices, the Ch'ol are also committed to maintaining the botanical diversity of the high-altitude forests that they occupy. Their agriculture is sufficiently versatile to be able to diversify crop options by planting, for example, chilli peppers, pumpkins, tomatoes, medicinal herbs and fruit. Another aspect that should be emphasized is their willingness to temporarily switch to other activities to supplement their income. During bad harvest years, several of the young farmers that were interviewed worked primarily as bricklayers in the area of the Mayan Riviera (Cancun, Playa del Carmen) or as employees at oil platforms in the Gulf of Mexico. The majority of them send money to their families but return to take up positions in religious ceremonies. This additional financial resource helps support community institutions and, in some ways, subsidizes family farming.

Agriculture and Rituals: Are They Part of the Same System?

The ritualization of Ch'ol farming is represented in different ritual moments through the mediation of shamans (*tlatuches*), which maintains a symbolic balance between nature and people. The agricultural calendar is made up of periods that coincide with Catholic festivities (2 February, 3 May, Corpus Christi, 12 December and patron saint celebrations), representing the circular time of the agricultural cycle.

Relationships between festivities and the rain cycles are intertwined in a framework of request and promise: one of the characteristics of rituals is to make a promise that will compensate for a request. Ritual time is divided into two periods that represent different agricultural activities: the time of the sun ($k`in$) and the time of the moon ($u`b$). The time of the sun is that of the growth of the cornfield. The time of the moon corresponds to the period of fertility (the harvest period), and it coincides with these celebrations of the Catholic Church: The Immaculate Conception, Our Lady of Guadalupe and Our Lady of Hope. There are numerous expressions of the time–ritual relationship; for example, the *quetzal* dance symbolizes the transition between the periods for harvesting and for sowing. The dance is performed with a representation of the four cardinal points (made with *aguardiente* liquor), and four men who dance with *quetzal* feathers representing the cornfield. The dance itself represents the growth of maize and is performed in December to represent the fall of *quetzal* feathers (April–May), and the ending of one cycle and the beginning of another.

The places of worship – mountains, caves and springs – represent different symbolic planes – celestial, terrestrial and subterranean. *Row Wan*, the deity to whom rain is requested on 3 May, lives in caves. For the Ch'ol, climate phenomena are codified through a ritual system that resolves any possible differences in the patterns of rain or drought. Every phenomenon has an explanation and consequently a ritual solution.

In some cases, *adjustment rituals* may be performed to manage climate uncertainty, such as exceptional ceremonies to petition for rain outside of the traditional calendar, which involves rethinking planting times. This practice shows that, for farmers, flexibility can be incorporated into their symbolic relations. However, these actions to adjust are not common and they engender a certain risk. In the event of a drought, for example, farmers who always plant on the same date following the traditional calendar may jeopardize their food security.

Conclusion

For the Ch'ol people, agricultural activities and rituals are part of the same system that provides for both social integration and the transmission of knowledge related to climate. This *ritualization* of agriculture goes hand in hand with practices such as environmental observations, experimentation and diversification of crops as a subsistence strategy.

The sense of flexibility of Ch'ol's knowledge of the environment provides an opportunity to analyse their ability to adapt to climate change, based on the fact that they have always faced up to the risks of seasonal climate variability. In this context, one can ask whether the farmers' practices of self-sufficiency, while being based on experience, social relations, production and cultural practices can serve as a form of adaptation to climate change. This requires thinking in timescales that are long term, rather than the seasonal periods that are more common for traditional knowledge.

In any case, the data obtained thus far from fieldwork allows us to conclude that the Ch'ol make decisions based on their accumulated experiences, which can be considered as *temporary adaptation* measures. However, these traditional knowledge and practices may function in specific temporal contexts and not necessarily in the long term. They constitute empirical initiatives that have been undertaken in response to manifestations of atmospheric irregularity identified by the Ch'ol themselves.

The case study confirms, as has been demonstrated historically, that the flexibility and diversity of traditional agricultural activities play an important role in subsistence. Paradoxically, the empirical evidence shows that these practices may favour conditions of risk if carried out in contexts of transformation, such as changes in the start date of the rainy season, thus compromising the success of harvests and food security.

References

Agrawal, A. 2002. Indigenous knowledge and the politics of classification. *International Social Science Journal*, 173: 287–97.

IPCC (Intergovernmental Panel on Climate Change). 2013. *Climate Change 2013: The Physical Science Basis. Contribution of Working Group I to the Fifth Assessment Report of the Intergovernmental Panel on Climate Change*. [Stocker, T. F., Qin, D., Plattner, G.-K. et al. (eds.)] Cambridge, UK and New York: Cambridge University Press.

Katz, E., Goloubinoff, M. and Lammel, A. 1997. *Antropología del Clima en el Mundo Hispanoamericano. Tomo 1*. Quito: Abya-Yala Editing.

Lammel, A., Goloubinoff, M. and Katz, E. 2008. *Aires y lluvias. Antropología del Clima en México*. México: CIESAS/CEMCA/IRD.

Lobell, D. and Burke, M. 2010. *Climate Change and Food Security Adapting Agriculture to a Warmer World*. Dordrecht and New York: Springer.

MacCabe, J. T. 2005. El impacto y la respuesta a la sequía entre los pastores turkanas: Implicaciones para la teoría antropológica y la investigación de riesgos. *Desacatos*, 19: 25–40.

Orlove, B. S., Chiang, J. C. H. and Cane, M. A. 2004. Etnoclimatología de los Andes. *Investigación y Ciencia*, 330: 77–85.

Orlove, B. S., Roncoli, C., Kabugo, M. and Majugu, A. 2010. Indigenous climate knowledge in southern Uganda: The multiple components of a dynamic regional system. *Climatic Change*, 100: 243–65.

Sillitoe, P., Bicker, A. and Pottier, J. (eds.) 2002. *Participating in Development: Approaches to Indigenous Knowledge*. London and New York: Routledge.

Part II
Our Changing Homelands

7

Indigenous Forest Management as a Means for Climate Change Adaptation and Mitigation

*Wilfredo V. Alangui, Victoria Tauli-Corpuz,
Kimaren Ole Riamit, Dennis Mairena, Edda Moreno,
Waldo Muller, Frans Lakon, Paulus Unjing, Vitalis Andi,
Elias Ngiuk, Sujarni Alloy and Benyamin Efraim*

Many of the world's remaining forests are found in indigenous peoples' territories. This is not surprising, considering that 60 million indigenous peoples are 'almost wholly dependent on forests' (World Bank, 2008). At the same time, the destruction of forests worldwide has proceeded at such an alarming rate that emissions from deforestation and land-use change have already accounted for around 17 per cent of all carbon dioxide emissions (IPCC, 2007). Destruction and degradation of these forests will thus impact the lives of indigenous peoples while also further contributing to global warming.

Fortunately, indigenous peoples in many parts of the world continue to practise traditional forest management and governance, which have been developed since time immemorial and have been handed down from one generation to the next. This chapter provides examples of highly evolved sustainable practices of forest management and governance by indigenous peoples from three countries. We discuss the centrality of the forest in the lives of indigenous peoples as reflected in their traditional knowledge and practices, even as they continue to struggle to keep control of this important resource. These sustainable practices are developed by indigenous peoples because of their unique relationship with the land that is antithetical to a utilitarian view of their territories.

In the context of the REDD+ initiative, these systems of forest management need to be supported and enhanced because they contribute to climate change adaptation and mitigation. For this reason, Tebtebba Foundation in 2010 embarked on the project 'Ensuring the Effective Participation of Indigenous Peoples in Global and National REDD+ Processes'. The general objective of the project is to ensure that indigenous peoples' rights and their effective participation in REDD+ processes, at all levels, are promoted and supported. With support from the Norwegian Agency for Development Cooperation (NORAD), the project aimed at documenting indigenous peoples' customary systems of forest management. This is in support of efforts under REDD+ to further enhance these sustainable ways of managing forests and provide incentives to the indigenous communities that practise them.

The project involved eight indigenous organizations working in their respective countries. The initial phase entailed documenting case studies of traditional forest management among indigenous peoples' communities in Kenya, Nicaragua and Indonesia.

Carried out by indigenous researchers, the studies aimed to document indigenous peoples' knowledge, systems and practices that are relevant in sustainable forest management, forest conservation, enhancement of carbon stocks and in the promotion of cultural and biological diversity in the context of climate change and REDD+.

This chapter draws on the results of the three case studies that highlight the communities' unique relationship with their forests as reflected in their traditional practices of forest governance. We argue that such a relationship provides the basis for complex but sustainable systems of community-based management of the forest and other natural resources, which may be considered as adaptive mechanisms and mitigation measures that have been developed in response to a continually changing environment.

In the context of efforts to reduce emissions from deforestation, threats and challenges that hinder the effective implementation of these diverse community-based practices of sustainable forest management need to be addressed.

Methodology

All three case studies were conducted from June 2009 to August 2010. As a research centre for indigenous peoples' concerns, part of Tebtebba's advocacy is to enhance the capacities of indigenous persons to carry out research and write up the results. An important assertion is that 'unless indigenous peoples hone their research capabilities, they will always be the objects of research' (Tauli-Corpuz in Alangui et al., 2010: ix). The Tebtebba partners had different levels of research exposure and experience. The more senior partners (who became the area focal persons for the research) already had extensive research experience, while others were doing it for the first time. Research workshops and trainings were held in each area by Tebtebba and the partner organizations in order to set the objectives of the research, agree on the research design, and train the research assistants and documentalists (who were all members of the local indigenous communities) on the methods to be used to gather data.

The partner organizations chose their own pilot areas. The local researchers, along with focal persons from the partner organizations, were responsible for the data-gathering process as well as writing the final report. Exploratory visits to the communities and region were carried out before the actual data-gathering process began. Free, prior and informed consent was strictly adhered to, and local researchers were identified. Focus group discussions (specifically with elders, women, farmers/producers), key informant interviews (community-recognized elders) and secondary research were the primary methods used in the data-gathering process.

The results of the three case studies were published in *Sustaining & Enhancing Forests through Traditional Resource Management* (Alangui et al., 2010). The book was launched during the UNFCCC Conference of Partners that was held in Cancun, Mexico in December 2010.

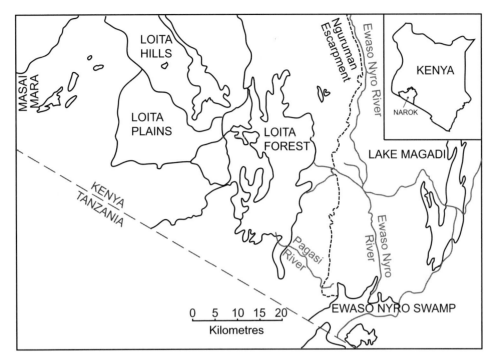

Figure 7.1 Map of the Loita Maasai Territory.

The Communities

The Loita Maasai People of Southern Kenya

The research 'Indigenous Peoples and the Naimina Enkiyio Forest in Southern Kenya: A Case Study' (Riamit, 2010) focuses on the Loita Maasai people who are found in the south-eastern part of the greater semi-arid district of Narok along the Kenya–Tanzania border (Figure 7.1). This community of 25,000 lives in the forest they call *Entim e Naimina Enkiyio* (literally, the Forest of the Lost Child), which is one of the few non-gazetted and largely undisturbed indigenous forests in Kenya (Riamit, 2010).

The Miskitu People of Tasba Pri Kuakuail II in Nicaragua

The case study entitled 'The Tasba Pri Kuakuail II Community's Relationship with the Forest in Nicaragua' (Mairena et al., 2010) highlights the Kuakuail II Miskitu People who belong to twenty-nine communities that make up the Tasba Pri indigenous territory (Figure 7.2). The forest in the study is a communal property covering 1,956 hectares.

The Kuakuail II territory borders on its north with the Sumubila community, on its south with the Akawas River and the Yatama group, on its east with the Naranjal community, and on its west with the Uriel Vanega y Altamira Collective.

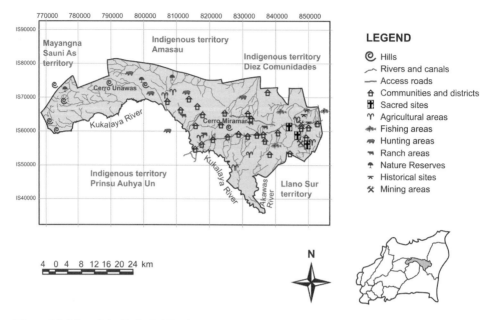

Figure 7.2 Map of the Tasba Pri Territory.

The Dayak Jalai People in Ketapangan District, West Kalimantan, Indonesia

The third study, entitled 'The Dayak Jalai People and their Concept of Dahas in Ketapangan District, West Kalimantan, Indonesia: A Case Study' (Lakon et al., 2010), documents the practice of integrated resource management by the Dayak Jalai people who live in the Jelai Hulu district of Ketapang Regency area. The Jalai are distributed in the three areas of Kampung Tanjung, Kampung Kusik Pakit and Kampung Pangkalan Pakit. In 2009, their total population was 2,714. The Dayak manage the Jelai Hulu forests, which are found in the three areas mentioned above (the encircled region in Figure 7.3).

Highlights of the Case Studies

Relationship With the Forest

For the indigenous peoples in the three case study areas, the forest is at the core of their existence. It is not only a source of sustenance and livelihoods but is the very basis of their identities, cultures, knowledge systems and social organizations. Living in subsistence economies, they exemplify those 'humans that maintain a relationship with land and water such that the meanings these humans find and make in them are constitutive of their identity and memory as a people' (Mendoza, 2007: 5).

The diverse ways in which the Loita Maasai describe the forest reveals a relationship that is linked to livelihoods and important community traditions and practices. The forest

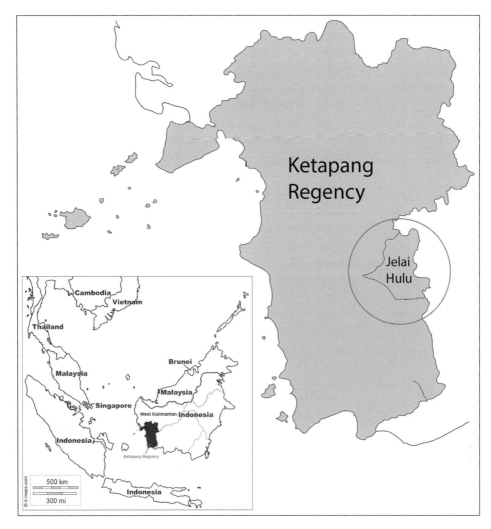

Figure 7.3 Map of the Ketapang Regency in 2006.

is at once a source of security and a safety net in times of drought when it is referred to as *saru-enkiteng* ('that which saves the cattle') and as a source of important food provisions, for example, when referred to as *entimoo Naishi* ('the forest of honey').

Other Maasai descriptions of the forest include *NooNkariakpusi* ('that of blue waters') to imply the pristine quality of the forest ecosystem, and *naigilakenyu* ('that which has two mornings or where the sun rises twice'), which to the people means minimal exposure to the scorching heat from the sun during the dry season.

Among these descriptions two stand out in terms of the message they convey: *SaruMaa/ SaruTungani* or 'that which saves/redeems the Maasai community or humankind' and *Mmenangiatua* or 'that which in it life is preserved'. On the other hand, the spiritual importance of the forest is reflected in the phrase *Osupuko le Mokompo* ('the highlands forest of Mr Mokompo') – it links the forest to Mokompo, the current chief prophet or seer of the entire Maasai community.

While change is inevitable in the Miskitu community in Kuakuail II, certain cultural and spiritual beliefs persist among the people. Like other communities in transition, the pull of tradition and the push for modernity reflect an ambivalence towards the values and norms that are used to define their identity as a people. Just as they declare the erosion of their sacred relationship with their ancestors and talk about the absence of any belief of supernatural beings in their forests (because of church teachings), they nevertheless make reference to bush and forest spirits, of *swinti/swinta* (elves) who might take away one's knowledge if this person entered the bush or the forest alone, of *Liwamairin* (mermaids) who are supernatural beings of their waters, and of *Liwa assure* who inhabit the swamps.

Despite the pervasive church presence, the Miskitu continue to perform forest-related rituals, which they impart to the younger members of the community through their stories. Storytelling is itself accompanied by a ritual. Before Miskitu children are told a story or lesson, the elders first place *kakamucwakia* (a woody vine that looks like the hands of an iguana) around each child's neck and waist to prevent them from seeing evil spirits.

Important community norms of the Miskitu, especially in relation to the forest, are integral to these stories. They speak of friendship with water and forest spirits, and the avoidance of acts that damage the natural habitat of these spirits.

Finally, in West Kalimantan, Indonesia, the Dayak Jalai people describe their relationship and interaction with nature as akin to the two sides of a coin: one depends on the other, and one's identity is determined by the other. People and nature both have roles to play to ensure their survival. *Hutanbajaluq, aria baikan, sasakbahundang* is the Dayak Jalai principle that speaks of ecological existence; a principle that has been passed down to them by their ancestors. According to the Dayak Jalai, the balance (of nature) is reflected in the way the *hehutan* (forest) is preserved, ensuring the availability of livelihood sources such as medicines and food. For them, the *hehutan* is also an important base of community beliefs and traditions. Their practice of utilizing the forest for agriculture is to them an important wisdom that continues to be practised for generations 'without destroying ecological balance' (Lakon et al., 2010: 144).

These indigenous practices and beliefs confirm what Mendoza (2007) refers to as a mutually implicated relationship between land and kinship, and to this may be added identity. A fundamental relationship exists between the indigenous communities and the forests making forest sustainability of strategic importance. It is a relationship that necessarily leads to the development of holistic and community-based sustainable forest management practices. Such practices are taught to the younger members of the community to ensure that such vital relationships with the forest shall continue through future generations.

The descriptions of the relationships of the Loita Maasai, Miskitu and the Dayak Jalai peoples with their forests stand in stark contrast to a relationship based on a utilitarian view of forest (and land) that is so prevalent in market-driven societies, especially by 'people who … are not tied to land' (Mendoza, 2007: 5). In this utilitarian view, a forest is a resource that should be exploited, and forest areas are easily converted to give way to various development programmes. In a utilitarian view of the forest, indigenous peoples eventually lose their relationship with the land, their kinship and their identity.

It can be argued that this utilitarian view has resulted in the massive destruction and degradation of the world's forests, generating negative social and ecological consequences, including climate change that is now disproportionately affecting indigenous peoples and other vulnerable communities (IPCC, 2007).

Traditional Forest Management and Governance

The relationship of mutual implication that is constitutive of indigenous peoples' identity and memory are manifested in the peoples' sustainable forest management practices that were developed by the communities and have evolved through the years.

If the forest is a place where life is preserved and humankind is redeemed, it is not hard to understand why it was necessary for the Loita Maasai to develop traditional forest management practices that ensure the sustainability of this important resource. These practices are reflected in guidelines adhered to by the community that prevent livestock grazing during the rainy season, the segregation of watering points by the elders according to their usage (e.g. for domestic use vs watering livestock), and the selective utilization of types of trees and other plants. Furthermore, the sacred sites and sacred trees for spiritual activities inside the forest including sites for rites of passage have served to regulate the utilization of this resource and its products. Aside from spiritual sites, there are sites for economic activities like firewood collection and the harvesting of honey, among others. The management of these sites could be under the control of individuals, groups or the entire community, determined by socially ascribed roles and duties in relation to forest sites and resources. The centrality of particular practices foster community-shared values and norms (Riamit, 2010).

In Nicaragua, the Miskitu people of Kuakuail II community possess a great deal of knowledge about the resources in the forest, and such knowledge is reflected on how they categorize the forest. In Miskitu language, there are three words related to the concept of forest based on its characteristics and use: *Unta, Unta Alal* and *Dus Ailal*. These concepts serve to guide the community's management of the forest and the resources found within. They serve to divide the forest into what Mairena, Moreno and Muller (Mairena et al., 2010) call forest zones.

The *Unta* is an area for cultivation of cereals by families from the community. Everybody knows where to work. Aside from planting, areas inside the *Unta* are also designated for pasturing animals. There are also areas that have not been utilized for at least twenty years which are assigned to the young people in the community for their own farming activities. Community members respect their neighbours' plots.

The *Unta Tara* is not used for planting and cultivation. Economic activities inside this forest zone include hunting, fishing and wood extraction in designated areas. There are strict guidelines for wood extraction, which is allowed only for house construction and to make *unuh* (pylons), *klasit* (latrines), *smalkaiawatla* (schools), *priaswatla* (churches) and *kapin* (coffins). Selling timber is allowed only when an emergency situation arises in the community, but this should not surpass 10 per cent of the total amount of wood used by the community. Community members are also required to reforest areas designated for wood extraction.

The third zone is *Dus Ailal*. This is the community's biodiversity reserve with minimal impact from human activity because of restricted access. The reserve abounds with wild animals, medicinal plants and precious woods. The whole community has been taking care of this zone for a long time, and individuals who 'damage' the reserve are dealt with accordingly. The community believes in the presence of spirits inside the forest that are capable of hurting careless and unsuspecting intruders. People are told to 'take caution to protect himself/herself' not only from snakes but also from spirits. This is one way of protecting the reserve.

Just like the Miskitu, the Dayak Jalai have an elaborate forest categorization scheme that guides the way they utilize the forest (Lakon et al., 2010). These include:

- *Rimbaq matuq*. This is a vast fully mature, virgin forest that is considered as a reserve and conservation area; flora and fauna are protected, although the area is occasionally exploited as a hunting ground and as a source of timber for building.
- *Jumpung/Pepulau*. Usually located on hilly land, the people also refer to this as jungle island; it is surrounded by farmlands, making the jungle resemble an island. It is also a conservation area reserved for future generations, although parts of it may be used for agriculture.
- *Pesapingan*. This forest is deliberately left intact to mark the border of one's farm from other land forms or from other people's farms.
- *Lakau mudaq*. Also called *bawas*, this forest fallow is utilized for shifting agriculture; as such, it goes through three stages: (a) *berahuk* (beginning stages of the farmland, around one to three years old; characterized by young plants, as well as the presence of sugar cane, ginger and sweet potato); (b) *mamat* (three to five years old, characterized by taller and bigger plants; not completely cleared because of the presence of bush or *mamat*); (c) *garas* (between five and twenty years old; vegetation has grown fully, cleared completely of bush).
- *Lakau humaq*. Farmland that eventually becomes 'the embryo for the establishment of dahas' (Lakon et al., 2010: 156).
- *Pengaraq*. This is reserved for farmland, but is never actually cultivated; it becomes a moderately dense forest after thirty years, and may eventually become a *rimbaq matuq*. The Dayak Jalai people have a saying, which goes: '*lakau mudaq* has passed, yet *rimbaq* has not arrived', which refers to the transition that a *panggarak* goes through into becoming a fully mature forest.
- *Pekampungan*. Forest containing various types of fruit trees.
- *Kabun presasaq*. Various commodity crops, notably rubber, are grown together in these forests, providing Dayak Jalai with a means to earn cash income.

- *Dahas*. Settlement area, containing houses, rice barns and various machinery for processing rice and rubber. The *dahas* area also contains *pekampungan* and *kabun presasaq*.
- *Pemaliq*. This forest houses particular things and objects considered sacred; these may take various forms (a forest patch, rock/s, wood, bay in the river and so forth). It is regarded as a holy place, a centre for prayers and offerings; it should be preserved and must not be cultivated or exploited in any way. People who do not obey these prohibitions are bound to meet tragedies and disasters.
- *Utung arai*. The uppermost part of a river where many springs occur. As this area is an abundant source of clean water for drinking, bathing, washing and fishing, it is forbidden to convert the surrounding forest into farmland. Doing so will 'disturb the ruler of heaven and earth' as well as the balance of nature (Lakon et al., 2010: 154).
- *Pandam pasaran*. Cemetery located near the settlement centre and characterized by the abundance of mature and big trees, *pandam pasaran* is a highly revered place for the Dayak Jalai.

According to Ellen (2008: 327) in his study on Nualu in Central Seram, Indonesia, the complex categorical constructions about the forest reflect an interaction 'inimical to a concept of forest as some kind of void or homogeneous entity'. Such complex categorical constructions are evident in the case studies featured in this chapter. The Loita Maasai, Miskitu and Dayak Jalai peoples certainly do not regard the forest as a homogeneous entity as shown by the multifaceted relationships that they have with the forest. Like the Nualu, they relate with different parts and resources of the forest in different ways (Ellen, 2008).

The three case studies also confirm the long-held view that the presence of sacred sites and adherence to other religious beliefs related to the forest help in the conservation of important water sources and plants that have medicinal value, as well as regulate the hunting of animals and wood extraction, especially of trees that have special uses in the community.

All these practices have been developed to ensure that the forest and other natural resources are utilized wisely by the community members. Doing otherwise would not only threaten a source of livelihood and security, but would also endanger the very basis of their identity as a people.

Dahas as an Integrated Resource Management

This section discusses the unique practice of the Dayak Jalai people called *dahas* as an important example of community-based integrated natural resource management.

There is a lengthy process in the formation of *dahas*, starting from the clearing of forestland for farming, building of a temporary cabin for shelter, expansion and building of a permanent house and a warehouse. Slowly, fruit plants and trees are planted and grown through the years; at this stage, the owner starts to settle and live in the area.

This area that evolves from the practice of *dahas* may be thought of as a village within a village. It is referred to by the Dayak Jalai as *pedahasan*, or centre of forest management and of agricultural and economic activities. If ownership remains unknown for many years, the *pedahasan* becomes a property of the entire village.

Once established, a *pedahasan* becomes a settlement area for the families who 'opened' it. Aside from the residential houses, a typical *pedahasan* would include rice barns, warehouses, animal pens, rice mill, rubber mill and a generator. Most of the Dayak Jalai villages in the research sites started from *pedahasan*. Within the *pedahasan* one may find other types of forests like *pekampungan* (fruit trees) and *kabun prasasaq* (commodity crops such as rubber trees). A mature *pedahasan* includes the settlement area as well as the various types of forest and agricultural zones, as shown by the schematic diagram in Figure 7.4.

The *pedahasan* exemplifies what Fairhead and Leach (1995) refer to as a vegetation-enriching local land use similar to what they saw in Kissidougou in West Africa. Based on 'local values and productivity criteria', the *dahas* owners develop a proportion of useful vegetation and forest forms in the landscape. The initial clearing of the forest for the establishment of the *dahas* does not harm the overall landscape. Instead, it becomes the impetus for more diverse and ecologically sound forms of resource management.

The Dayak Jalai people have also evolved clearly laid out rules on leadership within the *pedahasan*, including rules on access, control, use, possession, ownership and inheritance. A *dahas* is identified by the clan or family that first established it. Ownership is inherited, thus the *pedahasan* is passed on from one generation to the next. All rituals and customary laws for forest management of the Dayak Jalai are also performed and followed in the *pedahasan*. One example of customary law is *daradiumbungan*, which is imposed 'on someone who … does harmful acts on the vegetation such as felling the trees of *durian, tengkawang, kusik, mentawaq, langsat*, and other investment plants of the people' (Lakon et al., 2010: 172).

This complex system of management may be explained by the spatial and temporal importance of *dahas*. It becomes the nucleus of evolving villages that survive on the diverse vegetation of the *pedahasan* and on the various economic and agricultural activities that have been developed inside.

But it has another important function. The *dahas* also becomes an identity marker for the families who have put it up, as well as for the others who have established their own. This is because inheritance serves to perpetuate kinship relations. As a property that may be inherited by future generations, the *pedahasan* serves to identify kinship relations within villages that have evolved from it.

Thus, the linkage of land, kinship and identity is clearly operative in the Dayak Jalai people's practice of the *pedahasan*. From such relationships evolve community-based practices that help sustain the forest. By avoiding deforestation and other activities that are 'harmful', these community-based practices on forest management help the people adapt to climate change and mitigate its impacts at the local level.

Figure 7.4 Elements of the *pedahasan* forest management system.

Threats to Sustainable Resource Management

The case studies documented current challenges the indigenous peoples in the research sites face that might dispossess them from their communities and their forests. Losing their land and forests means losing the very basis of their knowledge on sustainable forest management. With worldwide efforts to reduce emissions from deforestation, the indigenous peoples and other stakeholders need to address such threats and challenges to ensure the continued protection of indigenous territories and allow the diverse community-based practices on sustainable forest management to flourish.

A common problem has to do with government control over their land and the forest. In the case of Loita Maasai, these resources are held in trust by the Narok County Council on behalf of the local indigenous community; trust which is often abused by the government (Riamit, 2010). For the indigenous peoples of Nicaragua, their use, control and access of natural resources is impacted by government norms and regulations. Another challenge, which the government has not done anything about, is the arrival of *mestizo* settlers at the outskirts of the territory, which has posed a serious threat to the forest. Mairena et al. (2010: 100) write that 'the banks of the river are already deforested ... These people are not settled in one population center but rather, [they are] spread out through the bush ... [causing] deforestation to take place over a broad geographic area'. Meanwhile, Dayak Jalai are faced with government-pushed expansion of palm plantations and continued operations of mining companies.

Also part of the problem, as in the case of the Loita Maasai, is the non-recognition of indigenous knowledge systems, including knowledge and practices in natural resource management at the local level. This experience shows that in some nations and states, the growing recognition at the international level of the important role played by indigenous peoples and indigenous knowledge in addressing climate-change-related issues do not get translated at the national and local levels.

Conclusion

The indigenous peoples' practices on forest management in the case studies involve strategies similar to what were practised in various Mexican communities as described by Klooster and Masera (2000): their strategies involve the setting aside of conservation, woodcutting and watershed management zones, which, in Mexico, helped reverse deforestation trends and enhanced forest density (despite having adopted such practices as recently as the mid-1980s). They argue that community-based forest management such as those documented in Mexico and in the three case studies 'has an important role to play in reversing process of deforestation, sequestering carbon, and promoting rural development' (Klooster and Masera, 2000: 259).

Unlike in Mexico, such strategies among the Loita Maasai, the Miskitu and the Dayak Jalai have been developed much earlier – knowledge that has been adapted to the environment, evolved and passed on to many generations. Moreover, these strategies are guided by principles and religious practices that, while not based on Western science, are viable, robust and consistent with the peoples' values and view of the world.

The three communities in the study areas are undergoing extreme pressures from the inside but even more so from the outside. As what is happening in many indigenous communities in other parts of the world, some residents start abandoning traditional practices for modern ways of land uses in order to meet their financial needs. Still, the majority of the community members continue to implement and practice their traditional practices of forest management and governance. There are several reasons for this. One is the presence of elders who continue to influence the affairs of these communities; the other is a continuing conviction that their community-based indigenous knowledge systems remain valid and relevant because these have helped them sustain their forests for all these years.

As peoples whose culture and identity are tied to the land, the protection of their natural environment, especially the forest, is of prime and strategic importance. These community-based forest management practices they have evolved ensure the sustainability for future generations. And, in the context of REDD+, they provide the world with viable and sustainable adaptation and mitigation practices that help reduce deforestation and forest degradation, thus needing further recognition and support in all scales – not only at the international level, but more so, at the local and national levels.

Acknowledgement

The editors and authors are indebted to John Bamba, Executive Director of Institut Dayakologi, for his review and comments on the *dahas* system of the Dayak Jalai.

References

Alangui, W. V., Subido, G. and Tinda-an, R. (eds.) 2010. *Sustaining & Enhancing Forests through Traditional Resource Management*. Baguio City: Tebtebba Foundation.

Ellen, R. 2008. Forest knowledge, forest transformation: Political contingency, historical ecology, and the renegotiation of nature in central Seram. In Dove, M. R. and Carpenter, C. (eds.) *Environmental Anthropology: A Historical Reader*. Oxford: Blackwell, pp. 321–38.

Fairhead, J. and Leach, M. 1995. False forest history, complicit social analysis: Rethinking some West African environmental narratives. *World Development*, 23(6): 1023–35.

IPCC (Intergovernmental Panel on Climate Change). 2007. *Climate Change 2007: Synthesis Report*, www.ipcc.ch/pdf/assessment-report/ar4/syr/ar4_syr.pdf

Klooster, D. and Masera, O. 2000. Community forest management in Mexico: Carbon mitigation and biodiversity conservation through rural development. *Global Environmental Change*, 10: 259–72.

Lakon, F., Unjing, P., Andi, V., Ngiuk, E. and Alloy, S. 2010. The Dayak Jalai peoples and their concept of dahas in Ketapang district, West Kalimantan, Indonesia: A case study. In Alangui, W., Subido, G. and Tinda-an, R. (eds.) *Sustaining & Enhancing Forests through Traditional Resource Management*. Baguio City: Tebtebba Foundation, pp. 119–78.

Mairena, D., Moreno, E. and Muller, W. 2010. The Tasba Pri Kuakuail II community's relationship with the forest. In Alangui, W., Subido, G. and Tinda-an, R. (eds.) *Sustaining & Enhancing Forests through Traditional Resource Management*. Baguio City: Tebtebba Foundation, pp. 59–117.

Mendoza, J. 2007. Alterity and cultural existence. In Tapang, B. (ed.) *Cordillera in June: Essays Celebrating June Prill-Brett, Anthropologist*. Quezon City: University of the Philippines Press.

Riamit, K. O. 2010. Indigenous peoples and the Naimina Enkiyio Forest in southern Kenya: A case study. In: Alangui, W., Subido, G. and Tinda-an, R. (eds.) *Sustaining & Enhancing Forests through Traditional Resource Management*. Baguio City: Tebtebba Foundation, pp. 1–58.

World Bank. 2008. *Bank Book is Practical Guide to Sustainable Development of Forests*, http://go.worldbank.org/61KD58WW70

8

Indigenous Knowledge, History and Environmental Change as Seen by Yolngu People of Blue Mud Bay, Northern Australia

Marcus Barber

This chapter examines some key aspects of the knowledge used by coastal-dwelling Yolngu people in North East Arnhem Land, Australia, to live successfully in the world they inhabit, and examines how that knowledge has adapted and grown in response to changing conditions.

Yolngu knowledge includes elements which might be classed as 'environmental' – climatic, geographic, ecological, meteorological and so on – and incorporates a sophisticated understanding of coastal water cycles (Buku-Larrnggay Mulka Centre, 1999; Magowan, 2001; Barber, 2005). However, for Yolngu people, a crucial aspect of knowledge derives from the understanding that their country is actively involved in human life, responding to human actions and historical events in myriad ways. Therefore, to live safely, people must not only have knowledge of the country, they must also be known by it and by the ancestral beings that created it and still live there. Colonial and capitalist processes have created ongoing challenges for Yolngu people, requiring the development of new ways of knowing and new kinds of accommodations with the ancestors and with the places they inhabit. The emerging issue of climate change represents an additional challenge, one that is interpretable in Yolngu terms as both a response by the country to inappropriate actions by human beings, and as requiring further adaptation by local people to ensure their continued survival.

Yolngu-speaking peoples living in the area that subsequently became known as Arnhem Land (Figure 8.1) were among the last diverse indigenous groups of Australia to experience the effects of hostile British colonization. The first significant colonial contact only occurred in the late nineteenth and early twentieth century, and Arnhem Land was designated as a protected reserve in 1928. The substantial ongoing national and international profile of Yolngu people emerges in part from the cultural continuities arising from such colonial circumstances, from the way non-indigenous people value such continuities in their assessments of indigeneity, and from the strategic way that Yolngu people themselves have promoted their own interests. Nevertheless, Yolngu ways of life have altered significantly over the last hundred years from the hunter-gatherer existence that dominated indigenous Australian occupation of the continent for many millennia. The patterns of these changes have important implications for understanding both contemporary Yolngu knowledge and Yolngu responses to ongoing social and environmental change.

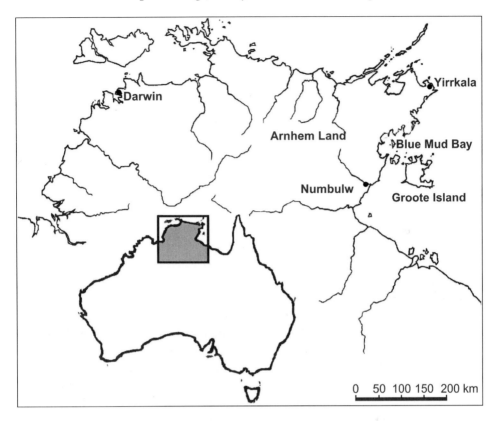

Figure 8.1 Location of Blue Mud Bay and former mission settlements.

History and Knowledge

Between the 1920s and the early 1950s, Christian missions were created on or adjacent to North East Arnhem Land – at Yirrkala, Groote Island and Numbulwar (Figure 8.1). The missionaries strongly encouraged residential centralization, inducing indigenous people to remain on the missions using incentives such as food and tobacco. This had an increasing impact on residence patterns, individual and communal dependencies, and knowledge-sharing and communication across a wide area. Blue Mud Bay, which lies between the missions, was depopulated at times during and after the Second World War. A change in government policy in the early 1970s and concerns about the impacts of newly established mines in the region led Yolngu people to leave such centralized missions and establish smaller independent communities, known as homelands, on their own clan territories. A number of these homelands now lie adjacent to the northern shores of Blue Mud Bay.

Material resources for these contemporary homelands are obtained in two main ways. Some food and other materials (such as those needed for art production) are sourced in customary ways from the surrounding landscape, while cash income is derived from a combination of social welfare payments, remittances, state-funded employment programmes and proceeds from art sales (Altman, 1987; Povinelli, 1992; Altman, 2002). High-value, small-scale cultural and ecotourist ventures are increasingly considered important for the future, but a particularly significant contemporary development for the homelands of Blue Mud Bay has been income and activity associated with formal, primarily state-funded natural and cultural resource management initiatives (Smyth et al., 2010; Altman and Kerins, 2012; Laynhapuy Homelands Aboriginal Corporation, 2013).

These lifestyle changes over the past century can be summarized as a shift from an independent hunter-gatherer existence to a centralized existence concentrated in the missions and dependent on mission rations and services, then to a more decentralized and autonomous way of life on the homelands, but with that way of life relying on support from government resources (state-supported employment, welfare income, housing and limited infrastructure, service provision, etc.). Within those broader patterns, high levels of personal mobility have remained, but this mobility follows particular kinship and residential ties, and occurs against a backdrop of deeply embedded customary relationships with particular places. Collectively, these circumstances mean that contemporary Yolngu communities are characterized by a complex mix of residential stability and mobility on the one hand, and of local autonomy and institutional dependency on the other. Therefore, depending on the frame of reference adopted, Yolngu people and communities can be depicted as highly vulnerable or as highly adaptive and resilient to social, political and environmental change. How Yolngu knowledge is understood and represented is crucial to generating such depictions.

The categorization of knowledge held by indigenous peoples remains the subject of considerable debate (Agrawal, 1995; Ellen, 2004; Sillitoe et al., 2005; Agrawal, 2009; Barber and Jackson, 2015). Each proposed alternative emphasizes important aspects, but in doing so also can elide or deprioritize alternate ways of knowing and practices undertaken by indigenous peoples. This account identifies key characteristics and dynamics of Yolngu knowledge as they are expressed by Yolngu elders' interactions with landscape and responses to recent historical and environmental changes. In this sense it can be depicted as either 'indigenous knowledge' or 'local environmental knowledge', yet neither of these terms individually convey the full range of knowledge that Yolngu people use in living in and relating to their country, and in observing, interpreting and adapting to changes within it.

Regardless of the terminology adopted to describe such knowledge, locally generated and continually evolving knowledge is crucial in enabling Yolngu people to retain the capacity for both resilient adaptation and constructive transformation. Understanding the historical and cultural development of local knowledge is crucial to understanding how its practitioners will adapt to new circumstances, including the still emerging consequences of climate change and sea-level rise. The Yolngu context described here combines a sophisticated local cosmology, deep awareness of millennia of continuous residence in the area, more recent colonial intrusions, and the implementation of contemporary natural and cultural resource management programmes that directly address the local management of ecological change. All of these factors are potentially significant in contemporary Yolngu peoples' ongoing thinking and learning about climate change and about human-environment relations more generally.

The Place of Learning

Teaching and learning occurs every day in every society in a wide variety of contexts, many of them informal. However, one characteristic of Yolngu societies, indeed of indigenous Australian societies more generally, has been the relatively low priority given to formal and institutional instruction of the young. Children are given a high degree of freedom to choose their own activities, and are encouraged to learn by watching older people performing daily tasks *in situ* and by attempting those tasks themselves, rather than through a heavy emphasis on formal schooling. Ceremonies were the contexts where the closest approximation to formal instruction occurred, but this still did not equate to the highly time-regulated institutional setting of the contemporary school that is now so crucial to securing future employment. The place-based nature of learning, the importance of regular movement between places and the fact that knowledge is transmitted by relatives are demonstrated by an elder, G, who recalled his early childhood years where he learned some of the skills that were critical to his later success in life:

AUTHOR: Who taught you spear-fishing?
G: Dhukal's father was teaching me.
AUTHOR: Did you stay for a long time at Wangurrarrikpa [an island in Blue Mud Bay]? Was he teaching you on Wangurrarrikpa?
G: Wangurrarrikpa, Garrapara, Baniyala, Djarrakpi [locations in Blue Mud Bay] and a few more places. He was teaching me about animals (*G pauses, thinking back*) Baykutji [an inland river area]. We stayed there, and then came back [to the coast]. In the wet season, we went to the beach. In the dry season, we went inland; Gangan, Baykutji, Dhuruputjpi, those places.

Later in this conversation, G recounted a well-known story of when he was a young man and how, without permission, he borrowed a newly built canoe to travel south to find a senior elder. This man, W, had been absent from the region for many years following a series of disputes, but his presence was deemed highly desirable at a major funeral that was soon to occur. G's journey involved moving through Blue Mud Bay and south to Numbulwar, and the staging points were determined by his knowledge of the wind, the weather, coastal geography and where freshwater could be found. When he stopped, he used his well-honed hunting skills, and some fifty years later he still remembers the kind of turtle he harpooned during the journey and the place he harpooned it. He located W and brought him back for the funeral. The renewed presence of W in the region thereafter had significant positive ramifications for several decades (Barber, 2008) and G's geographic knowledge and his ability to successfully hunt and travel independently were critical in his contribution to W's renewed presence.

G's journey had a specific purpose, but such journeys were made for a range of reasons, most significantly to maintain the complex extended set of kin relationships that structure Yolngu society. Such journeys resulted in people gaining regular experience of their surroundings under the guidance of those who had experienced them many times before, and the explicitly geographic and environmental knowledge emerging from that process is usually the easiest for those trained in the western natural sciences to comprehend, for it can be easily subsumed into the wider analytical framework characteristic of western natural sciences. Yet, a further exploration of Yolngu understandings of their world reveals that, in their terms, much more than this kind of knowledge is required for people to successfully travel through, live upon and understand the country.

Knowing Country

To live safely, it is not enough for Yolngu people to know how to travel between these places and what to find there. Rather, the places themselves must in their turn recognize that person, and must know them as someone who belongs. 'Knowing country' is not only an act and a state of being for Yolngu people, it is a state of being for the country itself (Rose, 2005).

Yolngu country is more than the backdrop to Yolngu life, more than a stage on which the action occurs; rather, it is an active participant in that life, constantly responding to people and to events. Strangers to a place must be wary of negative reactions, and there are protocols for introducing people to new places. In the Blue Mud Bay dialects, one way of expressing that the country does not recognize someone can be translated as 'does not know your sweat'. Being known means being known as a human body, and the country recognizes the bodies that belong to it. As one elder, WM, explained, 'if a stranger from another community comes along that has never been to that place, well we just put our sweat on them, and then ask the country for good luck'. Strange bodies are given the smell of familiar ones to keep them safe as they travel through places they have not been before.

The idea of a landscape that is an agent in the world, and that recognizes who belongs to it, has been noted before by other ethnographers of indigenous Australians (Povinelli, 1993; Rose, 1996; Bradley, 2001; Poirer, 2005; Bradley, 2010) and is regularly re-expressed in different forms by indigenous Australians themselves. It is a key point to establish when trying to reach a genuine understanding of what 'indigenous knowledge' might mean – the knowledge one holds is vital, but in relation to life that is grounded in place, it is equally important that one is known. The first stage in that process is being recognized as a body, as no longer smelling like a stranger. However, there are other ways of communicating to the country and to the potentially dangerous ancestral powers that inhabit it, that one has a legitimate right to be in that place:

DM: You talk to the country, talk to the [ancestral] Yolngu, so that he knows you. You talk to them: 'Give us something! We are the generation after you! Brrrr!' [vocalization designed to gain ancestral attention].

WM: At [the place called] Yathikpa, we would call Ngurrumula, Nimbarrki [less commonly used names identifying special features of Yathikpa].
DM: Nimbarrki, and we can call outside [more commonly known] names too, like Borrak – that [ancestral] man who was hunting in that area. He's a hunter, and his name [means] hunter so we always … have respect by calling his name out.
AUTHOR: So if you are in areas you know, do you worry about the *mokuy* [spirits of deceased ancestors] when you are out hunting?
DM AND WM: No!
AUTHOR: Why not?
DM: Well, we can talk to them. Because we can feel it in our life too, they are talking to us too. When you walk around you can have confidence.

Truly known and knowledgeable people not only possess familiar sweat, they are those who speak the language and know the names that come from those places. DM makes it clear above that it is not just the country but the spirits of those who have gone before and still live there in another form that are being addressed. Talking to the place means talking to both ancestral spirit and ancestral country simultaneously; and staying safe and hunting successfully means knowing how to talk and how to listen to the generations who came before. DM simply calls them Yolngu, making no distinction between the living and the dead, even if it is clear from the context that he is talking of the latter. The names that follow emphasize this point, for Borrak is simultaneously the name of the ancestral creator dugong hunter, of a long-deceased relative and of DM's younger brother. The calling out of such names is a mark of respect, but in areas where hunters are confident of their place, it is not a sign of fear. The ability to talk to the Yolngu of the past, to call their names and to feel them talking back to you, is the basis for the confidence that both men strongly express.

This calling of names is not just associated with hunting but with life in general – it can be a call for assistance or for a change in luck, and can therefore be applied to tasks such as starting a faulty boat engine, removing bark from a tree to be used for artworks or scoring a goal playing football. Knowing how to live successfully requires not just physical skills and knowledge of geography but familiarity with those who lived in those places in the past, and who continue to live there today in a different form. Knowing they are there, how to talk to them, and that they in turn know you, is a critical component of Yolngu knowledge. Such an attitude also suggests that knowledge claims about Yolngu country made by outsiders from elsewhere using non-local techniques (remote sensing, extrapolating from similar environments and so on) may be treated with some scepticism – how can one truly know about a place, if one is not also known by it? There are clear implications here for discussions with non-local experts of phenomena such as climate change.

Reading the Signs

Nevertheless, even for those who are known, smelling familiar and knowing the appropriate names are not guarantees of complete safety. One day, as DW and I were walking along the riverbank of a place for which he was a senior owner, he explained how crocodiles that recognize the sweat of Yolngu people are less likely to attack those people. Yet, as he searched for a fish to spear, he saw small bubbles rising up from deep in the water. It was, he said, the sign of a crocodile, and from then on he was much more wary near the water. People who are confident that their country knows them – that they are familiar bodies – are still aware of the risks and take precautions. This sense of confidence about their own place is augmented by practical knowledge borne of long experience. Moving through and living in a potentially dangerous environment requires both attitudes; the confidence that one has what is required to live there safely, but equally the knowledge to read the signs and be aware of the risks.

Such signs may be straightforward, like the bubbles signalling the presence of a crocodile, but if the country is engaged with life in the way that Yolngu people describe, then it is capable of communicating far more to those who are ready to listen and able to interpret the signs. One day, a senior man called N and I had begun to fish at a well-known and productive waterhole. Straightaway, he caught two fish and soon after I caught one. We eagerly prepared our lines again, expecting a good day, but slowly the time passed and the lines stayed slack. The wind dropped and it became much hotter. Three hours later, neither of us had caught another fish and we returned to the homeland disappointed. On our arrival, we learned that an invalid brother of N's had passed away early that afternoon. Later, N reflected on the day:

That fishing going quiet was the country telling us about that death, our brother passing away. Sometimes, the spirit of that person becomes an animal and follows you, a fish, a bird, a dog. Sometimes the sea sends a message, with a big wave rising up. Because the spirit goes back to the land and to the sea, to the country where it comes from.

Travelling and hunting safely requires knowledge and being known, but those who are sensitive, who feel they can read the signs, may also learn other things. As N describes, the country is involved in life, reacting to important events elsewhere, and communicating that news to those who are listening. Ingold (2000: 57) describes the hunter-gatherer landscape as 'infused with human meaning – that this meaning has not been "pinned on" but is there to be "picked up" by those with eyes to see and ears to hear'. Part of knowledge is being able to recognize what is unusual and reflecting on what it may mean. Full understanding may not be possible until later, but sensitivity to changes in a sentient world is another way of learning, of communicating and of knowing. From this perspective, interpretations of changing environmental conditions take on an additional aspect, being understood as a commentary or indirect response by the country to events in the human realm, as well as a direct material consequence of human activities (Petheram et al., 2010; Barber, 2011). A prominent Australian politician described climate change, and its effects, as 'the greatest moral, economic and social challenge of our time' (Van Onselen, 2008). From a Yolngu perspective, the 'moral' component of climate change-derived environmental degradation may encompass far more than a generalized sense of responsibility for carbon emissions. It may also be seen as a direct response by local places to the failure of their human custodians to manage their conduct and their responsibilities appropriately (Barber, 2011). This amplifies the significance of such changes even further.

Madayin-codified Knowledge

If you know where the *madayin* is, you know where the *gapu* [water] is, you know where you can find kangaroo or *miyapunu* [turtle and dugong]. Like in the sea, we call it *batpa* [reef and seagrass where *miyapunu* feed]. You know where you can find *batpa*. Or inland, you know where the water is and where the wallaby is and where the tortoise [is]. You can catch those ones ... (From DM).

Madayin is the word that Yolngu people use to describe their ancestral heritage as a whole, including art, song, myth, dance and ritual. The richness of this aspect of Yolngu society has been documented by many authors (Morphy, 1984; Williams, 1986; Morphy, 1991; Keen, 1994; Morphy, 1996). Morphy (1996: 177) describes the relationship between three key features of the ancestral world and Yolngu ritual life in terms of the *wangarr* or ancestral creator beings constituting the ancestral world, the *madayin* representing manifestations of the ancestral world, and *marr* (Thomson, 1975) as the power emanating from those manifestations of the ancestral world. DM makes clear in his above statement that, along with the mythological richness, knowledge of the *madayin* has a practical import – it tells you where to hunt and what you will find there. It has been argued that such sacred lore is not consulted by aboriginal people as westerners would consult a map (Ingold, 2000), but the dominant words in the quotation from DM concern location: 'you know where'. The song cycles contain place names, local features, environmental detail, the names of spirits and ancestors and so on. Furthermore, the *madayin* is a repository for more than just information about the wider environment: a significant number of the songs and paintings emphasize skills and processes, particularly hunting processes. DM's clan often sing about a group of fishers paddling out from shore to fish from the reefs, then returning to shore in rough weather, making a fire, cooking and eating their catch, and discarding the bones. DM, when out hunting on the boat one day, explicitly made the point that he was undertaking the same actions as the *wangarr* hunters had in the past, and this would be the basis for a successful hunt.

Yet, as the previous discussion showed, knowing the *madayin* means more than just knowing the names of the places and the animals that live there. It also means being known; being recognized by that place and the beings that inhabit it as somebody who belongs. This is far more than a person's knowledge of the environment; indeed, taken as a whole it is even more than knowledge of that person by the place. Rather the two are bound up together, constituted by the same ancestral power, pursuing relationships that are dynamic and ongoing, but which follow structures and processes laid down in the ancestral past. Knowledge as mutual recognition and relationship is much harder to place in terms of categories normally associated with the natural sciences and western thought more generally. There are times and places where it can be incorporated into such categories and the *madayin* itself represents a formal condensation or codification of that knowledge which lends itself more easily to such incorporation. Yet, as the previous examples show, knowledge is enacted constantly in everyday life in different ways – the *madayin* provides foundations for knowledge, but it is in practice that this knowledge is learned and actualized. This allows for a dynamic quality to the relationship with *madayin*, something that may be essential should the places and ecological relationships of which it speaks undergo significant and unprecedented changes. There is a precedent for such changes, at least in the social and historical realm, if not so strikingly in the ecological one, as the next section demonstrates.

Madayin and Mission Knowledge

Many of the skills required for an independent hunter-gatherer life were still relevant in the semi-supported context of the mission, particularly during travel between missions or shorter stays out in the bush. The creation of homeland communities from the early 1970s transformed the lives of Yolngu people throughout the region. The centralization process that had begun with the missions in the 1930s was reversed, and a significant number of people moved back to their traditional country. However, while the knowledge required to make this new form of life successful drew on the *madayin*, it also required adaptations of it and additional knowledge that people had gained during the mission era. Their goal now was to establish permanent settlements on their own country; smaller versions of the missions they were now leaving. This meant using the *madayin* to identify the safe and dangerous places (Biernoff, 1978), but also understanding the practicalities of community infrastructure which would support a way of life quite different from the nomadic life of the past.

The site of Baniyala – the largest of the homelands in Blue Mud Bay – was largely determined by W, the man G had gone to fetch in his borrowed canoe as described earlier. W was a senior elder and a lawman, a conduit for and an interpreter of the *madayin* of his clan. Yet he was also experienced in the logistics of establishing permanent settlements for he had been heavily involved in the creation of the mission at Numbulwar on the coastline south of the bay; building permanent houses, organizing the water supply, and identifying the range of food and other resources that needed to be within easy reach for the community to be sustainable. His clan homeland had to be sited on clan territory, and this clan owned two significant areas in the northern part of the bay.

However, this significance arose from the ancestral power of those places, making them potentially dangerous to the uninitiated. This was one of the problems that needed to be negotiated, and appropriate ceremonies would be necessary to minimize the risk. Yet one site had an added danger – it was close to a crocodile breeding ground and the seawater there was muddy and opaque. The other site looked out over a long, gently sloping beach with clear water – sharks and crocodiles could be seen well before they reached the shallows where children would play. Neither site had permanent water, but W knew that bores could be dug at either site to solve that problem. The clear beach site had food sources close by and flat land for an airstrip. The *madayin* told W what his people owned and where they could live, but new forms of knowledge derived from elsewhere were critical to establishing a new form of life for people who had been hunters and gatherers less than forty years before.

The mission era and the creation of the homelands altered the way that people related to the country on an everyday and experiential level. Quite simply, there are places which were formerly visited quite often but are now visited relatively rarely, and places that were once temporary residential sites which are now permanently inhabited. Ancestrally important areas remain highly significant to group and individual identity, but changes in the forms of residence have had an impact, for the homeland is no longer just one of a number of camping and hunting sites within the country: rather it is the place where people live constantly, a reference point, a destination, a home and a source of pride.

The significance of the homelands grows the longer people live there, as each of these old places acquires a new kind of history (Barber, 2008). Yolngu people have, at least in one sense, created a new world for themselves, a world that is an outcome of new knowledge and new accommodations with ancient powers. The homelands are, in this sense, an example of controlled change – change over which Yolngu people were able to exercise considerably more control than they had over other colonial processes such as the establishment of missions and mines in the region. The knowledge, skills and experience that young people now gain has been shaped by these circumstances: a hybrid of the bush skills of the past with the new demands of living in a permanent settlement. Local knowledge is changed and reformed to suit new circumstances, and negotiations with the powers that sustain life continue in the light of these new conditions. The ability to conduct such negotiations is a critical part of Yolngu knowledge. Yet such adaptations in turn develop new kinds of vulnerabilities – in Blue Mud Bay, human life is now concentrated around, invested in and dependent upon key infrastructure, significant parts of it sited on low-lying coastal areas in a cyclone zone. New adaptations will be required in such circumstances.

Recent Arrivals

If you know where the madayin is … (you know) swamps. You know those places. Only these days all the piggypiggy spoil them! Those pigs, they haven't got any story about that country.

DM's statement about the *madayin* and its role in successful hunting also contained the final comment above, in which he identified an element in the bay that had no prior place in the *madayin*. The wild pigs are an introduced pest species and, along with the cane toad and the buffalo, are the most prominent introduced animal species in the tropical north of Australia. The pigs can damage the swamps and other waterways in the region, and their numbers can increase rapidly in the right conditions. For Yolngu people, like DM, the ultimate cause of this damage is that these animals 'haven't got any story about that country' – they are strangers who do not carry the *madayin* required to interact with their surroundings appropriately, and who do not follow the law of the land. In addition, there is a general aversion among Blue Mud Bay people to hunting pigs for food: some years ago when Australian quarantine officers arrived to shoot pigs to test for disease, this was considered acceptable by people, when in the past the killing of a large animal without intending to eat or otherwise use it would have been considered wasteful and inappropriate.

Buffalo are another large introduced species which can substantially damage coastal plains and waterways; buffalo also carry no *madayin* – there are no major songs and ancestral stories about them in this area. However, when I asked DM some years ago about culling this introduced animal, he looked troubled and said 'some of those buffalo are the spirits of dead Yolngu, but we don't know which ones'. At a tense moment during a sad and upsetting funeral, the unexpected arrival after dusk of two buffalo caused consternation; there were suspicions that these creatures were the manifestations of spirits come to do harm to the funeral attendees. Compared with the pigs, these postcolonial arrivals are ambiguous within the landscape – potential spirit carriers but not *madayin* – making them harder to classify and to deal with. The ongoing rise of formalized, government-sponsored indigenous land management programmes has seen the introduction of buffalo and pig culls to the area (Marika et al., 2012), but the activity remains controversial (Altman, 2012; Marika et al., 2012).

From an indigenous perspective, the British colonizers did not follow 'the law of the land' (Rose, 2000) and so there is a certain irony in non-indigenous people attempting to manage feral pigs (or other recent arrivals): beings who by and large have still not understood and respected the *madayin* properly are now attempting to limit and control the activities of other animals who also 'have not got the story'. Yet the absence of pig *madayin* means that Yolngu people's attitudes to the animals are more consistent with the perspective that a natural scientist might adopt towards a pest species in the same circumstances. The complex Yolngu response to the presence of the buffalo indicates that while concerns about environmental management may intersect with those of western environmental science, the foundations on which Yolngu knowledge rests have a different emphasis, potentially leading to different responses.

Flows and Cycles

The pig and the buffalo are useful markers of the potential for diverse Yolngu responses to broader-scale environmental changes. Elements of the *madayin* of this coastal location embody and describe water flows, tidal cycles, clouds, winds and weather, as DM describes:

The saltwater country has names for each clan or tribe. For the sea country, there are people who know about their country, about the deep sea and over to where the clouds stand. Where the big clouds arise from, that explains it further. Inland it explains to us where the clouds stand and where that place is, where it will rain. Also the floodwaters, which become the rivers [from DM].

(Buku-Larrnggay Mulka Centre, 1999)

The patterns described in these *madayin* shape coastal life in diverse ways (Buku-Larrnggay Mulka Centre, 1999; Magowan, 2001; Barber, 2005; Morphy and Morphy, 2006), not least contemporary hunting and fishing (Figure 8.2). The implications of climate

Figure 8.2 Dugong hunting at Yathikpa. (A black and white version of this figure will appear in some formats. For the colour version, please refer to the plate section.)

change and sea-level rise for such *madayin* and for the people who hold it seem likely to be significant, and people have already begun to notice changes in the weather (Petheram et al., 2010; Barber, 2011). Yet the process of Yolngu analysis of these changes in terms of ancestral power represented by that *madayin* has barely begun, and must be properly explored and understood if support for local adaptation is to be appropriate and effective.

Changing Futures

Nothing ever changes. There have never, ever been any changes [BW when asked about the madayin he was singing].

We live in a different world now [W, as reported by DM in a conversation about changes over his lifetime].

This meeting is for the future [DM, opening a meeting to discuss commercial fishing in Blue Mud Bay].

Given that it is located in Australia, a wealthy developed country, Blue Mud Bay remains a genuinely remote location. Only in the last ten years have the small settlements in this area received commercial radio and television broadcasts, while poor literacy and a lack of supply have made printed media relatively scarce. As late as 2003 there was no mains power supply for even the largest of the communities, and the roads are regularly impassable for periods of the wet season (Morphy, 1991; Barber, 2005). However, in the last decade, mains power, satellite television and regular internet connection have all arrived in the area. BW clearly articulates a central tenet of Yolngu ancestral ideology: that there are no real changes, that everything happens according to the precepts of ancestral law. Yet W, the oldest man in the region at the time, commented that Yolngu people now live in a different world. I would strongly suspect that, if asked, each man would have endorsed the statement made by the other, and I have tried to show in the preceding examples of knowledge and change how they might feel able to do this. The Aboriginal Dreaming has been called an 'everywhen' (Stanner, 2009 [1953]), a place where past, present and future lose their distinctness, and there is a very real sense in which this is true for the people of Blue Mud Bay. There are ways of knowing and understanding the world, and of perceiving change, which minimize the discontinuities with the past and emphasize the ongoing relationships with beings who occupy past, present and future simultaneously.

However, the immense changes of recent decades can hardly be denied, and the abrupt brutality of colonization remains close to living memory (Williams, 1986; Barber, 2010). People are not blind to the dynamism in human life and to the changeability of their present circumstances; but the ancestors still live in Blue Mud Bay, and have made their own accommodations with the current generation of human residents. The *madayin* has not fundamentally changed: it is simply re-expressed and reinterpreted by each successive generation in response to the challenges they face and the opportunities they have been given. The establishment of the homelands was precisely this kind of step: a vision of a collective future with its origins in the knowledge of ancestry and the important connections between people and places, yet a vision which also took account of new possibilities and new opportunities. People emphasize their connections to past practices, but acknowledge the obvious differences that now exist.

Yet contemporary remote indigenous settlements and homelands are frequently characterized as vulnerable and dependent rather than resilient. Despite living in a wealthy developed country, indigenous Australians are a minority that have low socioeconomic status and suffer from high rates of crime, unemployment and ill-health. Federal and state authorities have implemented numerous policies and initiatives over previous decades to try to improve this situation, but their effects have been diminished by issues such as institutionalized discrimination, poor implementation, insufficient funding, lack of consultation and an absence of long-term continuity. More recently, poor indigenous socioeconomic status and past policy failures have been used as a justification for recent federal government interventions in the Northern Territory to reduce the powers of local indigenous peoples to manage their own administrative affairs, to introduce a system of tightly regulating welfare payments and to curtail other freedoms (Altman and Hinkson, 2007). The government also removed certain rights to land granted under previous acts of parliament. Poor socioeconomic indicators are correlated with policy and political vulnerabilities, creating ongoing challenges for indigenous people in places like Blue Mud Bay.

Conclusion

In terms of climate change, considerable recent work has gone into identifying potential climate-related vulnerabilities, impacts and adaptations for indigenous people across Australia (Green et al., 2010; Petheram et al., 2010; Langton et al., 2012). These identify existing challenges and provide guidance for future research and action, notably with respect to: sensitivity and exposure to climate risk; vulnerability and adaptive capacity; population movement and displacement; and the relationship with such issues as extreme weather events and biodiversity protection (Langton et al., 2012). The studies frequently highlight that sensitivities and vulnerabilities to climate change impacts are highly dependent on poor underlying socioeconomic and health indicators. Blue Mud Bay residents receive significant external income from state and other sources, and so a major shift in rainfall and weather patterns does not spell instant disaster for them. Yet these are places where income is comparatively low and subsistence hunting is correspondingly still significant (in social as well as economic terms). They are also places where people are highly sensitized to their surroundings and to the meanings that changes in these surroundings can convey. Such a situation would suggest that even gradual changes in local environments will be noted and important, and careful attention will need to be paid to ongoing (and potentially) increasing climate change vulnerabilities.

Yet the knowledge Yolngu people possess can also give them confidence about their capacity to negotiate possible futures:

Yolngu have been here for 50,000 years and we have survived many changes in the past. It is going to affect you guys, not me. Because I've done it in the past. If the store runs out of food, that will simply make people go back to the bush and start eating healthy again [JG commenting on the idea of human-induced climate change].

JG's comment eloquently demonstrates the knowledge people have of the duration of their own presence on the country and of how to survive on it unsupported, providing a basis for confidence in the face of future changes. Complementing the sophisticated conceptions of coastal ancestral flows and associations in the *madayin* has come Yolngu absorption of local archaeological studies, which have demonstrated evidence of human habitation of past environments when the sea level was different (Faulkner, 2011). In the early 2000s, climate change was not an explicit topic of conversation in this region. A decade and more later it has a higher profile, but remains peripheral when compared with both the short-term pressing issues that most remote indigenous Australians have to contend with (Altman and Hinkson, 2007; Petheram et al., 2010), and the long-term sense of confidence provided by Yolngu knowledge of their own country and of the duration of their collective residence.

Yet in the future, political and environmental challenges may intersect more strongly, as for example external support for ongoing homeland maintenance or for recovery following extreme events becomes harder to find for these remote, low-lying coastal settlements. At the moment it is the political and economic domains, rather than the usually slower processes of environmental or climate change, that have provided the most fundamental recent obstacle to the knowledge, practices and ways of being that are so important to the individual and collective identities of Blue Mud Bay people. Sustainable and resilient futures will require a stronger understanding of the bases of Yolngu knowledge, likely local responses to environmental change, and the consequences of intersecting political and environmental risks to these unusual and valuable places.

Acknowledgements

My thanks go first and foremost to the people of Blue Mud Bay, who generously gave their time and knowledge over the periods that I have lived among them. Material presented in this chapter was gathered while I was based at the Australian National University and at James Cook University, and was financially supported by the Marine Conservation Biology Institute via the Mia J. Tegner Memorial grant. I thank those institutions for their support. Further thanks go to Peter Bates for editorial comment on a previous version of this chapter, to UNESCO LINKS programme, and to anonymous reviewers supplied by UNESCO LINKS and the CSIRO.

References

Agrawal, A. 1995. Dismantling the divide between indigenous and scientific knowledge. *Development and change*, 26: 413–39.

Agrawal, A. 2009. Why 'indigenous' knowledge? *Journal of the Royal Society of New Zealand*, 39: 157–8.

Altman, J. C. 1987. *Hunter-Gatherers Today: An Aboriginal Economy in North Australia*. Canberra: Australian Institute of Aboriginal Studies.

Altman, J. C. 2002. *Sustainable Development Options on Aboriginal Land: The Hybrid Economy in the Twenty-First Century*. Canberra: Centre for Aboriginal Economic Policy Research (CAEPR).

Altman, J. C. 2012. Indigenous futures on country. In Altman, J. C. and Kerins, S. (eds.) *People on Country: Vital Landscapes, Indigenous Futures*. Sydney: Federation Press, pp. 213–31.

Altman, J. C. and Hinkson, M. (eds.) 2007. *Coercive Reconciliation: Stabilise, Normalise, Exit Aboriginal Australia*. Melbourne: Arena Publications.

Altman, J. C. and Kerins, S. (eds.) 2012. *People on Country: Vital Landscapes, Indigenous Futures*. Sydney: Federation Press, pp. 213–31.

Barber, M. 2005. *Where the Clouds Stand: Australian Aboriginal Attachments to Water, Place, and the Marine Environment in Northeast Arnhem Land*. Unpublished Ph.D. Thesis, Canberra, National University.

Barber, M. 2008. A place to rest: Dying, residence, and community stability in remote Arnhem Land. In Glaskin, K., Tonkinson, M., Musharbash, Y. and Burbank, V. (eds.) *Mortality, Mourning, and Mortuary Practices in Indigenous Australia*. London: Ashgate Publishing, pp. 153–70.

Barber, M. 2010. Coastal conflicts and reciprocal relations: Encounters between Yolngu people and commercial fishermen in Blue Mud Bay, Northeast Arnhem Land. *The Australian Journal of Anthropology*, 21: 298–314.

Barber, M. 2011. 'Nothing ever changes': Historical ecology, causality and climate change in Arnhem Land, Australia. In Musharbash, Y. and Barber, M. (eds.) *Ethnography and the Production of Anthropological Knowledge: Essays in Honour of Nicolas Peterson*. Canberra: ANU e-Press, pp. 89–100.

Barber, M. and Jackson, S. 2015. 'Knowledge making': Issues in modelling local and indigenous knowledges. *Human Ecology*, 43(1): 119–30.

Biernoff, D. 1978. Safe and dangerous places. In Hiatt, L. (ed.) *Australian Aboriginal Concepts.* Canberra: Australian Institute of Aboriginal Studies.

Bradley, J. 2001. Landscapes of the mind, landscapes of the spirit. In Baker, R., Davies, J. and Young, E. (eds.) *Working on Country: Contemporary Indigenous Management on Australia's Lands and Coastal Regions.* Melbourne: Oxford University Press, pp. 295–304.

Bradley, J. 2010. *Singing Saltwater Country: Journey to the Songlines of Carpentaria.* Melbourne: Allen and Unwin.

Buku-Larrnggay Mulka Centre (ed.) 1999. *Saltwater: Yirrkala bark paintings of sea country.* Neutral Bay, NSW: Buku-Larrnggay Mulka Centre in association with Jennifer Isaacs Publishing.

Ellen, R. 2004. From ethno-science to science, or what the indigenous knowledge debate tells us about how scientists define their project. *Journal of Cognition and Culture*, 4: 409–50.

Faulkner, P. 2011. Late Holocene mollusc exploitation and changing near-shore environments: A case study from the coastal margin of Blue Mud Bay, northern Australia. *Environmental Archaeology*, 16: 137–50.

Green, D., Jackson, S. and Morrison, J. (ed.) 2010. *Risks from Climate Change to Indigenous Communities in the Tropical North of Australia.* Canberra: Department of Climate Change and Energy Efficiency.

Ingold, T. 2000. *The Perception of the Environment: Essays on Livelihood, Dwelling, and Skill.* London: Routledge.

Keen, I. 1994. *Knowledge and Secrecy in an Aboriginal Religion.* Oxford: Oxford University Press.

Langton, M., Parsons, M., Leonard, S. et al. (ed.) 2012. *National Climate Change Adaptation Research Plan for Indigenous Communities.* Gold Coast: National Climate Change Adaptation Research Facility.

Laynhapuy Homelands Aboriginal Corporation. 2013. *Yirralka Rangers business plan 2013–2016*, http://docs.google.com/viewer?a=v&pid=sites&srcid=bGF5bmhhcHV5LmNvbS5hdXx3d3d8 Z3g6MTM0NGY5NmQwNzUzZWUz

Magowan, F. 2001. Waves of knowing: Polymorphism and co-substantive essences in Yolngu sea cosmology. *The Australian Journal of Indigenous Education.* 29: 22–35.

Marika, B., Munyarryun, B. Munyarryun, B., Marawili, N. and Marika, W., facilitated by Kerins, S. 2012. Ranger djama? Manymak! In Altman, J. and Kerins, S. (eds.) *People on Country: Vital Landscapes, Indigenous Futures.* Sydney: The Federation Press, pp. 132–45.

Morphy, H. 1984. *Journey to the Crocodile's Nest: An Accompanying Monograph to the Film 'Madarrpa Funeral at Gurka'wuy'.* Canberra: Australian Institute of Aboriginal Studies.

Morphy, H. 1991. *Ancestral Connections: Art and an Aboriginal System of Knowledge.* Chicago, IL: University of Chicago Press.

Morphy, H. 1996. Empiricism to metaphysics: In defense of the concept of the dreamtime. In Bonyhady, T. and Griffiths, T. (eds.) *Prehistory to Politics: John Mulvaney, the Humanities, and the Public Intellectual.* Melbourne: Melbourne University Press.

Morphy, H. and Morphy, F. 2006. Tasting the waters: Discriminating identities in the waters of Blue Mud Bay. *Journal of Material Culture*, 11: 67–85.

Petheram, L., Zander, K. K., Campbell, B. M., High, C. and Stacey, N. (2010). 'Strange changes': Indigenous perspectives of climate change and adaptation in NE Arnhem Land (Australia). *Global Environmental Change*, 20: 681–92.

Poirer, S. 2005. *A World of Relationships: Itineraries, Dreams, and Events in the Australian Western Desert.* Toronto: University of Toronto Press.

Povinelli, E. 1992. 'Where we gana go now?': Foraging practices and their meanings among the Belyuen Australian aborigines. *Human Ecology*, 20: 169–201.

Povinelli, E. 1993. 'Might be something': The language of indeterminacy in Australian aboriginal land use. *Man*, 28: 679–704.

Rose, D. B. 1996. *Nourishing Terrains: Australian Aboriginal Views of Landscape and Wilderness.* Canberra: Australian Heritage Commission.

Rose, D. B. 2000. *Dingo Makes us Human: Life and Land in an Australian Aboriginal Culture.* Cambridge: Cambridge University Press.

Rose, D. B. 2005. An indigenous philosophical ecology: Situating the human. *The Australian Journal of Anthropology*, 16: 294–305.

Sillitoe, P., Dixon, P. and Barr, J. 2005. *Indigenous Knowledge Inquiries*. Rugby: Practical Action Publishing.

Smyth, D., Yunupingu, D. and Roeger, S. 2010. Dhimurru indigenous protected area: A new approach to protected area management in Australia. In Paincmila Walker K., Rylands, A., Woofter A. and Hughes, C. (eds.) *Indigenous People and Conservation: From Rights to Resource Management*. Arlington, VA: Conservation International.

Stanner, W. 2009 [1953]. *The Dreaming and Other Essays*. Melbourne: Black Inc. Publishing.

Thomson, D. 1975. The concept of 'marr' in Arnhem Land. *Mankind*, 10: 1–10.

Van Onselen, P. 2008. *Year of Reckless Vows*. Sydney: Australian News.

Williams, N. 1986. *The Yolngu and Their Land: A System of Land Tenure and the Fight for its Recognition*. Canberra: Australian Institute of Aboriginal Studies.

9

Coping with Climate: Innovation and Adaptation in Tibetan Land Use and Agriculture

Jan Salick, Anja Byg, Katie Konchar and Robbie Hart

> Climate change is widely regarded as one of the greatest challenges of the 21st century
>
> (IPCC, 2013, 2014; Wuebbles et al., 2014)

In many areas of the world climate change impacts are already felt in the form of changed weather patterns, shifting seasons and extreme events (IPCC, 2012, 2013, 2014). Indigenous farmers in developing regions of the world are often left to interpret as well as to deal with its impacts by themselves (Salick and Byg, 2007; Salick and Ross, 2009). Here we report on a study carried out with Tibetan farmers on the impacts of climate change on their livelihoods. The study took place in six villages surrounding sacred Mt Khawa Karpo in the Tibetan Autonomous Prefecture (TAP) of Yunnan Province, China (Figure 9.1). Agropastoralism and Tibetan livelihoods across the eastern Tibetan Plateau and Himalaya are affected strongly by climatic changes (Salick et al., 2005; Yeh et al., 2013; Haynes et al., 2014; Klein et al., 2014; Xu and Grumbine, 2014). Increasing temperatures and water availability have impacts across the Tibetan agropastoral system (Lin et al., 2006; Xu et al., 2009). In order to survive, local Tibetan communities must adapt and adjust their livelihood systems to a daunting combination of milder winters, unpredictable monsoonal changes and shifts in local biodiversity. Informed by (1) participatory mapping and agricultural calendar building; (2) discussions with groups of farmers; and (3) interviews with local and regional professionals, this chapter outlines the impacts of climate change on Tibetan agriculture and natural resource use near Mt Khawa Karpo. It examines how farmers attempt to deal with these impacts and how climate-driven changes interact with other ongoing changes.

Methods

Study Site

The TAP of north-west Yunnan, China, borders on the Tibetan Autonomous Region (TAR). The upper reaches of four of Asia's great rivers – the Yangtze (Jinsha), Mekong (Lancang),

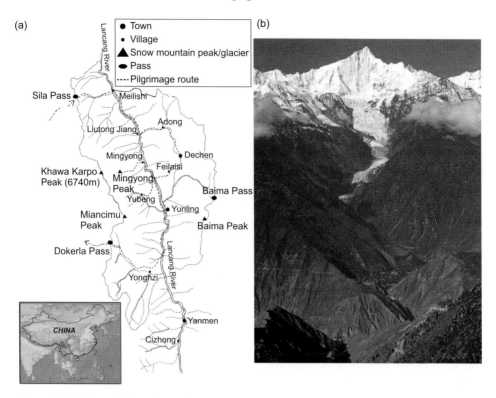

Figure 9.1 (a) Map of the study area in the easternmost Himalaya; (b) Mingyong village, Mingyong glacier and Tibetan sacred Mt Khawa Karpo. (A black and white version of this figure will appear in some formats. For the colour version, please refer to the plate section.)

Salween (Nu) and Irrawaddy (Dulong) flow through the area within 90 km of each other. This mountainous region of the eastern Himalaya is part of the Hengduan Mountains, a world biodiversity hot-spot (Mittermeier et al., 1998). Near Mt Khawa Karpo (Figure 9.1), the highest mountain in Yunnan and a Tibetan sacred peak, we conducted participatory rural appraisal (PRA) activities in six Tibetan villages chosen for livelihood variation from tourist villages to traditional agropastoralism. To provide broader perspectives and contexts, Tibetan religious, agricultural and environmental professionals also were interviewed in and around Dechen (county seat) and Shangri-la (TAP prefecture seat). After repeated referral, the primary author (JS) also interviewed Tibetan calendar makers in Lhasa (TAR provincial seat).

Participatory Rural Appraisal with Farmers

To study the effects of climate change on Tibetan agriculture and land use from the perspectives of Tibetan farmers, we used two PRA techniques (Chambers, 1994a, 1994b, 1994c): (1) village and farm mapping; and (2) diagramming yearly farm calendars and activity cycles (Figure 9.2a). First, groups of farmers (groups of between four and twelve people) mapped their farms and villages and diagrammed the farming cycle resulting in six maps and six calendars from six villages. We then asked the farmers to point out changes in these maps and calendars that had taken place over the last twenty plus years, along with causes of changes. Specifically, we inquired about long-term changes in weather patterns and their effects on farming and land-use patterns and on their yearly calendar as depicted. Finally, we opened the discussion, encouraging personal observations and reflections on agriculture and climate over the last twenty years.

Semi-structured/Open-ended Interviews with Local Tibetan Professionals

To provide a broader and more quantifiable context, forty personal interviews were conducted with local Tibetan agricultural, environmental and religious professionals (Tibetan Buddhist monks, Figure 9.2b), all we could contact in the towns of Dechen and Shangri-la. Tibetan calendar and astrological experts who make Tibetan almanacs were also interviewed in Lhasa after we were referred for explanations to them by many Tibetan farmers. Questions revolved around traditional agropastoral practices and the yearly calendar, agricultural changes over the last twenty years or more, changes in climate over this time period and agropastoral effects of these changes. Since little information on global climate change in this area was available to either professionals or farmers at the time of the study (summer 2007 and 2009), few Tibetans had systematically considered these issues to any great extent. Nonetheless, farmers and professionals all had many and detailed observations to relate, and many concerns to express.

In the following section, we present results by subject, along with brief introductory and comparative summaries and citations.

Climate Change in the Tibetan Autonomous Prefecture

For the Tibetan areas of the Himalaya, the Tibetan Plateau and the Hengduan Mountains, climate defines natural habitats and shapes traditional Tibetan livelihood practices (Tulachan, 2001; Goldstein and Beall, 2002; Aldenderfer and Zhang, 2004; Salick et al., 2005; Salick and Moseley, 2012). The parallel river valleys and north/south mountain ranges characteristic of north-west Yunnan Province include habitats ranging from subtropical to temperate to alpine (Salick et al., 2005; Ma et al., 2006). This area typically experiences two seasons each year, one wet (May–October) and one dry (November–April), due largely to the Indian monsoon. However, recent rises in temperature, irregular precipitation, glacial retreat and monsoonal changes have burdened Tibetans practising agropastoralism and increased threats to Tibetan cultural traditions (Byg and Salick, 2009; Klein et al., 2011; Salick and Moseley, 2012; Bai et al., 2013).

In the past several decades, annual temperatures have gradually increased across Yunnan Province and the Tibetan Plateau (Liu and Chen, 2000; He and Zhang, 2005; Xu et al., 2008). With observed increase in annual mean temperature of 0.16°C per decade and in

Figure 9.2 Participatory and interview methods: (a) villagers drafting maps using calendars; (b) interviewing Tibetan monks and professionals. (A black and white version of this figure will appear in some formats. For the colour version, please refer to the plate section.)

mean winter temperature of 0.32°C per decade (Meehl et al., 2007; Kang et al., 2010), temperatures across the high elevation Tibetan Plateau have increased earlier, faster and with larger variation in magnitude than most of the northern hemisphere outside the Arctic (Liu and Zhang, 1998; Jones et al., 2007; Meehl et al., 2007; Xu et al., 2009). According to IPCC reports, annual temperatures are projected to continually increase across the Tibetan Plateau region over this century (Christensen et al., 2007, 2013).

In recent years, northwest Yunnan has seen fluctuations in the timing and duration of the rainy season as well as an increase in the intensity of storms (Ding et al., 2006; Xu et al., 2009). Given the complex topography of the Himalaya, projecting future precipitation change is particularly difficult (Jones et al., 2007). Although the IPCC predicts that precipitation increases across the Tibetan Plateau (Meehl et al., 2007; Christensen et al., 2013), changes are likely to be spatially specific and exhibit greater annual variability (He and Zhang, 2005; Kang et al., 2010). Some Tibetan areas of the Himalaya and Plateau may experience periods of water stress as water availability is affected by multiple factors: fluctuating precipitation, changes in seasonal monsoon patterns (Christensen et al., 2007, 2013), and the loss of the buffering functions of glaciers and permanent snow cover (Xu et al., 2009).

The glaciers of the Himalaya form the largest body of ice outside of the polar ice caps and, as such, this region is often referred to as the third pole (Qiu, 2008). Increasing temperatures in the region have resulted in easily observable reductions in glacial size and extent in many areas of the Himalayas (Rai, 2005; Moseley, 2006; Baker and Moseley, 2007; Sheehan, 2008). In recent years, the rate of glacial retreat in western China has significantly increased (Lin et al., 2006). The lower latitude Mingyong glacier on Mt Khawa Karpo (Figure 9.1b) is the fastest retreating glacier in the region (Sheehan, 2008; Li et al., 2010; Salick and Moseley, 2012).

Based on results from the PRA in six Tibetan villages surrounding the sacred peak of Khawa Karpo and the forty interviews with Tibetan professionals, we saw that Tibetans were acutely aware of changes in the local climate despite their lack of access to outside information. Their observations were in line with similar studies in the region (Salick and Byg, 2007; Byg and Salick, 2009; Chaudhary and Bawa, 2011) and concur with scientific evidence for rapid climate changes in the eastern Himalaya (He and Zhang, 2005; Baker and Moseley, 2007; Christensen et al., 2007; Cruz et al., 2007; IPCC, 2013). Somewhat surprisingly, changes in precipitation were discussed by more Tibetans (75 per cent of respondents) than rising temperatures (50 per cent of respondents; Figure 9.3a). This seemed to be because, although temperatures are rising, this increase caused less hardship and was often welcomed in the high elevation villages that were previously snowed in for a very long winter. Changing and variable precipitation, on the other hand, caused many problems for farmers and were thus of great concern. We were told of many extreme events including floods, droughts, hail, extraordinary snows and landslides (29 per cent of respondents). Changing seasons and their unpredictability were discussed (29 per cent of respondents). In drawing seasonal calendars, Tibetans illustrated and explained that seasons were now running together, when before they were more distinct. Traditionally, there were six Tibetan seasons – early winter, late winter, early spring, late spring, summer and autumn – with winter being the longest. Recently, winter had been the shortest season and there was little predictable differentiation between early spring and late spring or between early autumn and late autumn. Rains and snows were increasingly unusual in timing, amount and variability. Thus, villagers said it had become more difficult to depend on the Tibetan agricultural calendar, and they continually referred us to the makers of the published calendar in Lhasa. For people who were in contact

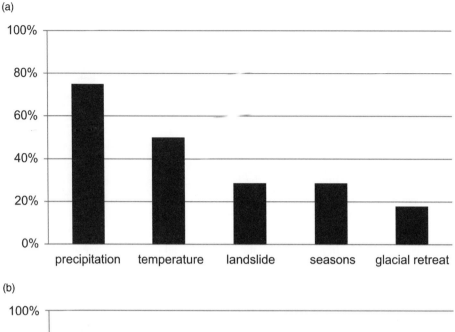
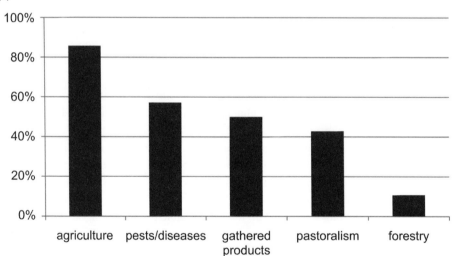

Figure 9.3 Percentage of respondents reporting (a) climate change perceptions; and (b) land-use impacts.

with glaciers (18 per cent of respondents), they were universally seen to be shrinking. However, many villagers and professionals who lived away from glaciers were unaware of glacial retreat. One elderly Tibetan woman recounted a trip to the glacier in her youth when it came almost all the way down to the village of Mingyong (village at base of Figure 9.1b); when we told her it was now almost a kilometre above the village, she was deeply disturbed and saddened.

Changing Tibetan Land Use

The Tibetan agricultural system is described as agropastoral transhumance: a mixed crop-livestock system that involves herding livestock across a range of elevations (Goldstein and Beall, 2002; Bauer, 2003; Yi et al., 2007). Field crops of the TAP include several cereals, such as Tibetan barley, wheat, maize and buckwheat; root crops including turnips and potatoes; vegetables such as cabbage and onions; and garden fruit and nut trees such as apples and walnut. Livestock include milk and draft animals, such as yak, cattle, sheep and goats; pack animals, such as yaks, horses, donkeys and mules, as well as pigs and chickens. Forest products, including timber for buildings and non-timber products – especially mushrooms, medicinal plants, firewood and soil amendments – also supplement the Tibetan livelihood system near Mt Khawa Karpo (Salick et al., 2005). Sacred space is a significant component of Tibetan landscape that preserves biodiversity, special natural products and old-growth forest (Anderson et al., 2005; Salick et al., 2007).

Tibetan Agriculture

The most frequently discussed effects of climate change in this study (Figure 9.3b) were agricultural challenges and changes (85 per cent of respondents). Seasons, water, soils and crops were discussed at length. In reviewing problems with changing seasons, Tibetan monks and religious experts related these to origin myths central to their identity and culture.

In the beginning, plants planted themselves and people gathered them. However with time, wild plants were over-harvested and people had to save seed and plant crops. In order to do this, the year – previously a random assortment of cold and warm days, rainy, snowy and clear days – was divided into seasons so that people would know when to plant and when to harvest

Local Tibetan cultural historian.

Recently, however, it seemed as though the seasonal organization provided to support agriculture was deteriorating and agriculture was suffering. Traditional Tibetan calendars published by specialized Tibetan astrologers were becoming difficult to use. Elders, who had traditional knowledge and experience about weather and planting, could no longer predict when to plant. Farmers were going increasingly to monasteries and temples to pray so that they would plant at a good time. Timing was made especially difficult as each crop had its own timing and weather requirements and new varieties of each crop were different. For some crops, flowering and harvesting were becoming earlier. With warmer temperatures and shorter winters, some villages could plant more rotations including winter grains, which in turn changed the traditional agricultural cycles and workloads. On the other hand, when seasons were bad, productivity declined and harvests were poor. Tibetans near Mt Khawa Karpo use many indicator species to tell when to conduct agricultural activities (Salick et al., 2005). Tibetans reported that these too were failing with climate change. Indicator flowers, such as rhododendrons, were blooming earlier (Hart et al., 2014), not in sync with agricultural activities (Hart and Salick, 2017). Indicator animals, such as cuckoos could be heard at unusual times of year; cranes were staying longer and eating ripened grain.

Precipitation and its unpredictability were also problematic for agriculture. Like changing temperature, changing precipitation affected planting and harvesting times, crop growth and productivity. Newly planted crops were shrivelling with unexpected droughts and matured fields of grain were rotting with extended monsoons. Lack of water for irrigation and evaporation of water from holding ponds were problems in some low, warm, dry Tibetan villages. Extreme events brought dreaded hail that demolished crops.

One reported change in Tibetan agriculture that we had not anticipated was the increasingly rapid breakdown of organic matter and the concomitant changes in soil management. Traditionally, Tibetans incorporate huge amounts of organic matter – leaves, sticks, detritus and manure – into their highly erodible mountain soils to retain both the soil itself and valuable soil nutrients, texture and water holding capacity (Salick et al., 2005; Salick and Moseley, 2012). Now, a few farmers complained that they needed to incorporate ever-increasing amounts of organic matter and that it was disappearing faster and faster. Soil organic matter is known to decay more rapidly in warmer climates (Sanchez, 1977); however the influence of climate change on soil organic matter decomposition rates in the Himalaya is a factor seldom discussed in the literature and would require investigation by agronomists and soil scientists.

Crops were changing rapidly and the number of crops and new varieties grown were increasing, making the Tibetan agricultural system more complex with much for traditional farmers to learn and to innovate. New, introduced varieties were replacing traditional Tibetan varieties developed locally. Tree crops such as walnuts and fruits were being grown at higher elevations. Cultivation of medicinal plants was promoted by the Chinese government. Some traditional crops such as rice were expanding their ranges into Tibetan areas. Maize, planted for centuries since its introduction into Asia at low elevations, was now being grown at higher and higher elevations. Some traditional crops were abandoned. This included culturally important buckwheat, a high elevation and very time-consuming crop, which now was being replaced by either lower elevation crops with expanding ranges or by other activities (e.g. tourist trade, gathered products and cash labour). Grapes were the outstandingly successful example of adaptation and innovation with a new crop (see below).

Pests and Diseases

Pests and diseases, of course, also affect agriculture but we present these separately because of the emphasis farmers and professionals (57 per cent of respondents) placed on pests and diseases with changing weather, and their inclusion of animal and human pests and diseases in these discussions. There were increasingly serious pest attacks attributed to changing weather that favoured diseases and insects. These were also associated with Tibetan beliefs that unclean and incorrect living (spiritual pollution) was punished by plagues. Mice were no longer killed by severe winters and were multiplying rapidly and out of control. Flies and mosquitoes were also increasing quickly and moving to ever higher elevations. The villagers had not yet experienced malaria but had heard that it was moving into Tibetan lands starting at the lowest altitudes (<2000m). Farm animals, particularly pigs, were succumbing to new and more virulent diseases. For crops generally, insects were bad when it was unusually hot and dry, while diseases flourished when rains persisted. Additionally, reductions of fallow periods with intensifying agriculture increased insects and weeds. New weeds were appearing that villagers had never seen previously and new insects were causing problems especially on new crops; for example, bees and wasps were reported as pests of the expanding grape crop.

Gathering

Half the Tibetans interviewed (50 per cent of respondents) were concerned with the effects of climate change on gathered products, however, it was acknowledged that over-harvesting also contributed to declining populations. Among products most frequently discussed were mushrooms (especially matsutake/*Tricholoma* and chung cao/*Ophiocordyceps*, Salick et al., 2005), medicinal plants (especially snow lotus/*Saussurea*, Law and Salick, 2005, and baimu/*Fritillaria*, Konchar et al., 2011) and pine nuts. Collectors were less certain of weather conditions favouring these products than farmers were of their crops. Nonetheless, these products provided a very profitable cash income to local Tibetans who were concerned about the health of these populations under the combined effects of climate change and over-harvesting.

Pastoralism

Yak herding along elevational gradients with the changing seasons has historically been an important part of Tibetan culture and this 'vertical transhumance' is still practised today (Goldstein and Beall, 2002; Bauer, 2003; Salick et al., 2005).

Grazing lands include high elevation grasslands and alpine meadows, typically grazed during the summer. Yaks and other grazing animals are let loose in fallowed fields in spring and autumn when agricultural crops are being planted and harvested. In winter, they are stabled with hay or allowed to pasture near villages (Xie et al., 2001). We were told that with climate change – earlier springs and later autumn – grazing extended over much longer seasons than previously. Families could not accomplish their agricultural tasks and simultaneously graze their animals (43 per cent of respondents). Some Tibetan families and communities were sharing the responsibility of herding livestock collectively, sending out one or more herders while other villagers remained to carry out agricultural work. Yaks were grazed on increasingly higher pastures for longer periods of the year with unpredictable weather conditions, including extreme events, which made herding much more arduous and lowered yak productivity.

Forestry

Only 11 per cent of participating Tibetans associated forestry with climate change. Those who did were unaware of carbon sequestration. They were worried that deforestation during the Cultural Revolution violated Buddhist bans on tree cutting: if too many trees were cut, resulting spiritual pollution would bring retribution in the form of extreme weather or plagues – as they were witnessing. At the same time, Tibetans linked spiritual gains with rising treelines and recent afforestation at higher elevations associated with climate change (Salick et al., 2005; Salick and Moseley, 2012). Thus with both negatives and positives, the relationship of forests to climate change seemed ambiguous to local Tibetans.

Tibetan Calendar

Repeatedly, Tibetan villagers and professionals referred us to Tibetan calendars published in Lhasa, Tibet. When we queried about climate changes and effects on land use and livelihoods, local Tibetans often deferred to specialists. So the primary author followed this advice and interviewed the Lhasa Tibetan astrologers who annually create Tibetan almanacs (Figure 9.4). The eldest of these astrologers descended from ancestors who had been astrologers continuously, dating back 1,200 years, since the first calendar was produced. He explained that climate change is integrated into calendars by the astrological readings as well as by pragmatic experimentations in villages.

Traditionally the astrologers spent months every year in villages to incorporate empirical observations into the calendars. The Tibetan monks who formulated the Tibetan calendar remarked on their struggle to interpret so many recent changes in the environment, along with traditional Tibetan astrology and pragmatic knowledge about the landscape and local ecology. Shifts in planting dates have been generally earlier.

One monk noted that three alternative auspicious planting dates were always provided – to account for uncertainty – and reported that in the recent past the earliest of the three dates has consistently been the best. It would be a fascinating study for a Tibetan scholar to test the direction and extent of shifts in planting dates through an intensive investigation of centuries of archived Tibetan calendars. The study could further inform agricultural adaptation strategies in the region as well as provide historical, cultural and spiritual contexts for climate change.

The Tibetan astrologers and monks detailed threats to Tibetan culture and health posed by climate change. Traditional yak fur robes were no longer practicable to wear in warming temperatures exceeding 30°C. The traditional Tibetan diet high in yak milk, cheese, red meat and fats was no longer metabolized to keep people warm; as a result Tibetans increasingly suffered from heart and respiratory diseases. Traditionally foods were never refrigerated, but now with a warming climate people suffered from intestinal bacteria as their foods spoiled. Plagues spawned by the warming climate threatened traditional livelihoods and caused suffering. Climate change, the venerable astrologer submitted, may deliver the final blow to Tibetan culture.

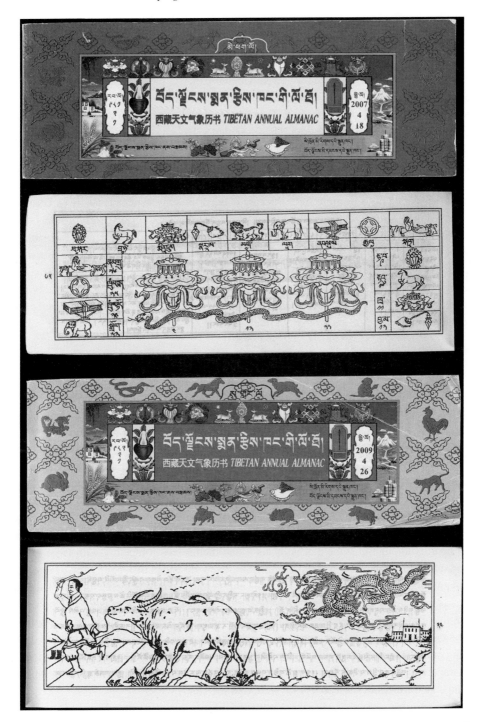

Figure 9.4 Tibetan calendars or almanacs. (A black and white version of this figure will appear in some formats. For the colour version, please refer to the plate section.)

Causation, Adaptation, Mitigation and Policy

Causation

With the exception of a few interviewed Tibetan experts, most of the people who took part in the study had no prior information about climate change. Instead they interpreted the changes they had observed from Buddhist perspectives of causality combined with other changes happening in the area. These seemed to have been shared among people and villages since we repeatedly heard the same explanations. One of the most popular was that during the Cultural Revolution too many trees had been cut and Buddhist treasures stolen, which broke the heart of Mt Khawa Karpo. Traditional Tibetan land use has taboos on cutting sacred trees or more trees than are needed for subsistence. Treasures are sacred objects of great value that have been hidden in the distant past to bring good fortune. Other people thought that there were too many non-Tibetans moving into the area (subsidized by the Chinese government) along with too many new buildings. Too many tourists (also supported by the Chinese government) and their pollution (plastic bags, defecation, feet washing, etc.) defiled the environment and caused changes in the climate. Electricity and dams, we were told, had caused the climate to change. In general, good deeds caused good weather, whereas bad deeds brought climate change.

Adaptation

One of the outstanding results of this research was the creative and varied adaptations and innovations that Tibetan villagers attempted in response to climate change. These were not passive victims. Tibetans were actively observing, experimenting and evaluating alternative land uses and livelihood strategies. New crops and varieties were continuously being tested in small plots and, if successful, planted to increasingly larger areas. Similarly, animal breeders attempted to counter new pests and diseases. Cash crops and gathered products (mushrooms, medicinal plants, orchids) and wage labour were increasingly attractive both as traditional crops faltered and cash became necessary (e.g. for school, clothing, tools). Expanding tourism provided cash through merchandizing, guide services, lodging and board; although in some places locals found themselves competing with large tourist conglomerates from Shanghai and Hong Kong. New divisions of labour were developing to maintain traditional agricultural activities and vertical transhumance while also guiding tourists to far-flung destinations, providing services to tourists in the village, and gathering lucrative non-timber products (e.g. *chung cao*). Amid this dynamic search for adaptations, Tibetans increasingly sought spiritual solutions to climate changes through prayers, pure living and offerings to religious spirits, shrines, monasteries, monks and living Buddhas.

Grape and wine production were the outstandingly successful adaptations to climate change near Mt Khawa Karpo. As we were told the story:

In the mid-nineteenth century, French missionaries established a lovely Catholic church in nearby Cizhong [Figure 9.5a; its arched chapel ceiling beautifully illuminated with colourful Himalayan wild

Figure 9.5 Tibetan wine production: (a) Catholic Church in Cizhong and (b) the award-winning 'Sun Spirit' ice wine. (A black and white version of this figure will appear in some formats. For the colour version, please refer to the plate section.)

flowers]. Within the high walled church-yard the French priests and monks grew Cabernet Sauvignon grapes that they had brought from France for their much beloved wine. In 1952 the Cizhong vicar Francis Goré finally abandoned the church and fled to Hong Kong as the Communists formed the People's Republic and exerted control over the borderlands [also see Moseley, 2011]. Nonetheless, the local Tibetans, trained by the French, continued to manage the small vineyard within the warm, walled microclimate. As climate change set in, the warmth outside the church walls became suitable for grape production. Tibetans distributed cuttings of the French grapes and vineyards thrived throughout the area.

Grapes were very successful cash crops, completely displacing traditional Tibetan agriculture in some particularly suitable villages. But Tibetans were not satisfied with this success; they had seen wine produced and consumed and they were determined to add value. Tibetan wineries were established and Tibetans went to Europe to learn more. In mountain areas where frosts and snows are rapid, they saw ice wines produced. Ice wine is a type of dessert wine produced from grapes that have been frozen while still on the vine. Ice wine takes delicate management and the grapes once frozen must be harvested and processed immediately. The Tibetans near Mt. Khawa Karpo developed ice wine to the point where they won second prize in an international ice wine contest with a 'Sun Spirit' label (Figure 9.4b). However, the success of wine making in the area attracted large-scale investors who have since taken over most of the added value part of production, leaving villagers to produce the grapes only.

Mitigation

Interviews with Tibetans in the area of Mt Khawa Karpo managing their natural resources through traditional ecological knowledge uncovered at least five major ways that they were mitigating climate change: biodiversity conservation, afforestation, agroforestry, soil amendment and extremely low carbon emissions. Tibetans were not aware that their traditional practices were mitigating climate change any more than they were aware of global climate change itself. Nonetheless, their practices should be noted and potentially upscaled.

- Throughout the Himalaya, sacred sites, sacred mountains and whole mountain ranges are conserved for their sanctity, which also has the ecologically beneficial effect of harbouring great biodiversity (Anderson et al., 2005; Salick et al., 2007; Shen et al., 2012) that sequesters carbon from the atmosphere by reducing land conversion and deforestation. Mt Khawa Karpo, the site of this research is a Tibetan sacred mountain (Salick and Moseley, 2012, Figure 9.1b).
- In many places where indigenous Tibetan practices are informed by traditional ecological knowledge, there was increased forest cover (afforestation). In contrast, where non-indigenous people enter sub-alpine environments, deforestation (through logging and agriculture) can be a serious problem (Xu and Wilkes, 2004; Willson, 2006).
- Taking advantage of warmer microclimates along elevational gradients, Tibetans traditionally cultivate and manage many tree crops, including walnuts, apples, pomegranates, pears, quince, citrus and sea buckthorn. As climatic conditions in the Himalaya change, tree crop cultivation, like the natural treeline, was extending to higher elevations. Such indigenous agroforestry sequesters carbon.
- The incorporation of mulch and manure into soils (Salick et al., 2005; Salick and Moseley, 2012) is a common practice in the Himalaya, where steep slopes and newly formed mineral soils are the norm. This addition of organic matter reduces soil erosion and is a seldom recognized method of carbon sequestration.
- Traditional Tibetan livelihoods are very low in carbon emissions. Himalayan farmers sequestered great amounts of carbon as described above while releasing very little carbon into the environment. They do burn wood and dung, and consume small amounts of electricity (for many in the area there was a single, hydroelectric-powered light bulb in their dwellings). However, they seldom pollute (either chemically or spiritually), do not consume more than they produce, rarely own or depend upon automobiles, and do not rely on manufactured goods, coal powered electricity, central heating or air-conditioning.

Policy

Chinese climate change policy was non-existent at the time of this study. Many Tibetans near Mt Khawa Karpo voiced concern that the government should lead climate amelioration and stabilization. In the governmental vacuum, villagers felt powerless to address climate change except through Buddhism and prayer.

Climate change, of course, was not a single change taking place alone. There were many other changes, including government policies that exacerbated climate change for Tibetans. Dealing with these confounding factors made Tibetan livelihoods ever more precarious. We heard about these policies at length.

The Chinese government encouraged farmers to increase herd size which caused degradation of alpine pastures. Traditionally, Tibetan herders actively managed grassland resources and practised controlled burning to prevent the encroachment of shrubs and other unpalatable vegetation in alpine meadows. Since 1998, government policy banned burning and alpine grazing lands have been severely reduced (Goldstein and Beall, 2002; Brandt et al., 2013).

To reduce poverty in Tibet, the Chinese government was promoting many activities with unintended consequences. For example, intensified medicinal plant gathering was encouraged to provide cash income. However, commercial collection of wild and threatened plants has led to over-harvesting products, such as the popular snow lotus (Law and Salick, 2005). Tourism also was subsidized by the Chinese government for development, encouraging tourist conglomerates from Shanghai and Hong Kong to move into the area. However, the Tibetans saw little profit from this brand of tourist development and were relegated to the most poorly paid service sector. The Chinese government also subsidized cash crops, such as grapes but the primary producers rarely saw profits from such development.

Tibetans featured prominently in the Sloping Land Conversion Programme (also known as the Grain to Green programme), introduced in 1999 as one of the largest land-use transition programmes in the developing world. Tibetan and other highland ethnic minorities of Yunnan were unduly burdened by this programme (Xu et al., 2004; Bennett, 2008; He and Sikor, 2015). In combination with other environmental policy (Willson, 2006), these programmes, along with climate change, threatened traditional Tibetan land use and livelihoods.

Globally, climate change policy must include indigenous perspectives, traditional knowledge and empirical observations if there is any hope for creative solutions, justice or equality within climate change responses. Indigenous people are on the frontline of climate change, experiencing the most devastating threats to traditional livelihoods and culture, and are the least integrated into national policy formation – in China and elsewhere (Salick and Byg, 2007; Salick and Ross, 2009; Nakashima et al., 2012).

With this and our previous research on Tibetan responses to climate change we clearly document that Tibetans perceived, adapted to and mitigated climate change. We have incorporated Tibetans directly in our climate change research. They have detailed knowledge about, creative ways of adapting to, and traditional ways of coping with climate change. Other international environmental and cultural efforts – such as the Convention on Biological Diversity (UN, 1992) and the United Nations Permanent Forum on Indigenous Issues (Economic and Social Council, 2000) – include indigenous peoples' perspectives and knowledge. These agreements have incorporated diverse peoples easily, productively and to the benefit of all. It is past time for new climate change agreements and policies to include indigenous peoples and traditional knowledge. In good faith and in the fulfilment of obligations assumed under the United Nations Declaration on the Rights of Indigenous Peoples (UN, 2007), we must incorporate indigenous peoples (and Nations) in the Intergovernmental Panel for Climate Change and traditional knowledge in its reports.

References

Aldenderfer, M. and Zhang, Y. N. 2004. The prehistory of the Tibetan Plateau to the seventh century AD: Perspectives and research from China and the West since 1950. *Journal of World Prehistory*, 18: 1–55.

Anderson, D. M., Salick, J., Moseley, R. K. and Xiaokun, O. 2005. Conserving the sacred medicine mountains: A vegetation analysis of Tibetan sacred sites in Northwest Yunnan. *Biodiversity & Conservation*, 14: 3065–91.

Bai, L., Woodward, A., Liu, X. et al. 2013. Rapid warming in Tibet, China: Public perception, response and coping resources in urban Lhasa. *Environmental Health*, 12: 71.

Baker, B. B. and Moseley, R. K. 2007. Advancing treeline and retreating glaciers: Implications for conservation in Yunnan, P.R. China. *Arctic, Antarctic, and Alpine Research*, 39: 200–9.

Bauer, K. M. 2003. *High Frontiers: Dolpo and the Changing World of Himalayan Pastoralists*. New York: Columbia University Press.

Bennett, M. T. 2008. China's sloping land conversion program: Institutional innovation or business as usual? *Ecological Economics*, 65: 699–711.

Brandt, J. S., Haynes, M. A., Kuemmerle, T., Waller, D. M. and Radeloff, V.C. 2013. Regime shift on the roof of the world: Alpine meadows converting to shrublands in the southern Himalayas. *Biological Conservation*, 158: 116–27.

Byg, A. and Salick, J. 2009. Local perspectives on a global phenomenon: Climate change in eastern Tibetan villages. *Global Environmental Change*, 19: 156–66.

Chambers, R. 1994a. Participatory rural appraisal (PRA): Analysis of experience. *World Development*, 22: 1253–68.

Chambers, R. 1994b. Participatory rural appraisal (PRA): Challenges, potentials and paradigm. *World Development*, 22: 1437–54.

Chambers, R. 1994c. The origins and practice of participatory rural appraisal. *World Development*, 22: 953–69.

Chaudhary, P. and Bawa, K. S. 2011. Local perceptions of climate change validated by scientific evidence in the Himalayas. *Biology Letters*, 7: 767–70, http://rsbl.royalsocietypublishing.org/content/roybiolett/early/2011/04/16/rsbl.2011.0269.full.pdf

Christensen, J. H., Hewitson, B., Busuioc, A. et al. 2007. Regional climate projections. In Solomon, S., Qin, D., Manning, M. et al. (eds.) *Climate Change 2007: The Physical Science Basis. Contribution of Working Group I to the Fourth Assessment Report of the Intergovernmental Panel on Climate Change*. Cambridge, UK and New York: Cambridge University Press, pp. 847–940.

Christensen, J. H., Kanikicharla, K. K., Marshall, G. J. and Turner, J. 2013. Climate phenomena and their relevance for future regional climate change. In Stocker, T. F., Qin, D., Plattner, G.-K. et al. (eds.) *Climate Change 2013: The Physical Science Basis. Contribution of Working Group I to* the Fifth Assessment of the Intergovernmental Panel on Climate Change. Cambridge, UK and New York: Cambridge University Press, pp. 1217–1308.

Cruz, R. V., Harasawa, H., Lal, M. S. et al. 2007. Asia. In Parry, M. L., Canziani, O. F., Palutikof, J. P. et al. (eds.) *Climate Change 2007: Impacts, Adaptation and Vulnerability. Contribution of Working Group II to the Fourth Assessment Report of the Intergovernmental Panel on Climate Change*, Cambridge, UK and New York: Cambridge University Press, pp. 469–506.

Ding, Y. H., Ren, G. Y., Shi, G. Y. et al. 2006. National assessment report of climate change (I): Climate change in China and its future trend. *Advances in Climate Change Research*, 2: 3–8.

Economic and Social Council. 2000. *Establishment of a Permanent Forum on Indigenous Issues.* UNOHC (United Nations Office of the High Commissioner) for Human Rights.

Goldstein, M. C. and Beall, C. M. 2002. Changing pattern of Tibetan nomadic pastoralism. In Leonard, W. R. and Crawford, M. H. (eds.) *Human Biology of Pastoral Populations*, Cambridge, UK and New York: Cambridge University Press, pp. 131–50.

Hart, R. and Salick, J. 2017. Dynamic ecological knowledge systems amid changing place and climate: Mt. Yulong rhododendrons. *Journal of Ethnobiology* 37: 21–36.

Hart, R., Salick, J., Ranjitkar, S. and Xu, J. 2014. Herbarium specimens show contrasting phenological responses to Himalayan climate. *Proceedings of the National Academy of Sciences of the United States of America*, 111(29): 10615–19.

Haynes, M. A., Kung, K.-J. S., Brandt, J. S., Yongping, Y. and Waller, D. M. 2014. Accelerated climate change and its potential impact on Yak herding livelihoods in the eastern Tibetan plateau. *Climatic change*, 123(2): 147–60.

He, J. and Sikor, T. 2015. Notions of justice in payments for ecosystem services: Insights from China's Sloping Land Conversion Program in Yunnan Province. *Land Use Policy*, 43: 207–16.

He, Y. and Zhang, Y. 2005. Climate change from 1960 to 2000 in the Lancang River Valley, China. *Mountain Research and Development*, 25(4): 341–8.

IPCC (Intergovenmental Panel on Climate Change). 2012. *Managing the Risks of Extreme Events and Disasters to Advance Climate Change Adaptation. A Special Report of Working Groups I and II of the Intergovernmental Panel on Climate Change*. [Field, C. B., Barros, V. R., Stocker, T. F. et al. (eds.)] Cambridge, UK and New York: Cambridge University Press.

IPCC. 2013. *Climate Change 2013: The Physical Science Basis. Contribution of Working Group I to the Fifth Assessment Report of the Intergovernmental Panel on Climate Change* [Stocker, T. F., Qin, D., Plattner, G.-K. et al. (eds.)]. Cambridge, UK and New York: Cambridge University Press.

IPCC. 2014. *Climate Change 2014: Impacts, Adaptation, and Vulnerability. Part B: Regional Aspects. Contribution of Working Group II to the Fifth Assessment Report of the Intergovernmental Panel on Climate Change* [Barros, V. R., Field, C. B., Dokken, D. J. et al. (eds.)]. Cambridge, UK and New York: Cambridge University Press.

Jones, P. D., Trenberth, K. E., Ambenje, P. et al. 2007. Observations: Surface and atmospheric climate change. In Solomon, S., Qin, D., Manning, M. et al. (eds.) *Climate Change 2007: The Physical Science Basis. Contribution of Working Group I to the Fourth Assessment Report of the Intergovernmental Panel on Climate Change, Miller*. Cambridge, UK and New York: Cambridge University Press, pp. 235–336.

Kang, S., Xu, Y., You, Q. et al. 2010. Review of climate and cryospheric change in the Tibetan Plateau. *Environmental Research Letters*, 5(1): 015101, www.researchgate.net/profile/Shichang_Kang/publication/230993878_Review_of_climate_and_cryospheric_change_in_the_Tibetan_Plateau/links/0912f51231582ac865000000.pdf

Klein, J. A., Hopping, K. A., Yeh, E. T. et al. 2014. Unexpected climate impacts on the Tibetan Plateau: Local and scientific knowledge in findings of delayed summer. *Global Environmental Change*, 28: 141–52.

Klein, J. A., Yeh, E., Bump, J., Nyima, Y. and Hopping, K. 2011. Coordinating environmental protection and climate change adaptation policy in resource-dependent communities: A case study from the Tibetan Plateau. In Ford, J. D. and Berrang-Ford, L. (eds.) *Climate Change Adaptation in Developed Nations. From Theory to Practice*, New York: Springer, pp. 423–38.

Konchar, K., Li, X.-L., Yang, Y.-P. and Emshwiller, E. 2011. Phytochemical Variation in *Fritillaria cirrhosa* D. Don (Chuan Bei Mu) in relation to plant reproductive stage and timing of harvest. *Economic Botany*, 65: 283–94.

Law, W. and Salick, J. 2005. Human-induced dwarfing of Himalayan snow lotus, *Saussurea laniceps* (Asteraceae). *Proceedings of the National Academy of Sciences of the United States of America*, 102(29): 10218–20.

Li, Z., He Y., Pu, T. et al. 2010. Changes of climate, glaciers and runoff in China's monsoonal temperate glacier region during the last several decades. *Quaternary International*, 218(1–2): 13–28.

Lin, E. D., Xu, Y. L., Jiang, J, H. et al. 2006. National assessment report of climate change (II): Climate change impacts and adaptation. *Advances in Climate Change Research*, 2(2): 51–6.

Liu, X. D. and Chen, B. D. 2000. Climatic warming in the Tibetan Plateau during recent decades. *International Journal of Climatology*, 20: 1729–42.

Liu, X. D. and Zhang, M. F. 1998. Contemporary climatic change of the Qinghai Plateau and its response to the greenhouse effect. *Chinese Geographical Science*, 8: 289–98.

Ma, C. L., Moseley, R. K., Chen, W. Y. and Zhou, Z. K. 2006. Plant diversity and priority conservation areas of northwestern Yunnan, China. *Biodiversity and Conservation*, 16(3): 757–74.

Meehl, G. A., Stocker, T. F., Collins, W. D. et al. 2007. Global climate projections. In Solomon, S., Qin, D., Manning, M. et al. (eds.) *Climate Change 2007: The Physical Science Basis. Contribution of Working Group I to the Fourth Assessment Report of the Intergovernmental Panel on Climate Change*. Cambridge, UK and New York: Cambridge University Press, pp.747–845.

Mittermeier, R. A., Myers, N., Thomsen, J. B., Da Fonseca, G. A. and Olivieri, S. 1998. Biodiversity hotspots and major tropical wilderness areas: Approaches to setting conservation priorities. *Conservation Biology*, 12: 516–20.

Moseley, R. K. 2006. Historical landscape change in northwestern Yunnan, China. *Mountain Research and Development*, 26(3): 214–9.

Moseley, R. K. 2011. *Revisiting Shangri-La: Photographing a Century of Environmental and Cultural Change in the Mountains of Southwest China*. Beijing: China Intercontinental Press.

Nakashima, D., Rubis, J., Ramos Casillo, A., Galloway McLean, K. and Thulstrup, H. 2012. *Weathering Uncertainty – Traditional Knowledge for Climate Change Assessment and Adaptation*. Paris and Darwin: UNESCO/UNU.

Qiu, J. 2008. China: The third pole. *Nature News*, 454(7203): 393–6.

Rai, S. C. 2005. *An Overview of Glaciers, Glacier Retreat, and Subsequent Impacts in Nepal, India and China*. WWF Nepal Program, www.wwf.or.jp/activities/lib/pdf_climate/environment/Overview_of_Glaciers.pdf

Salick, J., Anderson, J. D., Woo, J. et al. 2004. *Tibetan Ethnobotany and Gradient Analysis: Menri (Medicine Mountains), Eastern Himalayas*. Conference for The Millennium Ecosystem Assessment. Bridging scales and epistemologies: Linking local knowledge and global science in multi-scale assessments. 17–20 March 2004, Alexandria, Egypt.

Salick, J., Amend, A., Anderson, D. et al. 2007. Tibetan sacred sites conserve old growth trees and cover in the eastern Himalayas. *Biodiversity and Conservation*, 16: 693–706.

Salick, J. and Byg, A. 2007. *Indigenous Peoples and Climate Change. Report of Symposium 12–13 April 2007, Environmental Change Institute, Oxford*. Oxford: Tyndall Centre for Climate Change Research.

Salick, J. and Moseley, R. K. 2012. *Khawa Karpo: Tibetan Traditional Knowledge and Biodiversity Conservation*. Saint Louis, MO: Missouri Botanical Garden Press.

Salick, J. and Ross, N. 2009. Traditional peoples and climate change. *Global Environmental Change*, 19, 137–9.

Salick, J., Yongping, Y. and Amend, A. 2005. Tibetan land use and change near Khawa Karpo, eastern Himalayas. *Economic Botany*, 59(4): 312–25.

Sanchez, P. A. 1977. Properties and management of soils in the tropics. *Soil Science*, 124(3): 187.

Sheehan, M. F. 2008. China climate change: A sacred glacier recedes. *The Nature Conservancy*, www.nature.org/ourinitiatives/regions/asiaandthepacific/china/explore/mingyong-glacier-receding-in-northwest-yunnan.xml

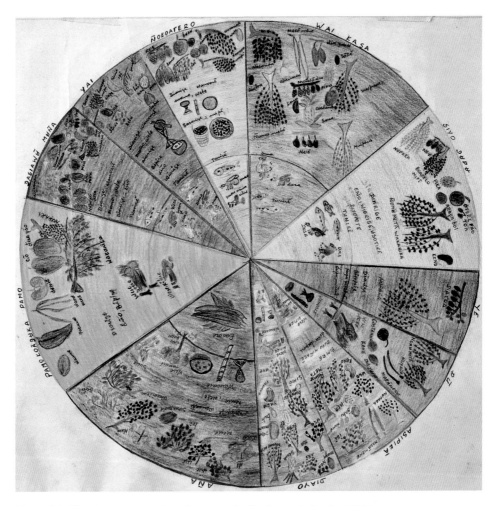

Figure 3.3 Circular representation of an annual calendar made by the AIMA.

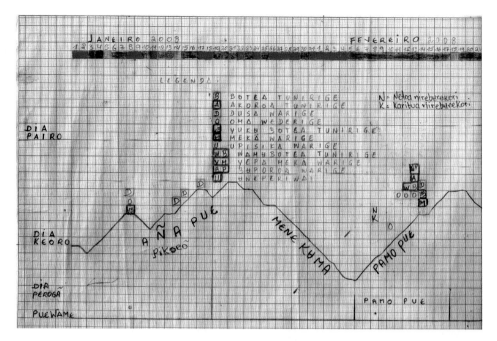

Figure 3.4 Excerpt of the timeline for 2008, showing season names along the water level line, as well as the timing of some phenological phenomena.

Figure 4.1 Moch Island, Mortlock Islands, Chuuk, FSM.

Figure 4.2 Land reclamation on Weno, Chuuk Lagoon, FSM.

Figure 4.3 Healthy taro patch. Moch Island, Chuuk, FSM.

Figure 4.4 Saltwater where there was once a taro patch, Moch Island, Chuuk, FSM.

Figure 8.2 Dugong hunting at Yathikpa.

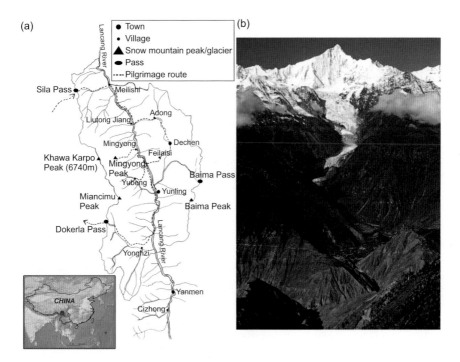

Figure 9.1 (a) Map of the study area in the easternmost Himalaya; (b) Mingyong village, Mingyong glacier and Tibetan sacred Mt Khawa Karpo.

Figure 9.5 Tibetan wine production: (a) Catholic Church in Cizhong and (b) the award-winning 'Sun Spirit' ice wine.

Figure 9.2 Participatory and interview methods: (a) villagers drafting maps and calendars; (b) interviewing Tibetan monks and professionals.

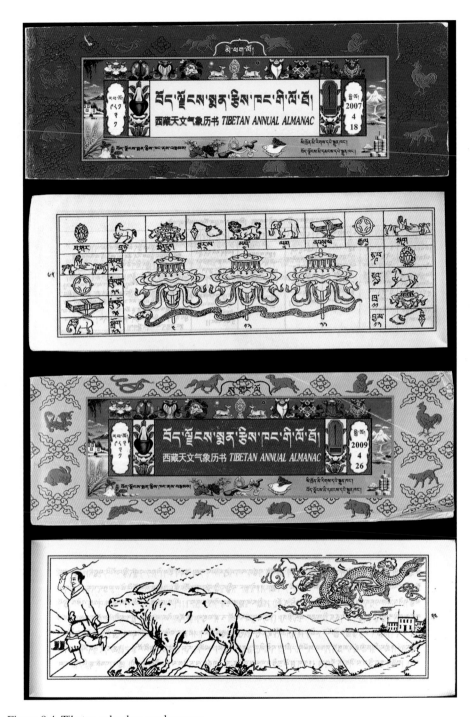

Figure 9.4 Tibetan calendars or almanacs.

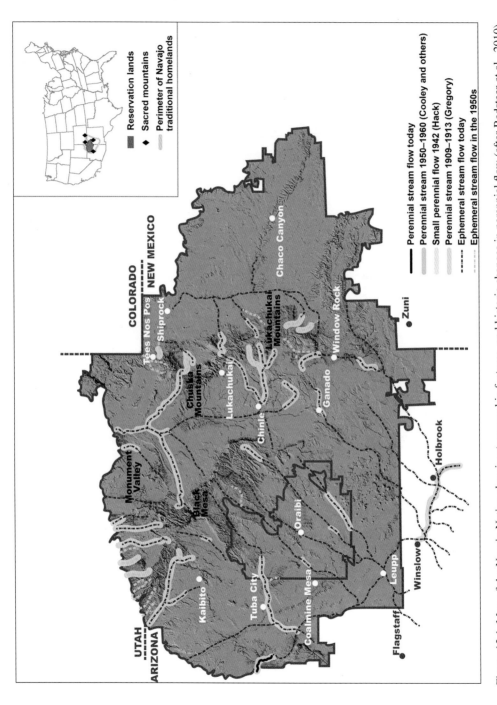

Figure 12.1 Map of the Navajo Nation showing topographic features, and historic changes in perennial flow (after Redsteer et al., 2010). Inset map shows location of reservation lands (red) black diamonds for locations of sacred mountains, and a pale line for approximate perimeter of Navajo traditional homelands.

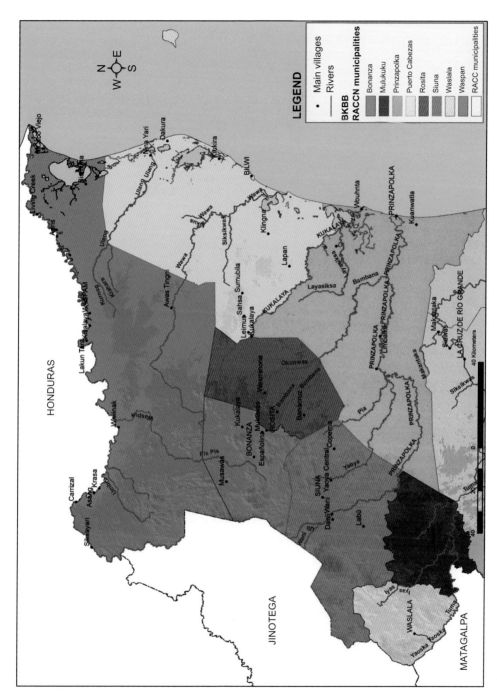

Figure 13.1 Map of the North Atlantic Autonomous Region (RAAN), Nicaragua. © Marcos Williamson.

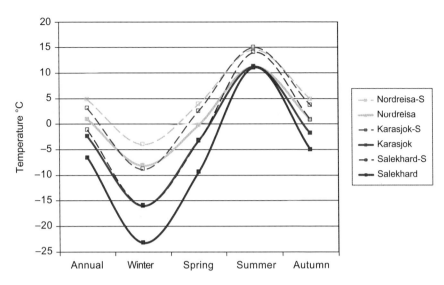

Figure 14.2 Annual and seasonal mean temperatures in coastal Finnmark, Norway (Nordreisa, yellow), inland Finnmark, Norway (Karasjok, red) and Yamal-Nenets AO, Russia (Salekhard, blue). Unbroken lines show 1961–1990 observed values. Dotted lines show similar average calculated scenarios (S) from fifty downscaled climate models for the year 2085 (Magga et al., 2011).

Figure 16.2 Hydropower infrastructure and reservoir in the Laponia World Heritage site.

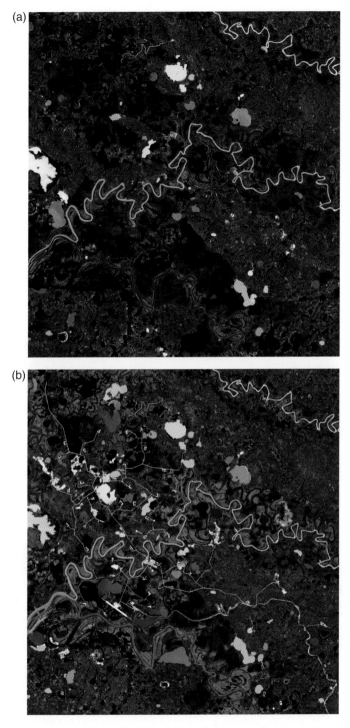

Figure 14.3 Landsat imagery from Yamal Peninsula, Yamal-Nenets AO, West Siberia, Russia: (a) in 1987, before industrial development; and (b) in 2011, after being heavily developed by the gas industry. Reindeer herders migrate through this region in midsummer to access the coastal summer pastures of Yamal Peninsula.

Figure 14.4 The 'rain-on-snow' circumpolar map for 8 November 2006 built on the Polarstereographic 12.5 km AMSR-E grid. A grey scale image of the AMSR-E 89 GHz polarization ratio is used for the background. The coincident ECMWF precipitation is shown in blue, with precipitation that occurs in the vicinity of subfreezing temperatures shown in cyan, and ECMWF snowfall shown in grey. The ROS categories are as follows: liquid layer (red), recent refreeze (purple), longer-term refreeze (pink), and temporal classification (beige). The extensive ROS event that was observed near the Yamal Peninsula in November 2006 can be clearly seen on the map.

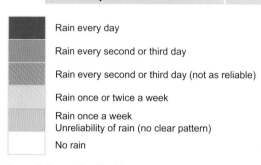

Figure 15.2 Calendar of observed changes in rain patterns.

Figure 15.3 The Nyangatom culture of sharing. © Julia Pfitzner.

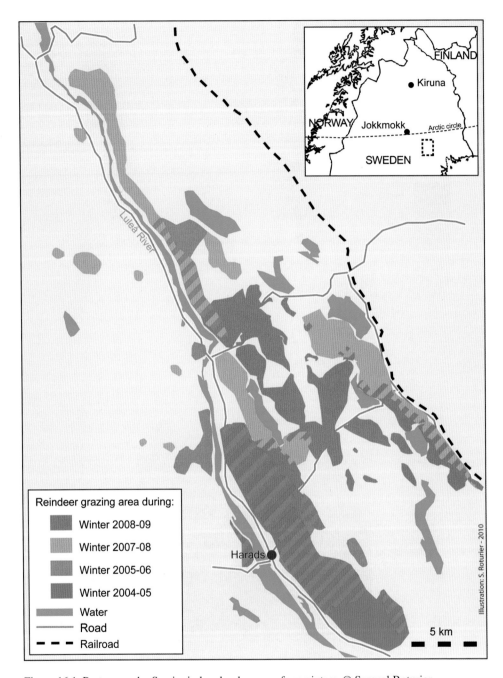

Figure 16.1 Pasture use by Sami reindeer herders over four winters. © Samuel Roturier.

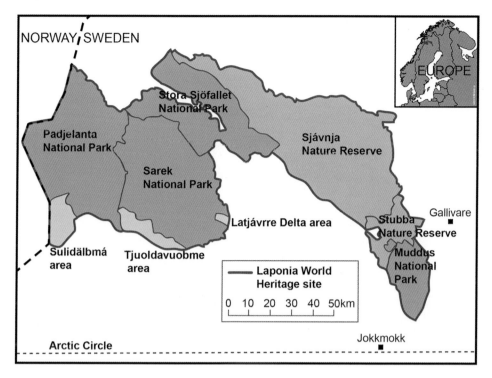

Figure 16.3 The Laponia World Heritage site, encompassing several national parks and nature reserves.

Figure 16.4 Sami from Tuorpon marking reindeer calves in the Laponia World Heritage site (July 2009). After a harsh winter, the Sami are particularly anxious to find out how many of their female reindeer are accompanied by calves.

Figure 19.1 The People of the Whales with a landed bowhead whale near Barrow, Alaska. © Jessica Jelacic.

Figure 19.2 Traditional Iñupiaq drum music performed at Nalukataq, the midsummer whale feast in Barrow.

Shen, X., Lu, Z., Li, S. and Chen, N. 2012. Tibetan sacred sites: Understanding the traditional management system and its role in modern conservation. *Ecology and Society*, 17(2): 13, www.ecologyandsociety.org/vol17/iss2/art13/

Tulachan, P. M. 2001. Mountain agriculture in the Hindu Kush-Himalaya: A regional comparative analysis. *Mountain Research and Development*, 21(3): 260–7.

United Nations. 1992. *Convention on Biological Diversity*, Rio de Janeiro: UN.

United Nations. 2007. *United Nations Declaration on the Rights of Indigenous Peoples (UNDRIP)*. UN General Assembly Resolution 61/295.

Willson, A. 2006. Forest conversion and land use change in rural northwest Yunnan, China. *Mountain Research and Development*, 26(3): 227–36.

Wuebbles, D. J., Chitkara, A. and Matheny, C. 2014. Potential effects of climate change on global security. *Environment Systems and Decisions*, 34: 564–77.

Xie, H., Richard, C., Xu, J. C. and Wang, J. H. 2001. Collective Management of Improved Forage in Zhongdian County, Deqin, Tibetan Autonomous Prefecture, Northwest Yunnan, P.R. China, www.eldis.org/fulltext/yakyunnan.pdf

Xu, J. and Grumbine, R. E. 2014. Building ecosystem resilience for climate change adaptation in the Asian highlands. *Wiley Interdisciplinary Reviews: Climate Change*, 5(6): 709–18.

Xu, J., Grumbine, R. E., Shrestha, A. et al. 2009. The melting Himalayas: Cascading effects of climate change on water, biodiversity, and livelihoods. *Conservation Biology*, 23(3): 520–30.

Xu, J. and Wilkes, A. 2004. Biodiversity impact analysis in northwest Yunnan, southwest China. *Biodiversity and Conservation*, 13(5): 959–83.

Xu, Z., Bennett, M. T., Tao, R. and Xu, J. 2004. China's sloping land conversion programme four years on: Current situation, pending issues. *International Forestry Review*, 6(4): 317–26.

Xu, Z. X., Gong, T. L. and Li, J. Y. 2008. Decadal trend of climate in the Tibetan Plateau-regional temperature and precipitation. *Hydrological Processes*, 22(16): 3056–65.

Yeh, E. T., Nyima, Y., Hopping, K. A. and Klein, J. A. 2013. Tibetan pastoralists' vulnerability to climate change: A political ecology analysis of snowstorm coping capacity. *Human Ecology*, 42: 61–74.

Yi, S. L., Wu, N., Luo, P. et al. 2007. Changes in livestock migration patterns in a Tibetan-style agropastoral system. *Mountain Research and Development*, 27(2): 138–45.

10

Seasonal Environmental Practices and Climate Fluctuations in Island Melanesia: Transformations in a Regional System in Eastern Papua New Guinea

Frederick H. Damon

This chapter introduces a case study of indigenous adaptation to various forms of environmental transformation in Island Melanesia: namely along the chain of small islands located in the Solomon Sea, to the east of the south-eastern tip of Papua New Guinea. The main argument is that approaches to environments and societies in small islands contexts require careful consideration of ecological relations, as locally understood, with special attention to existing long-term trends in the adaptive capacity of social and environmental milieu. The anthropogenic nature of key environmental spheres is critical to understanding the mutually constituted nature and resilience of social and physical worlds. Together with other contributions in this volume (see Mondragón; Henry and Pam; Resture; and Falanruw), this chapter outlines likely extreme versions of environmental dynamics that many if not all Pacific social systems face: climate variation and sea-level change on the one hand, and tectonic forces on the other. Evidence is building that Indo-Pacific societies have long been adapted to a, if not the, dominant climate variable in the region, El Niño Southern Oscillation (ENSO) (Davis, 2002: 280).

My study has been facilitated by close collaboration with specific local actors and repeated returns to the research areas in maritime Papua New Guinea where over three decades of fieldwork has been carried out since 1991 with a strong environmental focus (Damon, 1982; 1990; 1998; 2005a; 2005b; 2008).

Changing Scenes in Eastern Papua New Guinea

Background to Milne Bay Province, Papua New Guinea

From the vantage point of nearly four decades of research focused on the small island-offshore section of Papua New Guinea's Milne Bay Province, two matters stand out concerning traditional practices and knowledge, and experiences of climate change. One of these concerns experienced sea-level rise. For people who live by the sea – proximate to it and by means of it – these are significant observations. The second involves observed alterations in complex understandings of weather patterns. These entail perceived changes with respect to intra- and inter-annual oscillations between sun and rain. Without any question, people's lives are organized by the sea and by rain/sun dynamics. Their experience of change, as is the case with the sea level, is no small matter. Documenting these circumstances is the purpose of this chapter.

Muyuw (Muyua), or Woodlark Island, is in the north-east corner of one of anthropology's classic fields areas, the 'Kula Ring' which Bronislaw Malinowski made famous from his studies during the early twentieth century (e.g. Malinowski, 1984 [1922]). Matters of time were a principal focus of my initial work (Damon 1982; 1990); they re-emerged as important concerns in subsequent ethnobotanical research from 1991 to the present. Those inquiries led me to learn about indigenous peoples' interest in the climate/weather tension, and their experience of its shifts. I documented worried reports about weather – perhaps climate – changes for the first time in 1991. Other scholars reported similar things in Highland Papua New Guinea and in Bali. In Muyuw traditional intra-annual orientations followed sun and rain oscillations, modelling them on star behaviour. At the inter-annual level 'big suns', prolonged periods of drought that we call El Niño Southern Oscillation(ENSO), unpredictably disrupted the former pattern. Consequently, people customarily paid close attention to intra-annual expectations because they knew that they did not always unfold as anticipated. In 1991 people reported that sun/rain oscillations no longer regularly worked in accustomed ways. These early 1991 reports have been sustained over the last twenty and more years. Moreover, anomalous, prolonged wet periods in 1995/96 are becoming a new, recognized, though not appreciated, norm.

Two external temporal matters might be raised here by way of an analytical introduction. First, most of the inhabitants of Milne Bay Province derive from the Austronesian expansion into the South Pacific beginning some 4,000 years ago, an historical transformation which in this part of Melanesia overlaid a human history of at least 30,000 years. Because of its geology and climate variation, this region is one of the more dynamic – 'patchy' is the correct ecological term – ones in the world. Reciprocally, local human-controlled environmental systems were made possible by the collisions of the Indo-Australian and Pacific plates. All islands from New Zealand in the south going north through Vanuatu then bending west from, roughly, Fiji, and extending into island and mainland South-East Asia are on the proverbial 'ring of fire'. They regularly experience tectonic action, mild here, violent there. Furthermore, sea-level changes are a constant experience. A conclusion from this history is that for long the cultural forms of this region have probably been closely aligned with changing environmental relations, by the seas, on the lands and through the atmosphere.

The second temporal matter concerns the present: Milne Bay Province has been on the edge of Western colonial expansion since the first half of the nineteenth century. One recent transformation into the modern is the transition from colonial-era miners with shovels and picks (from 1895 to 1940) to massive petroleum-driven drills, earth removers and movers. Mass power rather than muscled dexterity is the new cultural reality. By Western standards the progressive views of the people in the island's recent mining company brought an entirely different calculus to their relationship to their environmental setting. Although its executives openly ridicule discussion of global warming, one of its on-site managers confirmed reports of coastal villages struggling with sea-level rises.

Changing Sea Levels

Many local people speak about gradual but observable rises in the sea level. Observations such as these are not to be trusted completely because coasts, in this part of the world in particular, are always and everywhere incredibly dynamic (Mei and Liu, 1993; Klein et al., 1998; Haslett, 2009: 131–54). Muyuw island is a mixed classic coral atoll subjected to many local forces and, according to some experienced scholars (William Dickerson, personal communication), intercontinental forces having to do with continuing adjustments to the last glaciation. If one model argues for gradual slow rising of the geological structure that governs the island, the Woodlark Rise (see Baldwin et al., 2008, 2012 for recent geological synthesizes), recent mining investigations suggest the island experiences periodic, catastrophic change. Nevertheless, when I told people in 2006 and 2009 that I never recalled seeing so many trees toppling into the water, they agreed with my observation. Young women who frequent shoreline coral reefs looking for shellfish report that the water always seems higher and some reefs never become exposed by the lowest of low tides, when in their recent past they did. Visits in 2012 and 2014 substantiated these reports. One south-eastern village experienced beach erosion from recent, anomalous winds unprecedented in anyone's life.

More alarming than these reports are those coming from small islands to the west of Muyuw; namely, Gawa, Kweywata and Iwa, and to the south, Nasikwabw. I have not visited the first three since the 1990s but reports from them say people are worried about their boat landings. Most of these islands appear like pillars raised up out of the ocean 30–150 m with flat, slightly concave tops. That is where people live, and they are not in danger. Yet life is managed on these islands by the water connections to elsewhere and higher water, and hence more wave action, is damaging their landings – the connections to the social system that has sustained them for millennia. The situation on Nasikwabw, a sailing village to the south, is different because for a hundred years or more the community lived not on the raised coral reef's high plain but rather nestled next to the sea along a semi-protected cove. By 2002, however, at least two families had moved on the top. Reports from 2006, 2009 and 2012 indicate more people are moving. Waves regularly washed under houses, and it was increasingly difficult to protect boats dragged on to large wooden rollers. These observations are consistent with the IPCC reports (Church et al., 2013).

Atop the island where people have their fields and to where some are now moving their houses, there are shattered and very old clay pots and stone tool trash. This suggests that people once lived there rather than along the shoreline. A significant pivot in the cultural history of this area is 1400; a debatable hypothesis is that the reputed drop of the sea due to the Little Ice Age would have enabled life on the beaches of Nasikwabw in a way perhaps not possible earlier (Nunn, 2000; Nunn and Britton, 2001; Dickinson, 2003).

The early colonial order sought to locate villages along shorelines, and many remain there. Before then, beach-hugging trees (e.g. *Casurina littorale*) defined coastlines with villages located back from and above the shoreline, probably a more intelligent policy. If the recent experience of rising sea levels is sustained, more villages may move to higher locations, perhaps reverting to a wiser, traditional orientation.

A Time Chiasmus – Star and Tree Times, and the New Confusions

The last two decades of ethnobotanical research revealed an astonishing adaptation between intricate cultural forms and climate/environment conditions. Space allows only for a sketch here (Damon, 1982, 1990, 2005a, 2017). These facts add new interest to the 'problem' of 'primitive calendars', for which this part of the ethnographic world has been very significant. From Leach's work (Leach, 1950) on the Trobriand material (e.g. Malinowski, 1927; Austen, 1939, 1950), Aveni created a synthesis that is suggestive for many of the world's calendars, including the forms underlying the contemporary Western construct (Aveni, 2002 [1989]).

The patterns are complex, not visible from a single research site nor evident over the customary year or two of anthropological research. A well-intentioned Australia National University (ANU)-based emergency trip through Milne Bay Province during the trough of the 1997/98 ENSO event arrived on the small island of Yemga only to discover that its gardens were not sufficient to sustain the population. This was falsely attributed to that ENSO event. Yemga is a sailing village that is never sustained by its gardens; its work was, and remains, to maintain a sailing fleet that enabled interisland trade. It does so in exchange for garden work and produce, i.e. vegetable food.

To begin, Muyuw has virtually the same set of lunar month names as found on the islands of Iw, Kitava and the Trobriands, stretching to the west for some 160 km. However, these same names cover different slots of the solar year, and new years are defined at different times on different islands. Traditionally Muyuw's new year started in the eastern end of the island in roughly March, clearly approximate to that equinox, but not exactly fixed to it. People watch relative motions of the sun, moon and tides very carefully. So looking over their expansive lagoon to the east Kavatan people watch the sun and full moon tracking more or less exactly due east and west and the observance of high and low diurnal tides at dawn and dusk at the March equinox. June low tides occur around noon, the full moon rising far to the south; December's low tide is near midnight, the full moon rising far to the north.

Although these movements are observed, they are not turned into explicit timing devices. And in theory new year declarations move progressively village by village as one moves west across the island; the organizing principle is the east to west movement, not a fixed time interval. By the western end of the island, and the next island over, Gawa, this paradigm dissipates, as if an orientation effectively dissolves. Further to the west still, people on Iwa Island, halfway between Muyuw and the Trobriands, think they start the calendar that defines the Trobriand sequence. They do so with a relatively precise observed coincidence. The new year comes with the first full moon after the heliacal rise of the Pleiades. This is thus after the 1st of June, proximate to the subsequent solstice. The progression moves on to the next island, and time districts in the Trobriands, by successive full moons so that while there is a spatial component to the sequence, the data argues that the form is fixed by the moon's temporal intervals. Vakuta, at the Trobriand's southern extremity and the last place that operates this calendrical system, declares its new year in September or early October. The whole set of sequences thus begins roughly with one equinox, becomes redefined with an intervening solstice, and seems to end proximate to the next equinox.

This is an inter-annual 'time system', mediated by a cultural transformation, which is part of a regional system organizing a set of sequential activities that transcend a solar year (Damon, 2005b).

These facts are part of the genius of this highly developed regional network. For now it is enough to observe that the positioning of the lunar month names about the solar year makes the eastern and western sides of this sequence almost mirror images of one another. It is important to note then that most of Muyuw and most of the Trobriands plant their main annual crops, yams, about the same time, in December, and harvest them after April and closer to June and July. The little islands in between invert this almost exactly, harvesting in the austral summer, planting in the winter.

Times and activities vary here because people expect intra-annual patterns to be unpredictably upset by inter-annual patterns, namely the ENSO droughts. The little islands are truly little. Consequently, they never absorb much rainwater, and they are high; hence the ground surface is far above the underground water lens. With a drought of any length they lose all of their crops and have to have them replaced by the new plants from people to their east. It turns out that the people on these small islands tend not to know very much about their root crops. While first puzzling, this situation finally made sense after I realized they regularly lost those crops, depended on others (with deeper knowledge) to replace what they lost, and concentrate instead on extraordinary sets of trees that, because of their deeper roots, more easily withstand all but the worst ENSO-generated droughts. Iwa Island people maintain hundreds of nut and fruit trees while those on Gawa and Kweywata maintain trees used to construct the highest class of outrigger canoe moving across the interisland system.

Trees mediate between the primary annual crops, yams and taro, and the inter-annual relations to a major degree effected by ENSO circumstances. This combination may synthesize the original inhabitants of Melanesia and eastern Indonesia – arguably more tree-focused – and the Austronesian expanse that brought the East Asian root crops to the region.

But let us return to the working model that runs as follows. A set of approximately twelve stars/constellations are watched for their heliacal risings and settings. Metaphorically they are conceived to be like 'Important Men' – which means men well established in interisland networks. And like these men, when they die, when they set in the west in the evening sky, they release a power which results in wind and rain. That period is then followed by a week or more of sun. The oscillation enables the swidden system, cutting a forest, burning it, then planting it. Sun is needed to power growth but also to dry a cut forest enabling burning, planting and harvesting. But that weather set has to be broken by the equally necessary rain. The annual movement of the stars *models* the oscillation between sun and rain.

Added to this patterning is a variance in ideal fallow periods. These realize the extremes between no gardens just forests, and gardens and forests grown together. Suffice it to say that the small islands between Muyuw and the Trobriands mix gardening and fallowing by planting trees in with their crops. One place on Muyuw is famous for the other extreme, no gardens at all, so of course no fallowing, just forests. In between these extremes, eastern Muyuw ideally uses a short fallow cycle – ten to fifteen year intervals; central Muyuw uses middle-length fallows – twenty to forty years; and western Muyuw ideally employs a long fallow.

These differences are regionally, not just locally, significant. Muyuw people believe that older forests keep the ground cooler and wetter. The convenient fact for these practices being located in western Muyuw is that it helps make their yam gardens impervious to ENSO extremes, and this is right next to the small islands for whom even the shortest droughts are disastrous for root crops. When I returned to Muyuw in 1998 after the severe ENSO event was over, all across Muyuw people were sending yams, as food and seed, to the small islands between Muyuw and the Trobriands. However, only in western Muyuw did people tell me that the ENSO had little effect on them, and only there did they say they were sending so many yams west that they couldn't count them. (These were not free gifts but explicitly informally tied to tree resources on the one hand and the interisland shell exchange, the Kula, on the other.)

This diverse set of fallowing generates the trees appropriate for producing the interisland outrigger canoes. As noted, the canoes are made in whole in Gawa and Kweywata. However, those islands lack the best trees for certain parts in general and some of the best replacement parts in particular. These come from Muyuw – therefore masts from the part of the island that does not garden; the best outrigger floats from central Muyuw by means of its ideal middle-length fallow; large numbers of really small trees useful for making outrigger platforms from the early fallowing practices in the east. The land and its products have been sculpted to generate the materials found necessary for the *region's* unique outrigger boat form. This is not coincidental because the chaotic conditions of sailing experiences (i.e. these boats) are models for the chaotic weather patterns these people customarily experience. A boat is supposed to be made in a certain way, but each boat has unique properties that have to be individually attended to, as do the relations between the islands. It is no accident that the idea of a boat is imposed over most forms – villages, gardens, whole islands – and social relations (Damon, 1998, 2008).

Superimposed over the annual movements modelled by the stars are inter-annual oscillations produced by what we call ENSO. These create oscillations between longer periods of drought and longer periods of rain. This pattern complements the annual cycle because with respect to it, sago orchards are managed and individual sago trees harvested. Sago is a palm-like tree domesticated somewhere between eastern Indonesia and eastern New Guinea; they are this part of the world's contribution to the human carbohydrate stock (Ellen, 2004, 2006). They reproduce by asexual reproduction, new trees growing off the sides of a mother tree. People harvest them by splitting the bole of a mature tree – which takes anywhere from seven to thirty years to mature – then pulverize the pith and by hand and with copious amounts of water separate carbohydrates from cellulose material – a flour-like material results. Although some parts of the larger Melanesian region use sago as their mainstay, in this part of Papua New Guinea the material forms a condiment to every ritual, exchange item for the interisland set of regional dependencies, a food supplement for the low spots in the annual yam and taro cycle – in the austral summer – and emergency food for severe ENSO droughts.

As part of the practice's management system, orchards are juxtaposed to meadows found in the high forest. These meadows are created by fire during ENSO droughts – otherwise trees would infill them. They produce important materials for tying (i.e. decorating) bodies and binding boat parts. Sago orchards are downstream from meadows therefore it is likely that run-off from them enhances nutrient availability in the orchards. Beyond that hypothesis, people coordinate sago orchard/meadow relations because 'wild pigs', avidly hunted and eaten, take refuge in the meadows and eat sago purposely left for them – the last metre or so of the top and bottom of the tree's bole – in the orchards.

The garden/fallow and sago orchard/meadow relation is a coordinated system partly fashioned together by means of the way social relations are represented in each form. Gardens contain models of the culture's gender and clan system; a set of continuous, complementary replacement. Sago orchards model subclan relations since the growth pattern of trees is equated with the way the culture's matrilineal units are reproduced by a daughter taking the place of her mother.

The relationships between the higher-order ENSO oscillations and the lower-order star patterns created the dynamic around which this cultural system organized itself over the last 2,000 or more years. But if it describes an oscillation of times around which the culture created itself, it is not what people experience, or talk about, now. Increasingly frequently, prolonged periods of rain all but eclipse the much shorter oscillation between sun and rain configured by the twelve or so stars. So increasingly frequently both burning and harvesting become extremely difficult, violating what is understood as a necessary order to gardening practices. Malaria outbreaks are more common; everyone consumes copious amounts of antimalarials and antibiotics. And it may be that ENSO dynamics enhanced by the effects of global warming will make the encompassing pattern truly dominant – long stretches of continuous rain broken by debilitating droughts. What cultural shape that will bring is for the future to determine. That future will also be determined in part by planned mining activities, such as too many other modern/Western intrusions, which are conceived independently of the natural patterns with respect to which this set of regional determinations organized itself over the millennia.

Conclusion

The case study presented here, from the eastern end of contemporary Papua New Guinea, offers one example of a diversity of knowledge, practices, adaptive capacities and environmental resilience in relation to an ecological milieu of the Western Pacific. From this short summary, it is possible to draw a number of important conclusions. First, as is the case for other regions, any comprehensive approach to the issue of adaptation to climate change in small island environments has to be particularly sensitive to local conditions – both natural and social (Cronin et al., 2004; Mercer et al., 2007). Only then can relevant insights and strategies be drawn in order to best address future forms of adaptation at a grass-roots level (Pernetta, 1992; Moran, 2006: 5). In this regard, there are clearly important differences, not only between the Melanesian environments discussed herein and those of other small island contexts –most notably Polynesia and Micronesia – but also between the societies and individual island groups.

On one hand, the islands with which Muyuw is associated in Papua New Guinea constituted an extremely dynamic climatological environment, one in which patterns of drought and rainfall have probably had the most profound impact on the human stewardship (and modification) of the islands' vegetation and soils. Tectonic activity is common on the island – earth tremors are called *nikw* and are conceived to result from the Creator's motions. Over my forty-year experience on the island only one tidal wave was noted, one from 2007, resulting from a quake north of the island in the Solomons. Paradoxically it made damage only on the island's south-eastern side. This experience, together with stories of the 2004 disasters associated with the tsunami stemming from the Sumatran quake and the creeping sense of rising sea levels and falling shoreline trees, pushed many people to rethinking the lovely beaches to which they were herded by initial colonial encounters a hundred and more years ago.

The forms of adaptation that we have reviewed here speak to a slightly more positive outlook for the possibilities of adaptation to sea-level rise in various small island locales. In this regard, I would draw attention to the fact that even such an apparently destructive and seemingly unstoppable process as the disappearance of current shoreline environments in many small island regions of the world need not necessarily fall prey to catastrophist projections (Barnett, 2001). The case study described here offers a powerful example of the resilience and adaptability of people and environments even in cases that seem extreme; indeed, we are tempted to proffer the additional argument that these Melanesian communities appear to be far more likely to adapt to abrupt climate change than the encompassing, globalized societies and institutions that are seeking to help them in this process of environmental transition. For, it might be said, these people's world was 'first' fashioned amid instability.

Acknowledgements

The original 2011 conference version of this chapter was written jointly with Carlos Mondragón, whose contribution to our joint effort from the island of Vanuatu goes separately in this volume (Mondrágon, Chapter 2). Many institutions have provided funds to support my research over the years that led to this publication. I thank most recently the University of Virginia for a Summer Research Grant (2002); American Philosophical Society for a Franklin Research Grant (2009); and a Research Grant from Dean of the College of Arts and Sciences and the Vice-President for Research and Graduate Studies, University of Virginia (2012).

References

Austen, L. 1939. The seasonal gardening calendar of Kiriwina, Trobriand Islands. *Oceania*, 9: 237–53.
Austen, L. 1950. A note on Dr Leach's 'Primitive Calendars'. *Oceania*, 20: 333–5.
Aveni, A. F. 2002. *Empires of Time: Calendars, Clocks, and Cultures*. Boulder, CO: University Press of Colorado.
Baldwin, S. L., Fitzgerald, P. G. and Webb, L. E. 2012. Tectonics of the New Guinea Region. *The Annual Review of Earth and Planetary Sciences*, 40: 495–520.
Baldwin, S. L., Webb, L. E. and Monteleone, B. D. 2008. Late Miocene coesite-eclogite exhumed in the Woodlark Rift. *Geology*, 36(9): 735–8.
Barnett, J. 2001. Adapting to climate change in Pacific Island countries: The problem of uncertainty. *World Development*, 29(6): 977–93.
Church, J. A., Clark, P.U., Cazenave, A. et al. 2013. Chapter 13: Sea level change. In Stocker, T. F., Qin, D., Plattner, G.-K. et al. (eds.) *Climate Change 2013: The Physical Science Basis. Contribution of Working Group I to the Fifth Assessment Report of the Intergovernmental Panel on Climate Change*. Cambridge, UK and New York: Cambridge University Press.
Cronin, S. J., Gaylord, D.R., Charley, D. et al. 2004. Participatory methods of incorporating scientific with traditional knowledge for volcanic hazard management on Ambrym Island, Vanuatu. *Bulletin of Vulcanology*, 66(7): 652–68.
Damon, F. H. 1982. Calendars and calendrical rites on the northern side of the Kula Ring. *Oceania*, 52(3): 221–39.
Damon, F. H. 1990. *From Muyuw to the Trobriands: Transformations Along the Northern Side of the Kula Ring*. Tucson, AZ: University of Arizona Press.
Damon, F. H. 1998. Selective anthropomorphization: Trees in the northeast Kula Ring. *Social Analysis*, 42(3): 67–99.
Damon, F. H. 2005a. The Woodlark Island calendar: Contexts for interpretation. In Chamberlain, V. D., Carlson, J. B. and Young, M. J (eds.) *Songs from the Indigenous Sky: Indigenous Astronomical and Cosmological Traditions of the World*. Bognor Regis, UK: Ocarina Books, pp. 348–57.
Damon, F. H. 2005b. 'Pity' and 'ecstasy': The problem of order and differentiated difference across Kula societies. In Mosko, M. and Damon, F. H. (eds.) *On The Order of 'Chaos'. Social Anthropology and the Science of 'Chaos'*. New York and Oxford: Berghahn Books, pp.79–107.
Damon, F. H. 2008. On the ideas of a boat. From forest patches to cybernetic structures in the outrigger sailing craft of the eastern Kula Ring, Papua New Guinea. In Sather, C. and Kaartinen, T. (eds.) *Beyond the Horizon: Essays on Myth, History, Travel and Society. In honour of Jukka Siikala*. Helsinki: Finnish Literature Society, pp. 123–44.
Damon, F. H. 2016. *Trees, Knots and Outriggers (Kaynen Muyuw): Environmental Research in the Northeast Kula Ring*. New York: Berghahn Press.
Davis, M. 2002. *Late Victorian Holocausts: El Niño Famines and the Making of the Third World*. London: Verso.
Dickinson, W. R. 2003. Impact of mid-holocene hydro-isostatic highstand in regional sea level on habitability of islands in Pacific Oceania. *Journal of Coastal Research*, 19(3): 489–502.
Ellen, R. F. 2004. The distribution of *Metroxylon* sagu and the historical diffusion of a complex traditional technology. In Boomgaard, P. and Henley, D. (eds.) *Smallholders and Stockbreeders: History of Foodcrop and Livestock Farming in Southeast Asia*. Leiden: KITLV Press, pp. 69–105.
Ellen, R. F. 2006. Local knowledge and management of sago palm (*Metroxylon sagu Rottbeoll*) diversity in south central Seram, Maluku, Eastern Indonesia. *Journal of Ethnobiology*, 26(2): 258–98.
Haslett, S. K. 2009. *Coastal Systems*. 2nd edn. London: Routledge.
Klein, R. J. T., Smit, M. J., Goosen, H. and Hulsbergen, C. H. 1998. Resilience and vulnerability: Coastal dynamics or Dutch dikes? *The Geographical Journal*, 164 (3): 259–68.
Leach, E. R. 1950. Primitive calendars. *Oceania*, 20(4): 245–62.
Malinowski, B. 1927. Lunar and seasonal calendar in the Trobriands. *Journal of the Royal Anthropological Institute*, 57: 203–15.

Malinowski, B. 1984. *Argonauts of the Western Pacific*. Prospect Heights, IL: Waveland Press.
Mei, C. C. and Liu, P. L. F. 1993. Surface waves and coastal dynamics. *Annual Review of Fluid Mechanics*, 25: 215–40.
Mercer, J., Dominey-Howes, D., Kelman, I. and Lloyd, K. 2007. The potential for combining indigenous and western knowledge in reducing vulnerability to environmental hazards in small island developing states. *Environmental Hazards*, 7(4): 245–56.
Moran, E. 2006. *People and Nature. An Introduction to Human Ecological Relations*. Oxford: Blackwell Publishing.
Nunn, P. D. 2000. Environmental catastrophe in the Pacific Islands about AD 1300. *Geoarchaeology*, 15: 715–40.
Nunn, P. D. and Britton, J. M. R. 2001. Human-environment relationships in the Pacific Islands around AD 1300. *Environment and History*, 7: 3–22.
Pernetta, J. C. 1992. Impacts of climate change and sea-level rise on small island states: National and international responses. *Global Environmental Change*, 2(1): 19–31.

11

Traditional Knowledge and Crop Varieties as Adaptation to Climate Change in South-West China, the Bolivian Andes and Coastal Kenya

Krystyna Swiderska, Hannah Reid, Yiching Song, Doris Mutta, Paul Ongugo, Mohamed Pakia, Rolando Oros and Sandra Barriga

Indigenous peoples and local communities often live in harsh natural environments, and have had to cope with extreme weather and adapt to environmental change for centuries in order to survive. They have done this using long-standing traditions and practices – or traditional knowledge (TK) – relating to adaptive ecosystem management and sustainable use of natural resources.

There is growing evidence of the role of traditional knowledge in responding to climate change (Berkes, 2009; Berkes et al., 2000; Gyampoh et al., 2009; Reid et al., 2009). The IPCC's Fifth Assessment highlighted the role of indigenous knowledge and crop varieties in adaptation (2007: 864–7). The Fifth IPCC report (2014) further recognizes that 'indigenous, local and traditional knowledge systems and practices, including indigenous peoples' holistic view of community and environment, are a major resource for adapting to climate change'. The United Nations University Traditional Knowledge Initiative (UNU-TKI) has identified over 400 cases of indigenous peoples' role in climate change monitoring, adaptation and mitigation, including a variety of successful strategies (Galloway McLean, 2010). The United Nations Framework Convention on Climate Change (UNFCCC) has also recognized the importance of traditional knowledge, in decisions on 'enhanced action on adaptation' and in the Cancun Adaptation Framework. However, because negotiating Parties at the UNFCCC COPs are countries rather than communities, TK is largely marginalized and has struggled to receive the acknowledgement it deserves.

This chapter provides evidence of the role of traditional knowledge and traditional crop varieties – which are the product of TK – in adaptation to climate change. It draws on case studies with indigenous farmers in the Karst mountains of SW China, coastal Kenya and the Bolivian Andes. These studies document farmer perceptions of climate change impacts and local TK-based responses. The China case also examines the use of TK in participatory plant breeding (PPB), and the threats to traditional varieties from modern agriculture and intellectual property rights (IPRs). The role of TK in weather forecasting is also considered in the Kenya and Bolivia cases. The studies adopt a holistic perspective on TK, as part of wider 'biocultural systems', building on previous research on TK protection by International Institute for Environment and Development (IIED) and partners (IIED et al., 2009). The chapter explores the role of wider biocultural systems in adaptation to climate change, and the drivers of change causing the loss of traditional knowledge and crop varieties.

The role of TK and ecosystems has tended to be under-represented in adaptation responses and economic evaluation of adaptation (World Bank, 2010). The main emphasis has been on structural approaches and agricultural intensification, although ecosystems or ecosystem services feature in about 56 per cent of National Adaptation Programmes of Action projects (Reid et al., 2009). In addition, the linkages between TK and genetic resources or ecosystems are not always adequately recognized, e.g. there has been limited attention to TK in work on ecosystem-based adaptation (Campbell et al., 2009).

This chapter highlights the close interlinkages and interdependence between TK and genetic resources, and their role in adaptation to climate variability and change. It also shows the need for climate change policymakers at national and international levels to address the multiple drivers of loss of TK and genetic resources, including the effects of modern agriculture and IPRs.

The Role of Traditional Knowledge and Crop Varieties in Adaptation to Climate Change

Traditional farmers have domesticated, improved and conserved thousands of crop species and varieties using their traditional knowledge (Andersen, 2006). The diversity of traditional varieties sustained by farmers around the world is increasingly valuable for adaptation as the climate changes, particularly as modern agriculture relies on a very limited number of crops and varieties (see FAO: Commission on Genetic Resources for Food and Agriculture). The Communities of the Potato Park, Cusco, Peru, for example, hold about 650 native potato varieties. This diversity has value in itself in the face of uncertainty.

In addition, traditional varieties or landraces are more genetically diverse than modern varieties and are therefore better able to withstand environmental stress, such as lack of water or nutrients (CBD, 2010). In SW China, laboratory analysis has shown that *in situ* varieties have much higher genetic diversity than those held *ex situ* for thirty years, which demonstrates the positive influence of environmental factors and farmers (Sihuang Zhang and Yiching Song, personal communication). In coastal Kenya, Mijikenda sacred forests (or *kaya*) conserve plant and animal biodiversity that represents a valuable source of germplasm for species that can tolerate the current extreme weather and soil conditions including drought, salt, pest and disease tolerance.

There are at least five types of TK useful for adaptation in agriculture:

i) *Knowledge about resilient properties, such as drought and pest resistance*: Traditional farmers often live on marginal lands where climate change impacts and selection pressures are greatest. Thus, they are well placed to identify resilient crop species and varieties for adaptation.

ii) *Knowledge about plant breeding*: Traditional farmers are also plant breeders actively engaged in farm experimentation. In SW China, local women's maize breeding experience and expertise has been developed over many years. For generations, farmers have carried out seed selection for preferred and adaptive characteristics and innovative farmers even cross some lines for crop improvement. Today, given the trend of male outmigration, local landraces are largely conserved by women and the old, most of whom are experienced in seed selection.

iii) *Knowledge about wild crop relatives*: Wild areas around farms provide wider gene pools for crop improvement and domestication (Jarvis et al., 2008), and communities use wild foods to supplement their diet. For the Mijikenda in Kenya, for example, some wild food plants are relied upon by farmers when crops fail.

iv) *Knowledge about resilient farming and resource management practices*: Traditional agriculture practices conserve key resources for resilience and adaptation – such as biodiversity, water, soil and nutrients. The Fifth IPCC assessment highlighted the potential of traditional water engineering, soil management, pest management, nutrient fixing, erosion control and land restoration (2007: 864–7).

v) *Knowledge about climate forecasting*: TK can help to forecast local weather, predict extreme events and provide accessible information to farmers, at a scale which can be more useful locally than sophisticated models. Climate models often provide information at a scale too large to be of use when planning what to plant and when at the farm scale. With the onset of climate change, the Mijikenda's traditional knowledge system has become a source of crucial information on weather forecasting for farmers in coastal Kenya (Mutta et al., 2009). TK can also monitor climate change in specific locations, and bridge the resolution gap of scientific models.

Farmers are already using traditional varieties and knowledge to adapt to climate change, particularly where traditional farming systems have been maintained. For example, farmers in the Eastern Himalayas have adopted varieties grown at lower altitudes in response to rising temperatures (Ruchi Pant, in Reid and Swiderska 2008).

Case Study 1: *The role of biodiversity, traditional knowledge and participatory plant breeding in climate change adaptation in Karst Mountain areas, south-west China*

Over the past fifteen years, PPB has been carried out in three provinces in south-west China, facilitated by the Centre for Chinese Agricultural Policy (CCAP). A systematic study on climate change impacts and adaptation was conducted in Guangxi, Guizhou and Yunan provinces in the Karst mountains, south-west China, a region inhabited by thirty-three ethnic groups and rich in plant genetic resources (a centre of maize diversity). The study involved a survey carried out in 2009 and 2010 in fifty-four villages and 162 households in the three provinces; semi-structured interviews in six PPB villages and six non-PPB villages in Guangxi in 2010; and use of qualitative findings from PPB action research.

Table 11.1 *Farmers' perceptions of the effects of climatic change (as percentages)*

Indicators	Perceptions				
	−2 (much less / lower)	−1 (less/ lower)	0 (normal)	+1 (more/ higher)	+2 (much more/ higher)
Drought		2	28	70	
Rainfall	4	77	17	2	
Temperature			13	87	
Wind force		4	34	62	
Sunshine		8	37	56	
Run-off	13	8	8	72	
New pests/disease			27	64	9

The main results of the study focused on the following aspects. First, the climatic changes and impacts in the region over ten years, i.e. from 1998 to 2008. Second, local adaptation tools and practices including TK, biodiversity, PPB, community group's collective actions. Third, adaptation results, by comparing PPB project villages with others, i.e. villages that used TK and local varieties with those that used modern technologies and varieties. Finally gender as a cross-cutting aspect integrated in the above three items. This is especially pertinent given the widespread phenomena of feminization of agriculture and ageing rural populations in China.

The study looked at changes in climate and in the socioeconomic situation of poor farmers, including male outmigration and increased rural poverty. It found that climatic and socioeconomic changes are interlinked and mutually affected complex processes. Farmers' perspectives of climate change revealed that they have been severely affected and most farmers felt the effect of increasing temperature and drought and lower rainfall (see Table 11.1). It is noticeable that: 70 per cent of the respondent villagers face the problem of increased drought; 77 per cent noted a lack of rainfall; 87 per cent remark that the temperature has increased; 62 per cent report stronger winds; 56 per cent report stronger sunshine; 72 per cent have more run-off; and 64 per cent of the villagers reported that new pests and diseases have emerged. Scientists from the Chinese Academy of Agricultural Sciences have reported new diseases such as banded leaf and sheath blight, grey spot, rust, ear rot and maize rough dwarf virus.

Farmers, mainly women and the elderly, are relying on their TK and genetic resources and on collaboration with public institutions and scientists for adaptation. Landrace conservation and PPB are two tools used.

Fifty-three per cent of respondent households still use landraces (especially of maize and rice) because they taste good, are better adapted to mountainous and barren land, are easy and cheap to obtain, and have drought and dislodging resistance (to protect from wind force). Most landraces come from the farmers' own saved seeds, whereas hybrids have to be bought and the old and poor cannot afford them. Farmers also noted that cultivating diverse varieties can help reduce risks. Evidence from the field in Guangxi has shown that most farmer-improved landraces and open pollinated varieties (OPVs) survived the big spring drought in SW China in 2010, while most of the hybrids were lost. The villages involved in PPB that have more landraces and OPVs survived and adapted to the drought. Other villages that had grown only hybrids lost all their production due to a shortage of hybrid seeds in the market. This event demonstrates the importance of farmers' seed systems for adaptation and resilience.

Faced with climate change, farmers require well-adapted crops and varieties with characteristics such as drought resistance, anti-dislodging and resistance to new pests and diseases. PPB is based on farmers' ancestral selection and breeding knowledge and genetic resources to better achieve locally adapted and preferred characteristics, while bringing in knowledge and expertise from formally trained plant breeders. The PPB initiative started in early 2000 in south-west provinces, focusing on maize initially and then also on rice. Its main aim is to establish cooperative and complementary relations between the previously distinct formal seed system and farmers' seed system. Cooperation is necessary to provide opportunities for empowerment of farmers who are mainly women, as most men have migrated to the cities. The farmers become active partners in plant breeding, on-farm biodiversity management and seed marketing (Ashby, 2009; Song and Vernooy, 2010).

Field experiments have targeted four types of maize OPVs and landraces: 'exotic' populations (from abroad), farmers' 'creolized' varieties (developed by breeders but further adapted by farmers, sometimes by crossing them with landraces), farmer-maintained landraces and formally conserved landraces. So far, more than 200 varieties have been used in trials at the Guangxi Maize Research Institute station and in the villages. Based on ten years of experimentation, six farmer-preferred PPB varieties have been selected and released in the research villages. They have also spread beyond these villages. In addition, five varieties from the International Maize and Wheat Improvement Centre that were showing increasingly poor results have been adapted locally; and five landraces from the trial villages have been improved. Agronomic traits, yields and palatability of all these varieties are satisfactory and they are better adapted to the local environment than modern hybrids (Song et al., 2006).

Participatory plant breeding and related activities have also led to enhanced crop and animal diversity, forest resources and herbal medicines, and related TK and culture system. PPB villages have higher rice and maize diversity and more cultural and collective activities. In addition, farmer incomes have increased by about 30 per cent in PPB project villages, compared to non-project villages, as a result of project supported activities like productivity increase, value addition to local speciality foods and women's seed production. PPB has also attracted some young people to work in their villages. Farmers are more confident and organized with better links to external markets, e.g. a women's group has won contracts for organic vegetable production. Women are also participating more in decision-making including as local leaders.

Evidence from field studies in SW China shows that traditional varieties face significant threats. The biggest reason for the loss of landraces and agricultural genetic resources is the extension of 'modern' varieties, mainly hybrids. For the three main traditional crops – maize, rice and soybean, between 1998 and 2008, the following findings are evident. First, varieties of both maize and rice, as staple food crops, have been highly commercialized with a large area planted with hybrids. Second, the cultivated areas of landraces of maize and rice have rapidly decreased, the reduction in maize being even more serious than rice. Finally, soybean landraces, as traditional intercropping and supplementary crops, have been well conserved and adopted, staying dominant and stable in the south-west region. This is in part because soybean is not a staple food crop (like rice, maize and wheat) and is not the target of hybrid breeding and extension by both public and private sectors. Yet, in the last few years soybean hybrid breeding has attracted increasing efforts from government and companies.

As a common good, agricultural germplasm resources should be enhanced by government, public research institutes and farmers. However, there is a lack of incentives and responsibilities among all these key stakeholders. There is an urgent need for policies and regulations to create incentives and encourage responsibility for plant genetic resources management. This includes changes in breeding, testing and release systems for 'modern' agriculture (such as distinctness, uniformity and stability criteria), because these standards lead to an increasingly narrow genetic base, less biodiversity and limited options for both breeders and farmers.

As the seed market is not yet well regulated in China, it is necessary for public agricultural service systems to provide seed access facilitation and regulation for farmers. However, the agricultural service system (previously called extension system) in SW China has proved dysfunctional in public service provision (Song and Vernooy, 2010). It needs to be strengthened, especially on (severe) weather broadcasting and warning, provision of information on varieties and market monitoring, among others.

The transformation of public breeding institutes from current mixed commercial and public roles to pure public functions, would not only enhance support for plant genetic resources and TK, but would also allow fairer market competition and bring more opportunities and resources, both technical and human, for seed enterprises. Having a clearly defined public role will also benefit fundamental research on issues such as meeting diversified varietal needs of farmers and others, conserving agrobiodiversity and broadening the genetic base in breeding. Currently, the public role played by these institutes is vague and weak (Song and Vernooy, 2010)

Another challenge is the negative impact of the big international seed companies, such as DuPont, Pionire, Monsanto, etc., and their use of IPRs. Their strategies are joint ventures with domestic companies, passing the national distinctness, uniformity and stability criteria and applying for IPRs. One hybrid maize, Zhenda 619, has wiped out half of the remaining maize landraces in Guangxi province since 2002 (CCAP report to IDRC, 2008). Another hybrid, Xianyu 335, by a joint venture company (Pionire and Denhai) has become the second biggest maize seed, in terms of extent of growing area in China, in 2010. Risks in terms of seed safety, genetic narrowing, resilience and adaptation to change have caused increasing concerns. The challenges of breeding in a climate change impacted world may require the widest possible circulation and sharing of germplasm to ensure effective, timely and adaptive breeding with farmers as a key partner.

Case study 2: *The role of traditional knowledge and biocultural systems in adaptation to climate change: The case of Kenya's coastal communities*

A preliminary study with the Mijikenda was conducted by the Kenya Forestry Research Institute (KEFRI) in June 2011. The study entailed interviews and focus group discussions around two kaya sacred forests (Kaya Fungo and Kaya Kinondo), coastal agroecosystems (crop and livestock) and fisheries in Kilifi and Kwale counties.

The findings show that changes in climate are impacting the Mijikenda's sociocultural, economic, environmental and food security situation. The communities are experiencing shifts in seasons including extreme temperatures, floods, prolonged drought, tidal changes and ocean storms. They have observed notably new and worrying unpredictability of the weather conditions including timing of onset of rains. The rains are either too low and unreliable or too much. Previously, droughts would occur once every ten years. Currently droughts are more frequent and prolonged. The communities are responding by using both traditional biocultural management approaches and contemporary government approaches.

Recurrent crop failures have been experienced in monoculture systems as crops receive inadequate moisture to grow to maturity. This leads to poor yields and food insecurity. Traditional cropping calendars have also been severely disrupted by changing weather patterns. Previously specific crop plants were cultivated at defined times of rain guided by traditional practices and rules of the elders. Following the Green Revolution and adoption of modern agriculture in an effort to improve food production and food security, a high proportion of farmers grow improved monoculture crop varieties and less traditional variants. However, most of the indigenous community members still maintain strong trust in indigenous varieties especially for maize, millet, simsim and cassava, and allocate a land parcel to grow these traditional variants.

In response to changes in climate, there has been a return to some of the traditional crop varieties that had been systematically overtaken by improved modern crop varieties. These include traditional maize varieties namely *mingawa* (white seeded, matures with extended rains), *mzihana* (black and white seeded, matures with medium rains), *kastoo/ njerenjere* (small yellow seeded, matures faster and is used when rains are not enough), and *bomba* (big seeded). There is an understanding that fast and slow maturing variants can be grown together to reduce the risk of crop loss. The indigenous communities believe that the traditional varieties are hardy and can cope better with the unpredictable weather conditions and local pests; easy to obtain from the last harvest or from friends and relatives; and known to establish without the need for inputs (fertilizer and pesticides) required by modern crop varieties.

The farmers share seeds of various varieties and hence enhance crop diversity on most farms (i.e. avoid concentrating specific varieties in certain areas only) in order to spread the risk and ensure the survival of the varieties even when there is total crop failure in some parts of the community area.

Farmers admit that old methods may not exclusively lead them to sustainable food security. But they insist that certain traditional crop varieties, practices and concepts must be maintained and synchronized with the modern knowledge system and practices. Owing to the high cost of modern farming practices, maintaining traditional crop varieties and farming practices forms an important survival strategy for the rural poor.

Some of the contemporary agricultural strategies work against the traditional systems (e.g. efforts to introduce improved seeds undermine traditional seeds). This may be very risky given that the latter are a possible life line for survival in the unpredictable future of changed climate. Government efforts should be directed towards the strengthening of traditional community seed banks of traditional crop varieties, as a complement to its current focus on modern crop varieties.

Changes in climate have also impacted on kaya forests, with significant consequences for food security (see Box 11.1). The current governance structures that reduce the role of traditional elders have not been effective in protecting the kaya forests. It is important that the traditional governance structures are strengthened to ensure that biocultural resources that could help in adaptation to climate change are more effectively protected and conserved.

Case study 3: Mountain agricultural systems, Andean people and decision-making processes under climate change realities

A study by Proyecto de Investigación de la Papa (PROINPA) with Quechua farmers in the highlands and valleys of the Cuchumuela community, Cochabamba, Bolivia, found that farmers are already experiencing changes in climate. These include unpredictable rainfall, more extreme weather events and higher temperatures. The farmers' perception is that the climate has considerably changed in recent years. This has had negative impacts especially on food security crops such as potatoes, owing to new pests causing crop losses. Farmers report reduced crop yields for all crops, water shortage, increased drought, increased pests (potato moth, Andean weevil) and diseases (chocolate spot), and emergence of new pests (black corn weevil). Because of increased moth attack, potatoes can only be stored for one to three months rather than seven months as previously, and farmers have to buy potatoes for the rest of the year.

The Andean potato weevil is a major potato pest in Bolivia, and the rise in temperature has led to a dramatic increase in the population of the more damaging *Rhigopsidius piercei* (from 9 to 69 per cent) which had previously been kept in check by low temperatures.

Rising temperatures have also enabled the potato moth to spread from the south of the country to the north and highlands. Changes in climate have led to changes in the planting

Box 11.1 Impacts of climatic changes and TK-based responses in Mijikenda forestry, livestock and fisheries

Forestry

Prolonged drought has led to a significant decrease in survival of vegetation leading to the scarcity and local extinction of certain species, in particular preferred animal and plant species including: small mammals and birds such as gazelles, rodents, doves and guinea fowls that provided common bush meat; plant species such as *Combretum schumannii* – an important house building plant; *Zanthoxylum chalybeum* – an important medicinal plant; and *Dialium orientale* (pepeta) – a plant that produces wild edible berries. This has led to a diminishing supply of forestry resources for subsistence. Coupled with increasing population, this has brought increasing pressure on the traditional kaya forests. This has led to the current situation where the kayas form islands of forest patches surrounded by wholly cleared farm or grazing land.

Long droughts result in shrinking swamps, drying rivers, diminished water supply and the formation of hard pan on the clay soils that lead to low survival of plant species and declining biodiversity. These events provide an opportunity to identify drought-tolerant plant species.

Response: These trends point to the need to strengthen traditional governance structures that would help ensure the protection of the sacred forests. The *kaya* elders feel that they should be enabled to control plant and animal resource harvesting using their customary laws and informal courts.

Other measures proposed by the government include the creation of buffer zones around the kaya forests, traditionally used as communal grazing grounds.

Livestock Farming

Unpredictable rains have led to rivers and water sources remaining dry most of the year and affecting grazing lands and availability of pasture for livestock. Owing to the extended drought, certain preferred grass species disappear causing livestock farmers to move to greener areas and causing soil erosion.

Response: Livestock farmers relocate to greener pastures and permanent rivers as a survival strategy. They follow the traditional practice of sharing their animals with friends and relatives with different sources of water and pasture until favourable conditions resume. This helps to minimize chances of losing the whole stock if put in one place where conditions may get worse leading to loss of livestock.

Traditional approaches also included setting aside a land parcel for communal grazing but these are difficult to maintain owing to privatized farms and increasing population pressure for agriculture expansion.

Fisheries

Marine biodiversity was once very rich but changes in climate have led to a declining population of some fish species, namely *paramanita* and *mkizi* (King fish) as suggested by substantially reduced landing statistics (except in marine protected areas). Fishing conditions have worsened due to rougher waters and erosion of beach lines. The destruction of coral reefs through bleaching has damaged breeding grounds and habitats for species such as *mkizi*, octopus and squid.

Response: The government has established Beach Management Units (BMUs) as a governance structure to manage fisheries activities. However, the leadership of the BMUs is comprised of youth and have left out traditional elders, which means useful TK and customary practices that can contribute to conservation may not be considered

Emphasis is being placed on environmental conservation including rehabilitation of mangroves to protect shorelines and breeding grounds for fish, placing control measures on endangered fish species, reintroducing the traditional practice of zoning off areas from foreign fishermen, and documenting traditional knowledge on fisheries.

season of many crops – and this has also favoured the potato moth. The moth causes about 30 per cent crop damage in the field and 50 per cent in stores when not controlled. Farmers in the highlands are not prepared to deal with this new pest and commonly use highly toxic chemicals to control tuber moths. As a result, production costs increase, farmer health is compromised and moths develop resistance to most treatments. However, to protect potatoes in storage, farmers also use native plants that have repellent properties (*muña* and eucalyptus leaves), and biological control – they bring ants that eat potato moth and Andean weevil larvae.

In response to new pests and low harvests of certain varieties, farmers are also specializing in more resistant and better adapted varieties. The diversity of local varieties has enabled them to select those that are best adapted to the new conditions. For example, 'Doble H' was not planted before in the community, but is now the most common variety in the region because it will always produce, even with little rainfall. The species grown have not changed, but the varieties have changed. Fewer varieties are now grown due to changes in climate and in the market.

The study also compared the predictions of a PRECIS (Providing Regional Climates for Impact Studies) model with actual meteorological observations and found approximate correlation in some areas. This suggests that the changes observed are owing to human-induced climate change. However, there are also significant differences in temperature and precipitation estimates in some areas, which suggests that climate change models may not be very precise in the Andean region, and should be complemented with assessments by local communities.

The research suggests that productive agroecosystems and related TK and practices are in danger of being severely impacted by climate change, and that the changes are happening too fast for local people to adapt their practices. Many natural indicators used by their ancestors to predict the weather have been lost with the loss of TK, and some signs are no longer useful because of recent changes in climate.

The study identified the following priorities for reducing the vulnerability of poor Andean farmers, and thus the need to use both science and TK:

- Developing crop varieties that are resistant to pests and diseases, and tolerant to abiotic impacts, through participatory plant breeding
- Local seed production and dissemination for timely response to natural disasters
- Promoting mass organic production because farmers prefer it and have limited or no access to chemical inputs
- Zoning microcentres of genetic diversity of cultivated species, semi-cultivated and wild relatives in Bolivia; and strengthening existing local conservation strategies in these microcentres
- Revaluation of local knowledge and linking it with scientific knowledge

The Role of Biocultural Systems in Adaptation to Climate Change

Research by IIED and partners with several indigenous communities in Peru, Panama, China, India and Kenya (IIED et al., 2009) found that traditional knowledge and traditional crop varieties are very closely linked and highly inter-dependent. The maintenance and transmission of TK depends on the use of diverse biological resources (wild and domesticated), and the reintroduction of traditional varieties can revive related traditional knowledge and practices. It also found that TK and traditional varieties are themselves dependent on other factors: traditional landscapes, cultural and spiritual values, and customary laws. These elements of biocultural systems are all inter-connected.

Traditional landscapes sustain genetic diversity and TK and contribute to adaptive capacity by:

- sustaining wild gene pools, diverse ecosystems, and ecosystem services
- sustaining sacred sites (e.g. forests, mountains, rivers) which promote TK transmission, cultural values and beliefs
- providing the physical space for sharing and exchange of landraces and TK across wide distances based on customary laws

Cultural values and preferences are a key reason why communities sustain traditional varieties despite the pressure to adopt modern hybrids. In SW China for example, landraces are sustained because their taste is preferred and for use in ceremonies (e.g. maize wine for weddings). For many communities, certain traditional varieties are sacred and used in rituals and ceremonies.

Our research found that customary laws promote TK use and transmission, conservation of natural resources and social equity. We identified three customary law principles that guide all aspects of life: Reciprocity (equal exchange), Duality (complementarity) and Equilibrium (balance in nature and society). We found that a wide sharing and exchange of seeds and knowledge (e.g. between villages) based on reciprocity enhances biodiversity, since receiving triggers the obligation to reciprocate in equal measure. Indigenous communities also have specific customary rules for conservation (e.g. taboos on harvesting at certain times). Much research has shown that recognizing existing customary laws is more effective than imposing external conservation rules (MA, 2005; Laird et al., 2009).

Climate change is likely to require exchanges over large distances for communities to access the range of crop varieties they need for adaptation, given the speed and scale of the changes predicted. The biggest constraint to such exchanges may not be differences in culture or language, but in farming and value systems, for example subsistence versus commercial. Strong sharing and reciprocity values are common to traditional farmers of different cultures, and exchanges can occur over large distances. In SW China, results of research on maize by CCAP (both in the field and in the lab) showed that variety exchange and gene flow occurs over large distances across the region (three provinces) and that women play a crucial role.

Drivers of Change Affecting Traditional Knowledge and Varieties for Adaptation

Traditional knowledge, crops, practices and forms of organization offer huge potential for resilience and adaptation to climate change. However, it is estimated that 50–90 per cent of all languages – an indicator of TK – will be extinct or threatened by 2100 (UNESCO, 2003). Despite being based on hundreds of years of accumulated practical experience, TK has been sidelined and undervalued for centuries, being considered unscientific, ungodly or anti-development. And climate change could accelerate the loss of TK.

Research by IIED and partners (2009) identified the following multiple drivers of loss of traditional knowledge and genetic diversity, which are often interlinked and mutually reinforcing:

- Agricultural policies, subsidies and research/extension which promote modern varieties and technologies, at the expense of local knowledge and biodiversity
- The existence of plant breeders' rights and patents to protect new varieties without commensurate protection of farmers' rights over traditional varieties, which means that farmers often have no economic incentive to sustain them
- Promotion of modern varieties/foods in the media, which influences consumer demand and reduces markets for traditional varieties
- Limited arable land and reduction in size of landholdings, which can force communities to adopt higher yielding modern varieties
- Erosion of cultural values and customary rules, due to modernization, weakening of traditional authorities, outmigration and changes in occupation

The findings of the China case study suggest that IPRs are facilitating the spread of commercial hybrids. The expansion of IPR regimes in agriculture tends to create a market for seeds that is dominated by a few large companies (Dhar, 2002). IPRs in agriculture also raise the price of seeds and have led to concerns over limiting access to seed by farmers and scientists (Tansey and Rajotte, 2008).

Conclusion

The three case studies presented above provide evidence of the crucial role of traditional crop varieties, knowledge and practices in enabling adaptation to changes in climate (Figure 11.1). The question is whether the climatic changes observed in these cases are human-induced climate change or just natural changes. The findings show that indigenous farmers in SW China, coastal Kenya and the Bolivian Andes are already severely impacted by changes in climate, including drought, with serious consequences for crop production and food security. The scale of the changes, and the fact that they have occurred quite recently (in the last ten or twenty years), suggests that they may be the result of human-induced climate change. In the Bolivian case, a number of observed changes in climate are backed up by scientific models, and hence likely to be human induced. Even if the changes are not human induced, the studies show that TK has clear potential to contribute to climate change adaptation, alongside modern science. This implies a revaluation of traditional knowledge, and a rethinking of mainstream agricultural policies that erode traditional varieties and knowledge.

In each case, the maintenance of diverse traditional crop varieties and access to seeds has been essential for adaptation and survival by poor farmers. Traditional varieties used include drought and wind resistant maize in SW China; maize resistant to unpredictable weather and new pests in coastal Kenya; and potato varieties in Bolivia that are more resistant to new pests and lack of rainfall. All three cases also found that traditional varieties have the advantage of being cheap and easily accessible as they come from farmers' own saved seeds, whereas modern varieties have to be bought, depend on market availability and require costly inputs. In SW China the old and poor cannot afford hybrids and hybrid farmers lost all their production to drought in 2010 due to the shortage of hybrid seed in the market. While modern agriculture and modern varieties may increase productivity, the findings show that under conditions of environmental stress and climatic variability, survival depends on more resilient and readily available traditional varieties.

In the China and Kenya cases, farmers also identified planting diverse traditional varieties as a means to reduce risk, and the importance of sharing and exchange of seeds to gain access to diverse varieties. In Bolivia, native plants and biological control is providing a less costly alternative to toxic chemical control which affects farmer health and leads to resistance in pests. In coastal Kenya, adaptation requires the strengthening of customary governance to restore kaya sacred forests since current governance structures have not been effective.

This study also underlines the need to look at climate change impacts alongside other socioeconomic issues and trends facing poor farmers. Often compounding the effects of climate change, these include outmigration and increased poverty in SW China and population pressure in coastal Kenya.

Interestingly, the three different case studies identified the need for very similar adaptation responses – conservation of traditional varieties/seed banks, community seed production and PPB. In the China case, these tools have been tested in practice and shown to be effective. The conservation of landraces and development of improved varieties

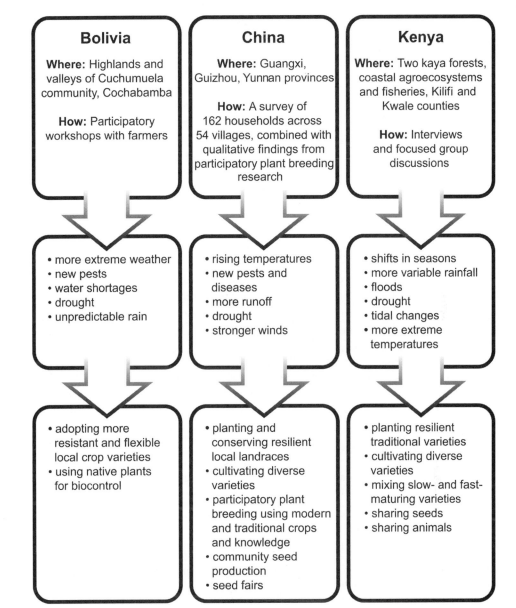

Figure 11.1 Community perspectives of climate change, and their responses based on traditional knowledge.

through PPB is not only enabling adaptation, but has enhanced biodiversity conservation and related TK, increased incomes of poor farmers by about 30 per cent over ten years, empowered poor farmers and women, and started to improve public sector support for farmer seed systems.

This suggests the need to support initiatives such as local landrace conservation, local seed production, seed fairs, community seed banks, and community-based conservation and adaptation. Adaptive capacity of the world's poorest and most affected communities depends not only on TK or on ecosystems but on both – on the interlinked biocultural systems from which new innovations can develop and spread; and on the landscapes, cultural and spiritual values and customary laws that sustain them.

Among the drivers of loss of indigenous knowledge and varieties perhaps the most significant are agricultural policies, laws and institutions that promote commercial varieties, and seed privatization at the expense of local varieties and seed systems. The challenge of climate change cannot be confronted by modern agriculture and intensification alone – indeed the spread of intensive agriculture to centres of genetic diversity could seriously undermine human adaptive capacity.

Yet, the current dominance of private sector objectives in government agriculture policies and institutions poses a serious threat to remaining centres of diversity as can be seen from the China case. IPR standards (plant breeders' rights and patents) can facilitate the rapid spread of hybrids, restrict access to genetic resources for adaptation by both farmers and scientists, and do not provide any incentives for *in situ* conservation by farmers. Reform of IPRs, protection of farmers' rights to traditional varieties (as required by the FAO Treaty on Plant Genetic Resources for Food and Agriculture), and improved market access for farmer varieties are urgently needed to create incentives for *in situ* conservation of thousands of traditional varieties that are in danger of being lost. These measures should be addressed as a priority in national adaptation actions and international climate change negotiations.

References

Anderson, R. 2006. *Farmers' Rights and Agrobiodiversity*. Issue Paper on People, Food and Biodiversity. Eschborn: Gesellschaft für Internationale Zusammenarbeit (GIZ).

ANDES (The Association for Nature and Sustainable Development), the Potato Park Communities and IIED (International Institute for Environmental and Development). 2011. *Community Biocultural Protocols: Building Mechanisms for Access and Benefit-sharing among the Communities of the Potato Park based on Quechua Customary Norms*. Full case study report. London: IIED.

Ashby, J. A. 2009. The impact of participatory plant breeding. In Ceccarelli, S., Guimaraes, E. P. and Weltzien, E. (eds.) *Plant Breeding and Farmer Participation*. Rome: FAO.

Berkes, F. 2009. Indigenous ways of knowing and the study of environmental change. *Journal of the Royal Society of New Zealand,* (39): 151–6.

Berkes, F., Colding J. and Folke, C. 2000. Rediscovery of traditional ecological knowledge as adaptive management. *Ecological Applications*, 10(5): 1251–62.

Campbell, A., Kapas, V., Chenery, A. et al. 2009. *The Linkages Between Biodiversity and Climate Change Adaptation. A Review of the Scientific Literature*. UNEP WCMC.

CBD (Convention on Biological Diversity). 2010. *Linking Biodiversity Conservation and Poverty Alleviation: A State of Knowledge Review*. CBD Technical Series No. 55, www.cbd.int/doc/publications/cbd-ts-55-en.pdf

Dhar, B. 2002. *Sui Generis Systems for Plant Variety Protection. Options under TRIPS. A Discussion Paper*. Geneva: Quaker United Nations Office, http://quakerservice.ca/wp-content/uploads/2011/07/SGcol1.pdf

FAO (Food and Agriculture Organization of the United Nations). nd. *Climate Change and Biodiversity for Food and Agriculture*. Natural Resource Management and Environment Department, ftp://ftp.fao.org/docrep/fao/010/i0142e/i0142e01.pdf

FAO. *Commission on Genetic Resources for Food and Agriculture. Plant Genetic Resources: Use Them or Lose Them*, www.fao.org/nr/cgrfa/cthemes/plants/en/

Galloway McLean, K. 2010. *Advance Guard: Climate Change Impacts, Adaptation, Mitigation and Indigenous Peoples. A Compendium of Case Studies*. UNU-IAS.

Gyampoh, B. A., Amisah, M, Idinoba, M. and Nkem, J. 2009. Using traditional knowledge to cope with climate change in rural Ghana. *Unasylva* 231/232, (60): 70–4.

IIED, ANDES, FDY, Ecoserve, CCAP, ICIPE and KEFRI. 2009. *Protecting Community Rights over Traditional Knowledge. Implications of Customary Laws and Practices*. London: IIED.

Jarvis, A., Lane, A. and Hijmans, R. J. 2008. The effect of climate change on crop wild relatives. *Agriculture, Ecosystems & Environment*, 126: 13–23.

Laird, S. A., Wynberg, R. and McLain, R. 2009. *Wild Product Governance: Laws and Policies for Sustainable and Equitable Non-Timber Forest Product Use. Policy Brief*. United Nations University, Center for International Forestry Research, People and Plants International, Institute for Culture and Ecology, and University of Cape Town.

Millennium Ecosystem Assessment (MA). 2005. *Ecosystems and Human Well-being: Biodiversity Synthesis*. Washington, DC: World Resources Institute.

Mutta, D., Chagala-Odera, E., Wairungu, S. and Nassoro, S. 2009. Traditional knowledge systems for management of Kaya forests in Coast Region of Kenya. In Parrotta, J.A., Oteng-Yeboah, A. and Cobbinah, J. (eds.) *Traditional Forest Related Knowledge and Sustainable Forest Management in Africa*. IUFRO World Series 23. Accra: IUFRO.

Parry, M. L., Canziani, O. F., Palutikof, J. P., van der Linden, P. J. and Hanson, C. E. (eds.) 2007. Cross-chapter case study. (C4). Indigenous knowledge for adaptation to climate change. *Climate Change 2007: Impacts, Adaptation and Vulnerability. Contribution of Working Group II to the Fourth Assessment Report of the Intergovernmental Panel on Climate Change*. Cambridge: Cambridge University Press, pp. 864–8.

Reid, H., Philips, J. and Heath, M. 2009. *Natural Resilience: Healthy Ecosystems as Climate Shock Insurance*. London: IIED.

Reid, H. and Swiderska, K. 2008. *Biodiversity, Climate Change and Poverty: Exploring the Links*. London: IIED.

Song, Y., Zhang, S., Hung, K. et al. 2006. Participatory Plant Breeding in Guangxi, South-West China. In Almekinders, C. and Hardon, J. (eds.) *Bringing Farmers Back into Breeding. Experiences with Participatory Plant Breeding and Challenges for Institutionalisation*. Agromisa Special 5, Wageningen: Agromisa.

Song, Y. and Vernooy, R. (eds.) 2010. *Seeds and Synergies. Innovating Rural Development in China*. Ottawa: IDRC.

Tansey, G. and Rajotte, T. (eds.) 2008. *The Future Control of Food. A Guide to International Negotiations and Rules on Intellectual Property, Biodiversity and Food Security*. London: Earthscan.

The World Bank/ The International Bank for Reconstruction and Development. 2010 *Economics of Adaptation to Climate Change: Synthesis Report*, https://openknowledge.worldbank.org/bitstream/handle/10986/12750/702670ESW0P10800EACCSynthesisReport.pdf?sequence=1

UNESCO (United Nations Educational, Scientific and Cultural Organization). 2003. *Language, Vitality and Endangerment*. Document submitted by Ad Hoc Expert Group on Endangered Languages to the International Expert Meeting on UNESCO Programme Safeguarding of Endangered Languages.

Part III
Confronting Extreme Events

12

Accounts from Tribal Elders: Increasing Vulnerability of the Navajo People to Drought and Climate Change in the Southwestern United States

Margaret Hiza Redsteer, Klara B. Kelley, Harris Francis and Debra Block

In recent years, many indigenous communities have engaged in studies that incorporate local knowledge in combination with conventional scientific data. These studies have provided ground truthing of scientific analyses and compelling information about changes to ecosystems that are already under way due to climate change (West et al., 2008; Kalanda-Joshua et. al., 2011; Mertz et al., 2012; Doyle et al., 2013; Leonard et al., 2013; Majule et al., 2013). These studies have also led to an increasing local awareness of climate change impacts, spurring community action and increasing adaptive capacity (Mimura et al., 2014). Although only a few of these studies have included Native Americans in the United States other than Alaska, we show that local knowledge from North American indigenous Navajo elders increases our understanding of climate change impacts to ecosystems and allows us a compelling glimpse at the interconnectivity and complexity of natural systems. Our findings also suggest that arid to semi-arid environments are extremely sensitive to small changes in precipitation type (rain vs snow) and small increases in temperature.

The Navajo Nation of northeastern Arizona, northwestern New Mexico and south-eastern Utah in the southwestern United States, is an ecologically sensitive semi-arid to arid area where rapid growth of the largest population of Native Americans is outstripping the capacity of the land to sustain them. Today, the Navajo land base is more than 65,700 km^2 (25,350 square miles) and is the largest reservation in the United States, encompassing 36 per cent of all reservation lands (Figure 12.1). People presently living on these Native lands are an exception in North American society as their traditional lifestyle requires intimate knowledge of the ecosystem, knowledge that has been passed on for generations through oral traditions. Many Navajo residents, especially the elderly, have relied on raising livestock as a significant part of their livelihood or as supplement to store-bought food supplies, and gather herbs for food and medicine (Iverson, 2002). Elderly Navajo people are usually not fluent in English, and in some cases have never attended school, in part because schools across the reservation were closed during the Second World War (Bailey and Bailey, 1986). Twenty per cent of the reservation population has not obtained an education past the ninth grade, and only 4 per cent have obtained a bachelor's degree or higher (American Rural Policy Institute, 2012).

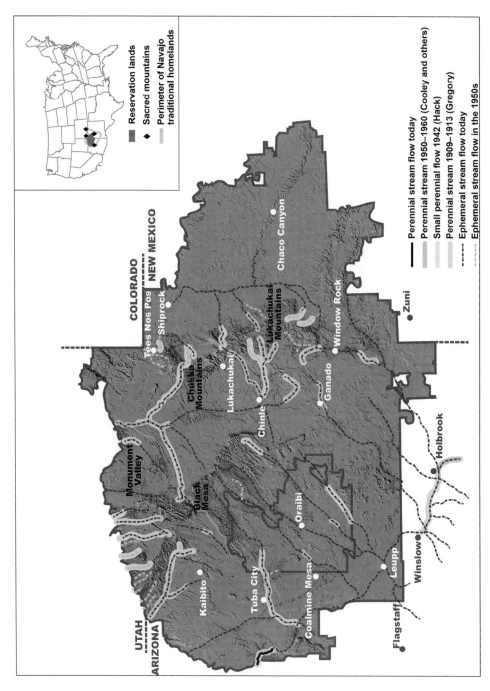

Figure 12.1 Map of the Navajo Nation showing topographic features, and historic changes in perennial flow (after Redsteer et al., 2010). Inset map shows location of reservation lands (red) black diamonds for locations of sacred mountains, and a pale line for approximate perimeter of Navajo traditional homelands. (A black and white version of this figure will appear in some formats. For the colour version, please refer to the plate section.)

Over 300,000 people are members of the Navajo Nation, but the proportion of that population living on the reservation has changed over time. Data from the 2010 US Census shows that the reservation population is declining. Although persons declaring Navajo ancestry or tribal affiliation continued to increase from about 298,000 in 2000, to over 332,000 in 2010, the reservation population decreased by 4 per cent. The changing demographics suggest that the younger segment of the population is emigrating from the reservation to urban areas such as Phoenix, most likely in search of employment.

Of those living on reservation lands, recent American Community Survey (2006–2010) data estimate that 38 per cent of the reservation population is severely poor, with 25 per cent of the total population reporting a wage income. Approximately 40 per cent of the reservation population lives in housing without indoor plumbing or electricity (Navajo Nation Department of Water Resources, 2000). Those without running water depend on springs, seeps and shallow alluvial aquifers that are sensitive to fluctuations in precipitation and threatened by drought (Breit and Redsteer, 2002).

Dryland regions, such as the Navajo Nation are highly sensitive to degradation from land use. In particular overgrazing has been a topic central in the discussions of local landscape conditions (Boyce, 1974; Graf, 1986; Weisiger, 2009). Ecosystem changes resulting from climate change have been slow in becoming part of the discussion about dryland regions (Orlove et al., 2014). Instead, those who are suffering from the deteriorating conditions are commonly blamed (Graf, 1986; West et al., 2008). In order to have a more complete understanding of the changes that have occurred in this poorly monitored region, we compiled the lifetime observations of seventy-three tribal elders from the Navajo Nation. Their accounts document changes in plants and animals, water availability and weather, as well as in the frequency of moving sand dunes and dust storms. We show that recent prolonged drought conditions that began in the 1990s combined with increasing temperatures are significantly altering the habitability of this region, a place already characterized by harsh living conditions. Detailed accounts of the changes that have occurred provide a clearer picture of how the local ecosystem has been altered over the past century. These observations help to refine our understanding of the impacts of climate change on Navajo traditions, culture and well-being, and how these impacts are magnified by historic changes in land tenure policies and economic conditions.

Regional Characteristics

The Navajo Nation is divided by the Lukachukai Mountains on the New Mexico–Arizona border with two-thirds of the Nation located to the west, and one-third to the east. In the centre of the Navajo Nation, at 2,000–2,200 m elevation, is the upland region of Black Mesa (Figure 12.1). Precipitation varies by subregion: lowlands on the western half of the Navajo Nation at 1,200–1,500 m in elevation are the hottest and driest, followed by the eastern Navajo Nation that lies below the Lukachukai Mountains in New Mexico. The Lukachukai Mountains, 3,000 m high at the crest, receive the greatest proportion of the regions' snowfall and rainfall, followed by Black Mesa. Annual precipitation totals in the more arid lowlands of the Navajo Nation average 100–150 mm, in contrast to wetter upland regions of Black Mesa and the Lukachukai Mountains that average 250–300 mm. There are also large seasonal and diurnal variations in temperature, with average annual temperatures from 11.0°C in areas of higher altitude to 14.5°C in the valleys and lowlands.

Navajo lands are characterized by a bimodal precipitation pattern. Winter precipitation typically occurs from November through March, and summer precipitation resulting from the North American monsoon, lasts from July through September. Approximately 45 per cent of total annual precipitation occurs during the monsoon season (Redsteer et al., 2010). These two wet seasons are separated by a dry, windy season in the spring, from April to June.

Documentation of changes within the region is difficult because it is topographically and climatically variable, and large portions of the region are poorly monitored. The density of operating National Weather Service COOP (Cooperative Observer Program) monitoring stations on the Navajo Nation is one for every 6,400 km^2. Moreover, a majority of these stations only record precipitation and are not well suited for climatic observations. The US Geological Survey Navajo Land Use Planning Project has four additional meteorological monitoring sites in the arid south-western Navajo Nation that record temperature, rainfall, wind speed and direction, and soil moisture.

Currently, the Navajo Nation is suffering from drought that has occurred during the years of 1994–2004, 2006–2009 and 2011 to the present (Figure 12.3). Although the winters of 2004 to 2005 and 2010 were wetter (close to average), and there was a wet summer monsoon in 2013, these periods were not sufficiently long to allow the vegetation in the region to recover (Draut et al., 2012). Moreover, this most recent drought has been accompanied by increasing temperatures, increasing 2°C in average temperature since the 1960s, causing a decrease in effective moisture. In the period of 2001–2002, Navajo officials reported that 30,000 cattle perished. The sky turned red with blowing dust. Local residents increasingly find themselves without water for livestock and domestic purposes, and seek additional water from local municipalities that sometimes turn them away, because municipal water supplies are not capable of handling the additional demand. Disputes over water supplies are becoming increasingly common (Redsteer et al., 2010).

Reservation History and Local Land Tenure

The boundary of the modern Navajo Nation was established over time, beginning with the Navajo Treaty of 1868, when land was set aside after the war between the United States and the Navajo ended. In the late 1800s, fierce competition among Anglo and Hispanic populations for the best rangelands surrounding this early reservation boundary precluded the retention of the more verdant traditional lands for Navajo use (Bailey and Bailey, 1986; Iverson, 2002). The reservation established by the 1868 Treaty occupies a relatively small area in the middle of the traditional Navajo homeland (Correll and Dehiya, 1978). The federal government subsequently was most successful in expanding the reservation into the drier country west and south-west of the treaty reservation in Arizona, where non-Indian ranchers did not oppose the extensions as strenuously as they did in the wetter country to the east in New Mexico. As a result, the reservation consists of the driest one-third of the traditional homeland of the Navajo people, defined by the Four Sacred Mountains. Overall, the average annual precipitation on the Navajo Nation of today is one-third that of the Sacred Mountain regions of the traditional Navajo homeland, and the average annual temperature is 7°C warmer.

Before 1868, Navajo families subsisted on raising sheep, goats and horses (as well as some cattle) mixed with farming, hunting and gathering, mainly for the family's direct consumption rather than for trade. This subsistence mix had required families to range widely over the vast area amid the sacred mountains (Kelley and Francis, 2004). After 1868, as non-Indian settlers cut them off from the wettest areas best for hunting, gathering and summer grazing, Navajo families were forced to depend more heavily on farming and especially stock-raising in the drier centre of their homeland, and to trade wool and other livestock products for mass-produced foods and other items. By the early twentieth century, government and other observers were warning about erosion and overgrazing on Navajo ranges (Kelley and Whiteley, 1989).

By 1930, federal government studies predicted that massive erosion from Navajo and other western ranges would silt up Boulder (Hoover) Dam, then under construction to provide electric power for Los Angeles. Therefore, in the 1930s the government began to regulate grazing on lands within its jurisdiction, including the Navajo reservation, where the goal was to reduce the number of livestock by half as well as to get more Navajos into wage jobs and to build water, erosion-control and irrigation facilities (White, 1983). The government required each Navajo family to have a permit to raise livestock, not to exceed a certain number, and to sell all their stock in excess of that number. The government also divided the reservation into twenty land management (grazing) districts, and grazing permits required each family to stay all year within a particular district (Fanale, 1982). These grazing regulations remain in force to this day. Before these regulations, families in normal years had moved their livestock around core customary grazing areas shared by networks of interrelated extended families. During droughts, they used other kinship ties to gain access to more distant places where conditions were better (Kelley and Francis, 2004). This land-use regime had helped families distribute their livestock over the range according to its quality at any given time. Encroachment by non-Indian ranchers into the best parts of the traditional Navajo homeland increasingly constrained this traditional range-conservation pattern (Kelley and Whiteley, 1989). The federal government's grazing permit system was even more restrictive, and especially hard on extended families that were split by district lines (Iverson, 2002).

The stock reduction programme left families in need of supplementing their livelihoods because the reduction in herd size, for families that did not initially have large herds, left people without adequate stock for a reliable food source. After 1940, labour recruiters from outside the reservation hired Navajo workers because of the labour shortage brought on by the Second World War, and in 1941 Navajo tribal members became eligible for financial assistance through federal welfare, temporarily alleviating the stress brought on by shrinking livestock subsistence (Boyce 1974: 130). However, after the Second World War, the US economy contracted, and most of the jobs created during the war ended. In an effort to stave off the poor conditions of Navajo families, the Navajo Tribal Council persuaded the Bureau of Indian Affairs to reduce enforcement of livestock restrictions (Kelley, 1986: 102). Even so, to meet the emergency needs of destitute Navajo families, the US Congress increased federal relief appropriations in 1948, and doubled it again in 1950 (Kelley, 1986: 102). These historical circumstances have resulted in the present Navajo family subsistence pattern, whereby wages and various forms of assistance (social security, etc.) provide most income (Choudhary, 2003). Stock-raising nevertheless remains important to family subsistence. Cattle and sheep are a major form of savings for many Navajo families, whose cash incomes are stretched to the limit. Livestock also is necessary, according to both custom and Navajo Nation law, to validate a family's use and occupancy rights on land that it has occupied for many generations (Young, 1961). Finally, sheep, especially, are essential to traditional Navajo ceremonies.

The Navajo-Hopi Land Settlement Act, passed by the US Congress in 1974, and amended in 1980, called for the forced removal of 14,000 Navajo residents from half of the 1882 reservation lands originally designated for 'Hopi Indians and such other Indians as the Secretary of the Interior may see fit to settle thereon' (Kammer, 1980; Brugge, 1994). When this section of the reservation, located in a zone that formed a perimeter around Hopi land was established in 1882, it was inhabited by approximately 1,800 Hopis and 3,000 Navajos, as well as members of the Paiute Tribe. This land, known as the 'Joint Use Area' as it was commonly called, was partitioned into areas designated as either Hopi or Navajo by the Navajo-Hopi Land Settlement Act, and most of the people forced to relocate from lands designated to another tribe were Navajo (Kammer, 1980; Brugge, 1994; Denetdale, 2011). As a result of Navajo relocation from these reservation lands, many families moved into adjoining Navajo lands.

This change has caused increasing population density and land-use pressure in remote areas of Navajo land adjacent to the lands now partitioned for the Hopi tribe. Although still members of the Navajo Nation, those who were forced to relocate have no grazing permits, though many relocatees continue to have some livestock. However, without livestock permits, those who relocated are not eligible for programmes designated to assist livestock owners, such as drought assistance and other US Department of Agriculture programmes (M. H. Redsteer, unpublished data). Even those who own no livestock are still in need of domestic water and other resources, thereby increasing pressure on scarce water supplies.

Accounts from Tribal Elders

In order to obtain a more complete history of the changes that have occurred on the Navajo Nation, tribal elders were interviewed between 2001 and 2004, regarding the availability and quality of water, observed weather patterns, and changes in farming and grazing practices, and local medicines. Interviews were conducted in the south-western part of the reservation in a region known as Tsezhin Tah ('within black rocks'). This region of the southern Navajo Nation has been continually occupied since AD 600 or earlier, including the well-known period of 'Anasazi abandonment' (Neff, 2003) and is characterized by numerous natural springs, some of which lie along routes of pre-Columbian travel that are documented in Navajo ceremonial songs (Kelley and Francis, 2003).

There were forty-two individual consultations with men and women, including ceremonialists, former tribal officials and those known locally as experts in Navajo place names and local history. In addition, there were three interviews that occurred in group settings at the Dilkon, Jeddito and White Cone Senior Citizen Centers, with fifteen, eight and eight individuals, respectively. The ages of the persons consulted were, at the time of the interviews, from sixty-two to eighty-seven. Interviews were conducted with a Navajo translator, and notes were taken in order to review and annotate discussions. Consultations were conducted under a permit from the Navajo Nation Historic Preservation Department. A standard checklist of questions was used to document observed changes in weather, location of water sources, land use (especially farming and livestock) and the landscape. These consultations were augmented by Navajo Land Claims records of tribal elder testimonies, the oldest of which took place in 1937, historic data on springs, stream flow, and the locations of other perennial and ephemeral water features, and other land-use archives in addition to National Weather Service data on rainfall and snowfall.

A long-term decrease in the amount of snowfall in the latter half of the twentieth century was commented on in every interview. Observations suggest that snowfall decreased markedly after the late 1960s, some of the oldest consultants remember amounts of rain and snow greater in the 1920s and 1930s, during years of their adolescence to young adulthood (Table 12.1). A decline in surface water features and water availability was also a frequent observation (noted in 85 per cent of the interviews). Several elders noted that a shift from wet to dry conditions began in the mid-1940s. Others noted a shift from wet to dry around 1937–1938, during anomalously dry years. A few commented that a drying trend began in the late 1960s, during the period that corresponds to the beginning of declining snowfall in the weather records. Wind and dust storms are noted as occurring around the time of the early 1950s drought, and were observed as frequent in the 1950s and increasing in the 1990s.

Observed changes in water availability include springs and lakes drying up, and washes and rivers flowing less often. The Little Colorado River and Jeddito Wash were specifically mentioned as flowing all year in the early 1900s. These water sources are almost entirely intermittent now, as are streams across the entire Navajo Nation (Figure 12.1). Twenty specific springs were named by elders as no longer flowing for an area that is roughly 60 km^2. Among these dry springs are sacred sites where offerings for rain were made by local medicine men (three were mentioned specifically). Noted changes in spring flow begin in the 1970s, but some have dried up as recently as 1999. One elder named six springs that his family had relied on for water that dried up during the 1990s. Some springs flow less often or only flow in the winter, while other springs were noted as becoming too salty to use. In places where lakes and streams have dried up and disappeared, elders noted that corn and squash were commonly farmed until the early 1970s.

Table 12.1 *Responses by Navajo elders to the questions: What changes in the weather have you noticed? Changes in rain, snow, wind, heat/cold, etc. How much and when?*

Responses	Number of persons
Today there is less rain and snow.	All
In the 1920s and 1930s, it rained a lot.	3
Rain and snow have gotten less since 1931.	2
In the 1930s, it snowed more and deeper [30 to 60 cm] and kept the ground moist down to 1.2–1.5 m.	1
In the 1930s, it would rain for a week at a time.	1
In 1934, there were heavy rains and good crops.	1
In the late 1930s and early 1940s, the climate shifted from wet to dry.	2 [the oldest consultants]
The last heavy rain was in 1937/was in 1938.	2/1
It has gotten drier since 1938.	8
In 1940, it rained a lot.	1
Rained a lot in 1940. All the washes ran, lots of sheep feed, lots of lambs.	2
In the 1940s, it snowed big every year, chest high on horses sometimes.	15
About 50+ years ago [1940s], it rained every day, good soaking summer rains, and snowed more – knee deep.	3
In 1942, there were deep snows and sheep starved and died. This was followed by rains and erosive flooding.	1
The climate has gotten drier in the last 57 years [since 1944]. Up until then, it rained in the afternoon every day [during the monsoon season] and snowed big. The ground stayed wet until the 4th of July.	8
In 1946/In 1948, there was heavy snow on the reservation and in Nevada, required National Guard help.	1/2
1950 to 1951: there was a drought that required moving or selling livestock.	1
In the 1950s, there was blowing sand and dust.	1
1954: violent storm of wind and rain.	1
During 1962–1965, 66, 67, there was snow, with little since.	1
In 1967/In 1968, there was deep snow and there were snow drifts [variously estimated at 0.6–1.2 m].	1/1
From 1965 to 1970, the weather shifted from wet to dry.	1
In the 1960s and early 1970s, Jeddito Wash ran all the time with high flows of 4–5 m, and lots of people farmed there.	1
In 1970, there were five days of heavy rain producing floods.	1
Until 1971, there was enough rain and people planted fields all along Jeddito Wash. In 1971, the rain started to decline.	1
Since the 1990s, there is drought and heat.	1
In 1962 and 1999, there were strong windstorms.	1
1999–2000: it is exceptionally dry.	15
Today it's hotter and windier with dust storms.	16

The lack of available water, in addition to changing socioeconomic conditions, was mentioned as a leading cause for the decline in the ability to grow corn and other crops (75 per cent of the interviews). Remarks about farming include

> People have quit farming because there is not enough rain; when corn is planted the wind starts blowing and the corn stops growing; people work for wages and don't depend on crops for food, younger generations don't know how to farm; the men who knew the songs for farming are gone; and the land is too crowded with people forced to relocate [resulting from Navajo-Hopi Resettlement Act of 1974].

Other noted changes include the disappearance of plant and animal populations. Changes in plant species include the disappearance of cottonwood, willow, reeds, medicinal plants and plants used as tobacco and for ceremonies. Greasewood, yucca, cliffrose and wild tea are now less plentiful. Both greasewood and yucca have ceremonial significance, and are also used for food. Plant species that are increasing include *Astragalus* spp. (locoweed) and *Salsola* spp. (tumbleweed). Locoweed is particularly problematic during droughts because it is very drought tolerant. Often, it is one of the few plants growing in dry years, but it is poisonous to foraging animals. Tumbleweed is another drought-tolerant plant that germinates earlier than most native plant species. It is also a poor substitute for livestock forage. This plant is plentiful on sand dune deposits that are commonly found in the region. Tumbleweed generally grows in disturbed areas, and is adapted to detach and blow during windy periods, leaving the ground bare and susceptible to wind erosion. Because of its characteristics, tumbleweed has added to problems of wind erosion and movement of sand dunes during the windy times of the year (M. H. Redsteer, unpublished data). Some currently abundant plants are so new that they do not have Navajo names.

Beavers were noted as being plentiful in the early 1900s but are now completely absent. Badgers, porcupines and horned lizards were once numerous in the region, as were antelope and deer, all of which are presently rare. Bird species that have disappeared include cranes which stopped in lakes and ponds during spring migrations. Other once common bird species, now rarely seen, include eagles, snowy egrets, heron, nighthawks, bluebirds and finches. Elders also noted the disappearance of bees and locusts. Locusts were once numerous, and are a traditional food source.

Discussion

Navajo elders agreed that the climate has become drier in their lifetimes. They tend to date the beginning of the trend to the late 1930s and early 1940s. This shift from wet to dry has been noted as the end of the early twentieth century pluvial, and is linked to multidecadal changes in precipitation linked to the Atlantic Multidecadal Oscillation and Pacific Decadal Oscillation (Hereford and Webb, 1992; Hereford, 2002; Gray et al., 2003).

Surprisingly, consultants did not mention the wet period that occurred from the late 1970s to mid-1980s in this region. However, it did not last as long as the wet period in the early half of the twentieth century, and was not characterized locally as anomalously wet as

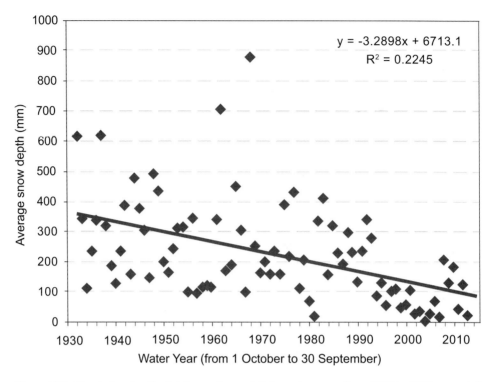

Figure 12.2 Graph showing the total annual snowfall for water years from 1930 to 2012, averaged over six sites in north-eastern Arizona near the region of the Navajo Nation known as Tsezhin Tah.

in the 1910–1930 event. The neglect of this observation may reflect the additional climatic influences of declining snowfall and warmer temperatures, and that many streams on the Navajo Nation had become ephemeral by the 1980s. In fact, some elders note the beginning of a drying trend when the amount of snowfall began to decline.

An examination of historical records and professional papers substantiate the changes noted by Navajo elders. Trends in records of total annual snowpack reported by Redsteer et al. (2010) corroborate a decrease in annual snowfall totals, and show that the trend may have begun in the 1930s, although two anomalously big snow years occurred in the 1960s (Figure 12.2). Weather stations across the Navajo Nation also show a long-term decrease in regional precipitation over the past century, with the years of more than 50 per cent above normal precipitation occurring before 1945 (Figure 12.3).

A compilation of historic observations of stream flow and surface water features show significant changes over the past century that are widespread, and lack spatial overlap with areas where water resource development has occurred. Some ephemeral water features of the past no longer exist (Figure 12.1). More than thirty major surface water features on the reservation are now dry year-round or ephemeral, and began to disappear sometime between the 1920s and 1950s. The number and length of stream reaches with perennial

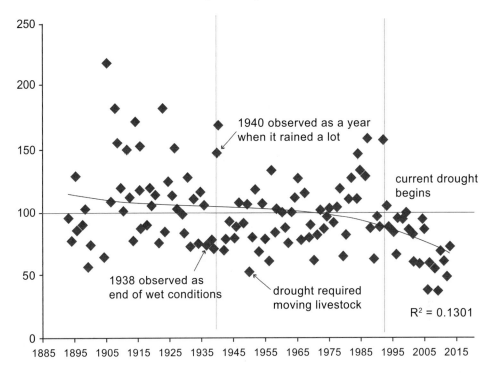

Figure 12.3 Per cent normal precipitation averaged over three meteorological monitoring stations closest to Tsezhin Tah on the Navajo Nation: Ganado, Holbrook and Winslow.

flow have declined so substantially, that out of all the past perennial streams on the Navajo Nation, only one section of the Little Colorado River, near its confluence with the Colorado River is now perennial. Coinciding with changes in stream discharge, large-scale changes to the riparian vegetation began in the mid-1940s, to more drought- and salt-tolerant species (Nagler, 2007). The declining trend in snow pack, and earlier stream discharge of snowmelt in the spring due to higher temperatures, are likely factors in the disappearance of perennial flow (and spring flow) during the last century (Stewart et al., 2005; Hidalgo et al., 2009). Other contributing factors possibly include the change in precipitation after 1940 and increasing aridity resulting from higher evapotranspiration rates (Weiss et al., 2009).

Important information that cannot be easily gleaned from meteorological and stream flow records are also recorded in our consultations. For example, there are observations of soil moisture and the description of disappearing migratory birds that rely on water sources. One elder noted that an older Navajo name for the nearby Sand Francisco Peaks, the highest mountain in Arizona, was 'the mountain with snow all year long'. Local monitoring of soil moisture conditions today indicate that a rapid decline occurs in the spring (M. H. Redsteer, unpublished data). Springtime conditions in this region are dry and windy from April through June, so that germinating plants must rely on whatever moisture was delivered as snow during the winter months to remain alive until monsoon season rainfall begins. The fact that soil moisture was noted to be present through the spring dry season, until the monsoon season that typically begins around 4 July, has profound implications for impacts to ecosystem viability that may have already occurred with climate change and drought. Drier sand dune deposits during the windy season can also lead to more severe sand and dust storms. The noted changes in soil moisture also help us understand the influence of seasonality contributing to current drought severity.

Although the accounts of elders provide information for most of the twentieth century, questions remain about how typical the dry conditions today might have been in the early 1900s, late 1800s or earlier. Navajos from the same region were consulted in 1937, the late 1950s and early 1960s for the Navajo Land Claim before the Indian Claims Commission. The 1960s records of these interviews named water sources that their forebears utilized before 1864–1868, when the US Army interned a large portion of the Navajo Nation at Fort Sumner, New Mexico. Navajos continued to use these water sources after the Fort Sumner internment. These water sources include shallow lakes, washes and springs that are now dry. Springs that were named in this land-use assessment include springs where offerings for rain were made. Records from Gregory (1917) also include observations from the early 1900s (1909–1915) that recount measurements of spring flow and the location of lakes and ponds that are currently dry. Information from federal land-use studies in the early 1930s also indicates plentiful sources of water: 'In almost every case, permanent water is available within one-quarter to two miles of hogan locations' (Russell, 1988).

We cannot attribute all ecosystem impacts to changes in snowpack and increasing temperatures. However, changes to religious practices, farming, plants, animals and water supplies have certainly been effected by a drying climate. Already dire conditions of increasing population pressures, poor socioeconomic conditions, and a limited resource base have acted in combination with climatic change to push the viability of living on Navajo land to its limit.

It is hard to quantify the magnitude of impacts to Navajo culture, traditions and well-being from climate changes. Ceremonialists are at a loss as to what can be done to ask for rain, when the offering places for rain are no longer there. Many blame the lack of prayers and offerings for the current drought, and for many other problems facing Navajo people today. As one consultant described,

My Nali (grandfather) was born two years after his family returned from the Long Walk [1870] and lived to be 107. He had a garden that was watered by rainfall, and it was his life. He grew lots of different fruits and vegetables and never had to go to the store to buy food. When he felt it was needed, he would paint his entire body with clay, and went to make offerings for rain, and it would rain. In this way, he lived a long and healthy life, as many people from his generation did.

Planting corn and other crops is central to Navajo culture and distinguished Navajos from Apache tribes in Spanish accounts from the 1600s to 1700s (Kelley and Whiteley, 1989). Many ethnologists believe that farming was even more central to Navajo culture prior to the years of war and conflict that occurred in colonial times (Hendricks and Wilson, 1996). Corn is traditionally cooked in a variety of ways and was a staple of the traditional

Navajo diet. Different kinds of ground corn are used for prayers and ceremonies as well. Traditional foods made from corn and other wild plants still play an important role in Navajo ceremonies. Corn pollen is used extensively in virtually all ceremonies and prayers, and custom dictates that it should be corn pollen gathered from one's own crops. This, of course, is impossible if it is too dry to farm. Additionally, all of the plants noted in our study as rare serve important traditional uses. The knowledge of plants for medicines and foods has been passed down from generation to generation. A wide area may be travelled to gather the right plants for a particular ceremony or to treat a specific illness. Currently, ceremonialists are travelling greater distances to find the proper herbs, to locations at higher elevations with more moisture. When the proper plants cannot be found, a ceremony may no longer be practised.

Animals also play a central role in traditional Navajo culture, and are endowed with well-defined qualities that are important, including having the powers of healing and protection. Each animal, bird and insect has its place, and teachings that are associated with it. Some animals play important roles in how conditions are interpreted. As an example, rats and mice are a bad sign because they live in graves and feed on the dead, so things that they have been eating are often burned. In contrast, porcupines have many uses and are also a source of meat (McPherson, 1992). Although the eagle is well known as having a place in Native American cultures and spiritual practices, other birds, such as the bluebird, are also used in specific ceremonies.

Consultants expressed anguish over many of the changes that have eroded Navajo culture, and mentioned most often the shift away from traditional stock-raising. Changes in traditional grazing practices resulted in the replacement of kin-based, reciprocal range-sharing to conditions where grazing disputes erupt among individuals who claim that their permits grant exclusive use or property rights. Disputes over land keep children from establishing homes near their parents and grandparents, and hinders the close cooperation in ceremonial and other traditional family activities that are essential to maintain traditional language and culture. Another result is the shift away from traditional Navajo cosmological knowledge and practices, since people who no longer depend on livestock are insensitive to the needs of the land and the deities to whom their elders and forebears have offered prayers for rain, vegetation and other blessings (Kelley and Francis, 2001).

Conclusion

Increasing aridity combined with drought threaten the very existence of Navajo culture and the survival of traditional Navajo communities. Climate change impacts have contributed significantly to poor living conditions on Navajo reservation lands. This region is characterized by harsh, dry conditions and sparse water supplies, even during normal conditions, and therefore is more vulnerable. The relocation of Navajo families from land allocated to the Hopi tribe has placed additional strain on local resources by increasing population pressures. Dire economic conditions and cultural ties to livestock add land-use stresses that create greater risk and vulnerability from drought impacts and climate change.

In addition, lack of available water has undermined the ability of Navajo people to grow corn and other crops, and to collect corn pollen. Corn pollen is used for blessings and is central to every Navajo prayer and ceremony. Changes in springs, and to plant and animal species have left ceremonialists without many of the resources necessary for traditional Navajo prayers, ceremonies and offerings.

The observations that traditional elders have contributed to our studies provide a clear and consistent picture of the changes resulting from increasing aridity after the middle of the twentieth century. Although land use and changes to traditional practices were not entirely the result of climate change, it has been a contributing factor. Moreover, we cannot expect to fully understand the significance that climate change impacts will have in societies by anticipating those changes to occur in isolation from economic, social and cultural conditions. It is difficult, if not impossible, to fully quantify the impacts of climate change to the cultural, spiritual and physical well-being of Navajo people, although the impacts are significant.

Increasing aridity and ecosystem stresses from increasing temperatures and decreasing snowfall are trends we can expect to continue with climate change (Seagar et al., 2007). A continuation of these trends, without addressing the economic and cultural needs of Navajo communities, will result in increasingly dire circumstances for the Navajo people. It is likely that younger generations of Navajo people, who cannot live off the land as their forbears did, will continue to emigrate from the reservation lands in response to dwindling water supplies and increasingly harsh surroundings. This alternative is difficult, if not impossible, for the more traditional elderly who tend to be the poorest of the Navajo people.

Acknowledgements

We would like to express our appreciation to our seventy-three anonymous co-authors, whose names are not listed at the beginning of this report although they contributed greatly to our understanding of the local changes in climate on the Navajo Nation, and the impacts these changes have had in local communities. Permission was granted for this study by Navajo Historic Preservation Office, NNHPD permit C0204-E, issued to Klara Kelley.

References

American Rural Policy Institute. 2012. *Demographic Analysis of the Navajo Nation Using the 2010 Census and 2010 American Community Survey Estimates*. Window Rock, AZ: Navajo Nation Planning and Development.

Bailey, G. and Bailey, R. G. 1986. *A History of the Navajos: The Reservation Years* (2cd ed.). Santa Fe, NM: School of American Research Press.

Boyce, G. 1974. *When the Navajos had too Many Sheep: The 1940s*. San Francisco: Indian Historical Press.

Breit, G. N. and Redsteer, M. H. 2002. Variations in water composition in the Hopi Buttes (Tsezhin Bii) area of the Navajo Nation, Northeastern Arizona. *Geological Society of America Abstracts with Programs*, 34: 395–6.

Brugge, D. M. 1994. *The Navajo-Hopi Land Dispute, an American Tragedy.* Albuquerque, NM: University of New Mexico Press.

Choudhary, T. 2003. *Navajo Nation Data from US Census 2000.* Window Rock, AZ: Navajo Nation Division of Economic Development.

Correll, J. L. and Dehiya, A. 1978. *The Navajo Nation: How It Grew.* Window Rock, AZ: Navajo Times Publishing Co.

Denetdale, J. N. 2011. *Bitter Water, Dine Oral Histories of the Navajo-Hopi Land Dispute.* Tucson, AZ: University of Arizona Press.

Doyle, J. T., Redsteer, M. H. and Eggers, M. J. 2013. Exploring the effects of climate change on Northern Plains American Indian Health. *Climatic Change*, 120: 135–48.

Draut, A. E., Redsteer, M. H. and Amoroso, L. 2012. *Recent Seasonal Variations in Arid Landscape Cover and Aeolian Sand Mobility, Navajo Nation, 2009 to 2012.* Flagstaff, AZ: Geological Survey Scientific Investigations Report.

Fanale, R. A. 1982. *Navajo Land and Land Management: A Century of Change (Doctoral dissertation).* Retrieved from ProQuest Dissertations and Theses (Accession Order No. 8208779).

Graf, W. L. 1986. Fluvial erosion and federal public policy in the Navajo Nation. *Physical Geography*, 7: 97–115.

Gray, S. T., Betancourt, J. L., Fastie, C. L. and Jackson, S. T. 2003. Patterns and sources of multidecadal oscillations in drought sensitive tree-ring records from the central and southern Rocky Mountains. *Geophysical Research Letters*, 30: 1316–19.

Gregory, H. E. 1917. *Geology of the Navajo Country, A Reconnaissance of Parts of Arizona, New Mexico, and Utah (Professional Paper 93).* Flagstaff, AZ: Geological Survey.

Hack, J.T. 1942. *The Changing Physical Environment of the Hopi Indians of Arizona. Awatovi Expedition Report No. 1.* Cambridge, MA, Peabody Museum paper.

Hendricks, R. and Wilson, J. P. 1996. *The Navajos in 1705, Roque Madrid's Campaign Journal.* Albuquerque, NM: University of New Mexico Press.

Hereford, R. and Webb, R. H. 1992. Historic variation of warm-season rainfall, southern Colorado Plateau, southwestern USA. *Climatic Change*, 22: 239–56.

Hereford, R., Webb, R. H. and Graham, S. 2002. *Precipitation History of the Colorado Plateau Region, 1900–2000 (Fact Sheet 119–02).* Flagstaff, AZ: US Geological Survey.

Hidalgo, H. G., Das, T., Dettinger, M. D. et al. 2009. Detection and attribution of streamflow timing changes to climate change in the Western United States. *Journal of Climate*, 22: 3838–55.

Iverson, P. 2002. *For Our Navajo People: Diné Letters, Speeches, and Petitions, 1900–1960.* Albuquerque, NM: University of New Mexico Press.

Kammer, J. 1980. *The Second Long Walk: The Navajo-Hopi Land Dispute.* Albuquerque, NM: University of New Mexico Press.

Kalanda-Joshua, M., Ngondgondo, C., Chipeta, L. and Mpembeka, F. 2011. Integrating indigenous knowledge with conventional science: Enhancing localized climate and weather forecasts in Nessa, Mulanje, Malawi. *Physics and Chemistry of the Earth*, 36: 996–1003.

Kelley, K. B. 1986. *Navajo Land Use: An Ethnoarchaeological Study.* London: Academic Press, Inc.

Kelley, K. B. and Francis, H. 2001. *Diné Land Use and Weather: Cause and Effect.* Paper presented in the session on USGS Interdisciplinary Study of Weather and Land Use Change in the 'Hopi Buttes' Region at the 13th Annual Navajo Studies Conference, Flagstaff, Northern Arizona University.

Kelley, K. B. and Francis, H. 2003. Abalone shell Buffalo People: Navajo narrated routes and pre-Columbian archaeological sites. *New Mexico Historical Review*, 78: 29–58.

Kelley, K. B. and Francis, H. 2004. Navajo land use and climate in Chezhin Bii. In Redsteer, M. H. (ed.) *The Arid Region of Hopi Buttes on the Navajo and Hopi Reservations, Arizona* (manuscript in preparation). Flagstaff, AZ: US Geological Survey Bulletin.

Kelley, K. B. and Whiteley, P. 1989. *Navajoland: Family Settlement and Land Use.* Tsaile, AZ: Navajo Community College Press.

Leonard, S., Parsons, M., Olawsky, K. and Kofod, F. 2013. The role of culture and traditional knowledge in climate change adaptation: Insights from East Kimberley, Australia. *Global Environmental Change*, 23: 623–32.

Majule, A.E., Stathers, T., Lamboll, R. et al. 2013. Enhancing capacities of individuals, institutions and organizations to adapt to climate change in agricultural sector using innovative approaches in Tanzania and Malawi. *World Journal of Agricultural Sciences*, 1(6): 220–31.

McPherson, R. S. 1992. *Sacred Land, Sacred View, Navajo Perceptions of the Four Corners Region, (Charles Redd Monographs in Western History No. 19)*. Provo, UT: Charles Redd Center for Western Studies.

Mertz, O., D'haen, S., Maiga, A. et al. 2012. Climate variability and environmental stress in the Sudan-Sahel zone of West Africa. *AMBIO*, 41: 380–92.

Mimura, N., Pulwarty, R. S., Duc, D. M. et al. 2014. Adaptation planning and implementation. In Field, C. B., Barros, V. R., Dokken, D. J. et al. (eds.) *Climate Change 2014: Impacts, Adaptation, and Vulnerability. Part A: Global and Sectoral Aspects. Contribution of Working Group II to the Fifth Assessment Report of the Intergovernmental Panel on Climate Change*. Cambridge, New York: Cambridge University Press, pp. 869–98.

Nagler, P. L., Glenn, E.P., Hinojosa-Huerta, O., Zamora, F. and Howard, K. 2007. Riparian vegetation dynamics and evapotranspiration for the riparian corridor in the delta of the Colorado River, Mexico: Implications for conservation and management. *Environmental Management*, 88: 864–74.

Navajo Land Claim Collection. *Navajo Statements*. 3. Star Butte, 1960; 7. Dilkon, 1960; 143. White Cone, 1961; 147. Winslow, 1961; 175. Indian Wells, 1961; 192. Dilkon, 1937; 198. Bidahochee, 1961; 202. Indian Wells, 1954; 203. Seba Dalkai, 1961; 205. Bidahochee, 1961; 206. Bidahochee, 1961; 207. Bidahochee, 1961; 246. Jeddito Wash, 1961; 271. Comar Spring, 1960; 273. White Cone, 1961; 289. Jeddito, 1961 and 1964; 412. Castle Butte, 1961. Window Rock, AZ, Navajo Nation Library.

Navajo Nation Department of Water Resources. 2000. *Water Resource Development Strategy for the Navajo Nation*. Window Rock, AZ: author.

Neff, L.T., Tsosie, N., Tsosie, C. Begay, R. and Sandoval, H. 2003. *Past pattern settlement in the southeastern Tsezhin Bii' (Hopi Buttes area): Results of an archaeological survey for the US Geological Survey* (Navajo Nation Archaeology Report NNAD 02-168). Window Rock, AZ: Navajo Nation Archaeology Department.

Orlove, B., Lazrus, H., Hovelsrud, G. K. and Giannini, A. 2014. Recognitions and responsibilities: On the origins and consequences of the uneven attention to climate change around the world. *Current Anthropology*, 55: 249–75.

Redsteer, M. H. (n.d.) (ed.) *The Arid Region of Hopi Buttes on the Navajo and Hopi Reservations, Arizona* (manuscript in preparation). Flagstaff, AZ: US Geological Survey Bulletin.

Redsteer, M. H., Kelley, K. B., Francis, H. and Block, D. 2010. *Disaster Risk Assessment Case Study: Recent Drought on the Navajo Nation, United States*. Contributing case study to the 2011 Global Assessment Report on Disaster Risk Reduction, annexes and papers, www.preventionweb.net/english/hyogo/gar/2011/en/home/annexes.html

Russell, S. C. 1988. *Supplemental Analysis: Navajo Use of Areas in Arizona Outside Land Management Unit No. 3 in 1934, vol. 1*. Expert witness report on behalf of the Navajo Nation for Masayesva V. Zah before the US District Court, Phoenix AZ. Ms. on file at Navajo Nation Justice Department, Window Rock, AZ.

Seager, R., Ting, M., Held, I. et al. 2007. Model projections of an imminent transition to a more arid climate in southwestern North America. *Science*, 316: 1181–4.

Stewart, I. R., Cayan, D. R. and Dettinger, M. D. 2005. Changes toward earlier streamflow timing across western North America. *Journal of Climate*, 18: 1136–55.

United States Census Bureau. 2012. *The American Indian and Alaskan Native Population: 2010; 2010 Census Briefs*. Washington, DC: Author.

Weisiger, M. 2009. *Dreaming of Sheep in Navajo Country*. Seattle, WA: University of Washington Press.

Weiss, J. L., Castro, C. L. and Overpeck, J. T. 2009. Distinguishing pronounced droughts in the southwestern United States: Seasonality and effects of warmer temperatures. *Journal of Climate*, 22: 5188–32.

West, C. T., Roncoli, C. and Outtara, F. 2008. Local perceptions and regional climate trends on the central plateau of Burkina Faso. *Land Degradation and Development*, 19: 289–304.

White, R. 1983. *The Roots of Dependency: Subsistence, Environment, and Social Change among the Choctaws, Pawnees, and Navajos*. Lincoln, NE: University of Nebraska Press.

Young, R. W. 1961. *The Navajo Yearbook, Report no. 8, 1951–1961: A Decade of Progress*. Window Rock, AZ: Navajo Agency.

13

The Spirits Are Leaving: Adaptation and the Indigenous Peoples of the Caribbean Coast of Nicaragua

Mirna Cunningham Kain

Throughout their lives, every Miskitu indigenous person from Honduras to Nicaragua has heard myths, tales and anecdotes and received counsel about spirits. Therefore, for researchers of the Center for Indigenous Peoples' Autonomy and Development, who carried out a study on adaptation measures and traditional knowledge of their peoples, it was not strange to hear that:

> The water channel is dangerous early in the morning, before the sun comes up and after it goes down. Our family says that no one should go there to bathe, wash, or make noise, because they will disturb the spirits of the children that bathe early in the morning and in the late afternoon before the sun goes down. If they are washing, they could be using the stone that the spirit uses to sleep. It is said that those who disobey will receive messages from spirits in their dreams.
>
> (Cunningham Kain, 2010: 17)

The worldview that human beings coexist with spirits, animals, plants and stones, among others, thus complementing each other, is a given for our peoples. Indigenous peoples still possess a rich and diverse cultural heritage of knowledge, languages, values, traditions, practices, modes of organization, norms of coexistence, symbolisms, spiritualities, cosmovisions and conceptions of development, that constitute the foundation of our cultural patrimony. This allows us to interact and influence positively the economic, social and political spheres, as well as the developmental dynamics, of the countries and of the region in which we live.

This knowledge is part of the collective memory and it is expressed through the activities carried out as part of daily community life by women and men of all ages. It is observed in stories, songs, folklore, proverbs, dances, myths, rituals, community laws, local languages and taxonomies, agricultural practices, instruments, plant varieties and animal breeds. Languages and oral traditions have served as collective mechanisms for the transmission of knowledge and the configuration of diverse cultures.

As our peoples' cultures are linked to their territories and surroundings, the natural environment makes a peoples' knowledge unique and different from others. Climate change has an impact on traditional ways of life and the ecosystems in which they are rooted, altering the practice of rotational planting, hunting, fruit gathering, herding, stock breeding, agricultural production, fishing and agroforestry. As a result, climate change threatens traditional and innovative forms of knowledge and the practices associated with them.

With the loss of economic opportunities and profits, as well as traditional cultural practices, it has been observed that, due to social and cultural pressure in the indigenous population, youth migration increases, with young people leaving in search of opportunities to earn income abroad. This situation erodes indigenous economies and cultures. It has caused changes in the roles of women and the importance given to them in territories and communities. Along with cultural migration, the loss of important plants and animals makes it more and more difficult for the elderly to practise and transmit their traditional ecological knowledge to subsequent generations.

The United Nations Framework Convention on Climate Change (UNFCCC) considers adaptation as modes of adjustment to climate change. Through adaptation, people seek ways to respond to changes that pose major threats to life and forms of life. Therefore, adaptation to climate change refers to any adjustment that occurs naturally in ecosystems or in the human system as a response to climate change, whether it counters the damage incurred or takes advantage of beneficial opportunities. This chapter presents the main findings of a study on climate change and adaptation methods of the indigenous peoples of the Nicaraguan Caribbean. It was conducted in 2009 by the Institute of Investigation and Development Nitlapan-UCA and the Center for Indigenous Peoples' Autonomy and Development, CADPI.

The study was carried out in twelve communities of the Miskitu, Suma-Mayangna, Mestizos and Garifunas of Nicaragua's Autonomous Regions of the North Atlantic and of the South (Figure 13.1), that are located in various ecological zones from the coasts and plains to tropical forests, and along the agricultural frontier. The initial thesis of the study is that, well before the global community began talking about climate change, the indigenous peoples and ethnic communities of these regions have been developing adaptation measures, based on their traditional knowledge.

The Study Area

The regional system of multiethnic autonomy was established in 1987 in two regions that make up approximately 50 per cent of the national territory of Nicaragua. This system is expressed in different juridical instruments such as the Constitution of Nicaragua, the Statute for the Autonomy of the Autonomous Regions of the Atlantic Coast of Nicaragua (National Assembly of the Republic of Nicaragua, 1987, Law No. 28) and its regulations, among others. They establish that the responsibilities of the autonomous authorities include, among others, the promotion of the correct use and enjoyment of waters, forests, communal lands and the defence of their ecological systems. In the Autonomous Regions, autonomous indigenous communal and territorial jurisdictions are complemented with the multi-ethnic municipal and regional authorities.

With the approval of Law 445 in 2003 (National Assembly of the Republic of Nicaragua, 2003), the regional system of autonomy was consolidated and the process of demarcation and entitlement of collective indigenous territories set in motion. Up until the present this covers 22 per cent of the national territory, and the goal is to cover 36 per cent, once the process concludes in the Autonomous Regions of Indio Maíz and Bocay.

For indigenous peoples in the Autonomous Regions, the territory is the comprehensive spatial area where the life of the people develops. It is where cultural patterns are reproduced through socialization between spirits, human beings and nature. Their own institutions come to life in the territory. It is where the governing methods of internal matters are defined and practised, mainly with collective and sustainable use of economic, natural, human and cultural resources, and in favour of socioeconomic livelihoods and self-development. The territory is also where coordination mechanisms are defined with external stakeholders, including governments and businesses.

The process of demarcation and entitlement has strengthened the indigenous institutions and the indigenous territorial governments are exploring options for territorial governance that may allow them to fulfil the principle of 'the common good' or *Laman Laka*. This principle functions as a linchpin for the cooperation system in communities. It includes all persons and families, and it is founded on social equity. It contributes to strengthening associative relationships which necessarily require trust, solidarity, reciprocity and ethnic and territorial belonging. The principle of 'the common good' is linked to the protection and appropriate use of the natural and cultural heritage of the community. They encompass the territory, natural resources, language, production, knowledge and practices, health, food and lifestyles. Therefore, 'the common good' contributes to ensuring the economic, social and cultural reproduction of the people.

Natural phenomena and their effects are a constant in the collective memory of the inhabitants of the Autonomous Regions. The force of winds and the frequency of storms have increased. From 1982 until 2004, Nicaragua has been struck by forty-one cyclones: nineteen hurricanes, twenty tropical storms and two tropical depressions. The torrential rains caused by hurricane Irene in 1971 and by hurricane Fifi in 1974 are incommensurable with the strength of and damage caused by hurricane Juana in 1988; and by hurricane Mitch, in October 1998.

In September 2007, hurricane Felix directly struck the coast and the mountain areas, destroying more than 300 Miskitu and Suma-Mayangna indigenous communities. It also destroyed the foundation of local subsistence: the forests and the coastal fishing zone (see Box 13.1). It affected a million acres of broadleaf and conifer forests as well as marine and mangrove ecosystems. Hurricane Felix was followed by hurricane Ida in November 2009. The impact of these natural phenomena on the lives of the indigenous peoples and on ethnic communities has been pervasive, also affecting traditional knowledge.

One limitation of analyses of climate change impacts is their narrow focus. They generally emphasize only impacts on natural resources, and do not consider the consequences for the peoples whose way of life depend on these resources and the spaces for cultural reproduction, including places for ancestral rites.

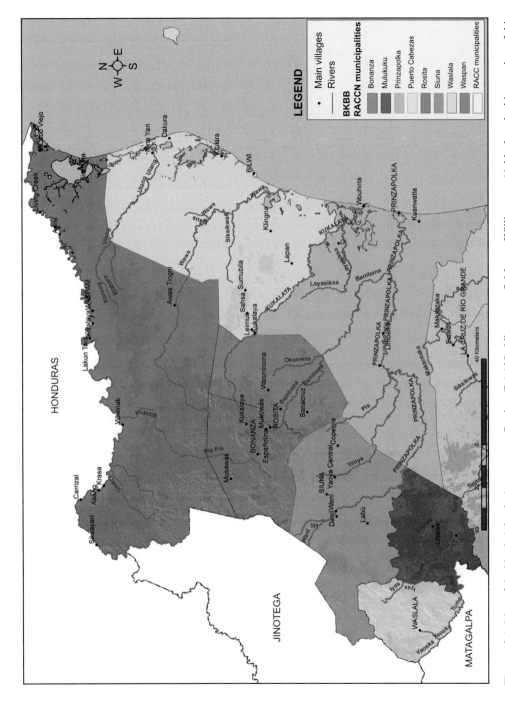

Figure 13.1 Map of the North Atlantic Autonomous Region (RAAN), Nicaragua. © Marcos Williamson. (A black and white version of this figure will appear in some formats. For the colour version, please refer to the plate section.)

Table 13.1 *Some protective spirits and their habitats*

Spirit or *Dawan*	Habitat	Protects
Duhindu	Swamps with Papta in the valley	Animals and water
Liwa Mairin	River, sea, creeks, lagoons	Water, fish, turtles, lobsters, others
Niki Niki	Pine	Land
Unta Tara Dawanka	Forest	Forests
Sisin Tara Dawanka	Tree	Specific trees
Hiltara dawanka	Hill	Hills

Perceptions of Climate Change Impacts and Traditional Knowledge

The perceptions of community members basically reflect two tensions. On the one hand, they reveal the pressures they face due to the ongoing loss of their territories and natural resources as a result of high rates of deforestation and the degradation of soils and forests. On the other hand, they express concern about the natural areas where the spirits or *Dawan* live. These are the powerful beings that protect resources. As a result, the balance between people and the beings that are central to their worldview has been altered, contributing to a loss of cultural and spiritual values.

The Miskitu explain that the events of daily life and the norms that guide social and economic interaction, are anchored within a worldview in which cultural, natural and political spheres are intertwined. The interrelationships between these elements that live in specific areas assure the survival of the Miskitu as individuals and as a people. The loss of balance between these elements or between these spaces is the cause of illness and death.

The cultural sphere includes the space inhabited by human beings, as well as all the practices that enable the physical and social reproduction of peoples. The sphere of nature includes plants, animals, bodies of water and other natural elements that remain untouched or undomesticated by human beings. For the Miskitu, the sphere of nature is also the space inhabited by spirits or *Dawan*. In this context, their explanations about climate change impacts refer to imbalances between these spheres and its implications for the life of communities. Therefore, one of the most common observations on climate change is related precisely to the fact that the spirits – *Dawan* – have lost their habitat due to the destruction of natural resources. Some outcomes of the spirits' discontent are seen in nature: in the low capture rates of lobsters and turtles.

The small river that goes through the community was the habitat of the *Liwa Mairin* that protected the river. With time, the river's water level decreased. We know she needed deeper water. This made her move away in search of other places. She left! The same happened with the other spirits, they left. But they still watch us because when we disobey, they punish us with their power [illness]. We have also contributed to the changes happening in nature. For instance, we make them leave when we cut down trees and cause rivers and lagoons to dry up [through sedimentation] and we make unnecessary forest and agricultural fires.

Some manifestations of this loss of balance identified by the Miskitu are:

1 **Loss of practices of community solidarity and reciprocity,** accelerated by the monetization of the indigenous economy. Some productive activities like hunting, planting, harvesting (*mano vuelta, pana-pana or bakahnu*), were carried out collectively by the communities. They were also associated with the communal redistribution of resources. Today, the workforce is paid, hunting is no longer a collective activity, and therefore sharing is not as widespread.
2 **Community institutions' loss of capacity to resolve conflicts.** New forms of organization, property and labour have been imposed. All this leads to the endangerment of institutions – such as the Counsel of Elders, the Communal Assembly, the Wihta and the Representative.
3 **Risks in food production.** With the spread of the culture of the agricultural frontier, large areas of land are being converted to pasture or for raising livestock for the sale of meat or milk. This situation is affecting food security because it alters the ecosystem and the use of the soil, which then reduces the harvest of tubers, rice, plantains and bananas, among other agricultural products.
4 **Shifts in traditional knowledge for reading weather and climate.** The loss of traditional knowledge on climate and weather is another climate change impact observed in the communities. In the past, the signs in the constellations, thunder and lightning, shapes of clouds, movements of insects, wind direction, among others, were elements in nature that were interpreted to understand the ecosystem's events and its dynamics. With climate change, these signals are no longer as reliable and this is creating confusion. Those signs, which have been the foundation of the cultural calendar, are no longer applicable due to climate change.

Adaptation Measures and Traditional Knowledge in the Face of Global Change

Adaptation is not new for the indigenous peoples and for afro-descendant communities in the Autonomous Regions of Nicaragua. They remember that they have transformed their families' traditional diets on several occasions. First, when banana plantations and logging companies arrived, their traditional diet was replaced with a diet largely dependent upon imported canned goods that came to the commissaries of the companies. Later, once those companies left, flour tortillas replaced the banana.

While communities are aware that they have to adapt to changes today, we did not identify any effort to systematize these experiences in order to identify which measures are the most successful. In most cases, community members suggested that the research gave them for the first time an opportunity to discuss the issue of climate change, and to analyse the impact it has on their lives.

Box 13.1 Forests and the indigenous women of Tuapi: Experiences with promoting food security by Nadezhda Fenly

Elders from the community of Tuapi, a Miskitu community in the municipality of Puerto Cabezas, North Atlantic Autonomous Region (RANN), Nicaragua, have been finding it more and more difficult to read the weather. They say that the weather is now irregular and operates outside its normal cycle. For this reason, hurricane Felix caught the population by surprise, leaving them unprepared and largely defenceless.

The Category 5 hurricane Felix struck the RANN hard. Flora and fauna were severely damaged over more than a million hectares with considerable impact on the capacities of the local ecosystems to sustain livelihoods. Agricultural production and fish catches sharply decreased. The estimated total losses amount to US$716.31 million, which was about 14 per cent of the Nicaraguan gross domestic product. In the context of this extreme event, the resilience of the Tuapi community offers an interesting case study to assess adaptation capacities in the face of increasing climate variability experienced over the last thirty years. Furthermore, this note provides a gender-oriented light, illustrating how gender-disaggregated understandings of responses to climate variability can benefit climate adaptation policies.

Tuapi communities have a subsistence economy that relies on agricultural activities (shifting cultivation) and fishing to produce resources mainly for family consumption. The main economic activities of the women are related to agriculture or the collection of fruit for household consumption. Some fruit may also be sold at the nearby municipal market of Bilwil. Of particular interest, Miskitu women from Tuapi community are responsible for the emotional, spiritual and physical health of the members of the group. For that purpose, they possess an extensive botanical knowledge relating to medicinal plants.

Gender relations and the division of labour in Miskitu cultures must be informed by recent history, in particular the expansion of the market-based economy. For more than a hundred years, coastal peoples were the auxiliaries of British foreign investors that ran the region with the economic logic of an enclave. Indigenous Miskitu men have participated in both subsistence-based and trade economies since the colonial era. Periodically, when boom times have made foreign money and goods plentiful, Miskitu men seemed destined to become incorporated into the larger economy and society. But ensuing busts have always triggered a return to dependence on livelihoods based on the adjacent forest and sea. While men were adsorbed into outside cultures and became increasingly acculturated, Miskitu women stayed behind as the society's conservative cultural core.

The severe impacts of hurricane Felix on local livelihoods, especially on the local farming systems, led to increased food insecurity but also drew women's attention to the current loss of biodiversity. Over-exploitation and deterioration of ecosystems result in an impoverishment of medicinal resources. Increased climate variability also put increased stress on the traditional way of cultivating their land (shifting cultivation). Strategies developed by Miskitu women from the Tuapi community to guarantee food security without over-exploiting the forest-based natural resources are no longer viable under the new environmental conditions. This loss of traditional products has led to an increased dependency on imported substitution products, which has in turn increased vulnerability and a diminished interest in preserving traditional knowledge. The women also note impacts on the social life of the community with erosion of the traditional practice of reciprocity embodied in the decrease of exchange of labour and collective work on the plantation (*pana-pana*).

Hurricane Felix triggered a reflection on the recent evolution of Tuapi traditional livelihoods in relation to current increases in climate variability. The distinctive relationship of the women to their environment predisposed them for this reflection. Questioned about the most valuable forest resources, the observations of Mairena et al. (2012) are indicative:

> Agricultural crops were cited more in the focus groups of women than in those of men, who cited precious woods more frequently. This could be related to the more timber-oriented vision of forest resources held by the male community members, who have been more involved in forestry projects or forestry extraction and who enjoy the benefits of those activities.

Women from Tuapi proposed adaptation measures intended to face the challenge of climate change such as the implementation of traditional recovery therapy for the plant, reinstatement of traditional collective practices such as the *pana-pana*, as well as a change in the diet and way of life. Investigations of this kind that shed light on the fundamental role of indigenous women in the evolution of their community are very important and further research should be encouraged. The women of Tuapi have shown themselves to be a driving force behind climate change adaptation. Women's roles in family dynamics have begun to filter into decision-making about forest resource use, along with other aspects of community life.

Reference

Mairena, E., Lorio, G., Hernández, X. et al. 2012. *Gender and Forests in Nicaragua's Indigenous Territories: From National Policy to Local Practice*. Bogor, Indonesia: CIFOR.

Some of the adaptation measures identified were:

1. The struggle for the legalization of ancestral territories:
   - The configuration and strengthening of territorial governments
   - Enactment of coordination mechanisms, negotiation or alliances with neighbouring communities for the joint protection of the territory and its resources
   - Formulation of norms of coexistence with third parties to facilitate sanitation in ancestral territories
2. Improvement of the mechanisms of social control in the communities:
   - Application of measures of community control, strengthening communitarian structures and the communitarian institutions themselves
   - Respect the role of the elders who control community management
3. Recreate community rules of protection of natural resources:
   - Develop the rules of use of natural resources in the community and territory in written form, including them as part of the Territorial Statutes
   - Include spiritual elements in the community rules, including traditional prohibitions
4. Comprehensive cultural revitalization:
   - Reinstate the roles of the elders
   - Return to the *Auhbi Piakan*: revival of traditional food

- Bring back practices of reciprocity and community solidarity: *pana-pana bakahnu*
- Use of traditional modes of transport
- Use of new fishing techniques
- Reinstate the traditional practice of combining marine-coastal production and subsistence farming
- Establish community areas dedicated to forest conservation
- Promote productive diversity and commercial interaction with neighbouring communities
- Changes in housing styles
- Revitalize traditional medicine
- Promote appropriate forms of waste management
- Encourage environmental education

Conclusion

The economic models used in the Autonomous Regions have contributed not only to excluding the population, but also in creating conditions that increase the vulnerability of the indigenous peoples and ethnic communities. This vulnerability is exacerbated by the current global climate crisis.

During the last decades, the continuing advance of the agricultural frontier has imposed a demographic pattern of instability that hinders the process of building communities, identity and belonging, especially in the *mestizo* population. Until recently, indigenous communities had been able to lessen the impact of those movements related to the agricultural frontier by strengthening their ancestral claims. The policy of appeasement and adjustment implemented since the 1990s, together with the processes of impoverishment and disinvestment that the farmers suffered in the zone of conflict (the northern centre), unleashed a new, massive and aggressive cycle of movement into indigenous lands. The livestock and timber sectors from other parts of the country are also participating in this movement. This new wave of displacement is fraught with violence accumulated since the war. It is heightened by the despair that arises when there are no options for fulfilling basic needs. Commercial exploitation and new settlers are practically devouring protected areas and ancestral indigenous territories. As the agricultural frontier advances, new conflicts are occurring, exacerbated once more by the impacts of climate change.

In the face of these global changes, the adaptation measures identified during the study coincide with the comprehensive perspective provided by the human rights of these individuals and peoples. These rights have been classified in the following areas:

- **Reconstitution of the natural and collective heritage of indigenous peoples**. This involves the reinstatement of their rights, territorial spaces and the management of their lands, natural resources and institutions.
- **Promotion and implementation of educational processes.** Centred and focused mainly on strengthening traditional practices, directed especially towards children and youth with the knowledge and practices of the elders as the point of departure.

- **Development of organizational capacities within different levels of autonomy.**
- **Strengthening coordination capacities.**
- **Revitalization and rescuing cultural practices.** Indigenous peoples have developed and adapted a series of knowledge and experiences as mechanisms to respond to change based on their cultural and ancestral principles (mutual cooperation, reciprocity, harmonious coexistence with others and with nature, a sense of solidarity and of collectivity).

Communities perceive the legalization of their ancestral territories as a fundamental step so that adaptation measures may contribute to improving their living conditions. The effective implementation of Law 445 is thus perceived as an urgent measure and as a strategy that would facilitate autonomous relationships among indigenous communities, and between them and the neighbouring peasants and *mestizo* farmers.

References

CICA (Consejo Indígena de Centro América) 2008. *Declaration of Tolupán on Climate Change*. 22 May 2008, Honduras.

Cunningham Kain, M., Mairena Aráuz, D. and Pacheco Sebola, M. 2010. *Cambio Climático: Medidas de Adaptación en Comunidades de las Regiones Autónomas de la Costa Caribe de Nicaragua*. Cuadernos de investigación Nitlapan no. 34 Managua: Nitlapan, http://biblioteca.clacso.edu.ar/Nicaragua/iid-uca/20170417025144/pdf_166.pdf

National Assembly of the Republic of Nicaragua. 1987. *Law 28. Autonomy Statute for the Regions of the Atlantic Coast of Nicaragua*, http://calpi.nativeweb.org/doc_3.html

National Assembly of the Republic of Nicaragua. 2003. Law 445: Law of Communal Property Regime of Indigenous Peoples and Ethnic Communities of the Autonomous Regions of the Atlantic Coast of Nicaragua and of the Rivers Bocay, Coco, Indio and Maiz. *La Gaceta, Official Daily*, 16: 74–83, www.calpi-nicaragua.org/wp-content/uploads/2014/12/Law-445.Chpt-1-to-5.pdf

14

Indigenous Reindeer Herding and Adaptation to New Hazards in the Arctic

Svein D. Mathiesen, Mathis P. Bongo, Philip Burgess, Robert W. Corell, Anna Degteva, Inger Marie G. Eira, Inger Hanssen-Bauer, Alvaro Ivanoff, Ole Henrik Magga, Nancy G. Maynard, Anders Oskal, Mikhail Pogodaev, Mikkel N. Sara, Ellen Inga Turi and Dagrun Vikhamar-Schuler

> Remember, it is not us reindeer herders who have been the cause of climate change. The reindeer know what paths to take. Many people have lost their connection with Nature, but the animals maintain this connection and that is why we follow the reindeer.
> *Vassily Namchaivyn, Chukchi herder, Community of Kanchalan, Chukotka, Russia.*

> I am not too concerned about climate change if it is due to nature itself. But if it is due to people, that people have been destroying nature, then I am worried.
> *Karen Anna Logje Gaup, Sami reindeer herder, Kautokeino, Norway.*

There is abundant evidence across the Arctic, coming both from climate scientists (e.g. Larsen et al., 2014) and polar indigenous peoples, that points to the advent of rapid climate change. Global and regional scenarios project significant changes in temperature, precipitation and snow conditions in the key areas for reindeer-herding communities. Reindeer herding is the primary livelihood for over twenty indigenous peoples in the Arctic and Subarctic, involving close to 100,000 herders and 2.5 million semi-domesticated reindeer (Turi, 2002; McCarthy et al., 2005 – Figure 14.1) Reindeer-herding peoples have lived and worked across wide areas of Eurasia since time immemorial. On the Yamal Peninsula in West Siberia, the earliest documented evidence shows that domesticated reindeer husbandry existed more than 2,000 years ago (Fedorova, 2003). This chapter explores the responses to various new hazards associated with climate change by two groups of indigenous herders in northern Eurasia – the Sami people of northern Norway and the Nenets people in the Yamal-Nenets Area of northern Russia.

In Norway, there are about 200,000 reindeer and about 3,000 Sami people working in reindeer husbandry. The Yamal Nenets in Western Siberia constitutes one of the largest

Figure 14.1 Map of indigenous reindeer herding peoples in Eurasia.

reindeer husbandry regions in the world, with more than 700,000 reindeer and over 14,500 Nenets people who practise nomadic reindeer husbandry (Oskal et al., 2009).

Reindeer economy in the Sami region of northern Norway is characterized by large herds and a high degree of mechanization and state control (Reinert, 2006; Reinert et al., 2009). Reindeer are primarily used for meat production, although hides, bones and antlers are an important source of material for clothing and handicrafts for the domestic and tourist industry. Recruitment to reindeer husbandry has been limited in Norway and Sweden by legislation, and the lack of pastures and economic opportunities has limited the expansion of this livelihood. In the Yamal area, reindeer herding is based on strong family networks that couple ecosystems and nomadic use of pastures (Turi, 2008; Degteva and Nelleman, 2013). The migrations in Yamal from winter to summer pastures are up to 700 km one way.

In contrast, Sami reindeer herders in Finnmark may migrate up to 350 km from inland winter pastures to the coastal summer grounds.

This article reports on some results from the multidisciplinary vulnerability study called EALÁT (Reindeer Herders Vulnerability Networks Study) that was implemented during the International Polar Year (IPY) 2007–2008. The first outcomes of the project were presented at the Seventh Arctic Council Ministerial Meeting in Nuuk, Greenland (Magga et al., 2011). The EALÁT project has, on the one hand, stimulated reindeer herders in their thinking about future strategies and responses to changes in the Arctic; and, on the other hand, generated interest in and understanding of indigenous reindeer herding across many institutions, both nationally and internationally. Reindeer herders in the Circumpolar North are on the frontlines of monitoring the rapid ongoing changes in the Arctic. Their knowledge and skills are the key for their future existence in their home territories.

Finnmark, Norway: Projected Climate Variability and Change

Temperature projections for the town of Kautokeino, the hub of Sami reindeer herding in northern Norway, signal a dramatic change in the mean seasonal temperatures over the next hundred years. Meteorological records for Kautokeino come from a weather station located just outside the town. The mean seasonal temperatures in the preceding decades (1961–1990) were $-16.0°C$ for winter, $-5.2°C$ for spring, $+10.0°C$ for summer and $-1.0°C$ for fall (Oskal et al., 2009). Current climate change scenarios indicate that summer temperatures in Finnmark may increase by 2 to 4° C in the next hundred years, while winter temperatures may increase by as much as 7 to 8°C (Benestad, 2011; Magga et al., 2011). This represents a significant shift and it is likely that rapid and variable oscillations between freezing and thawing conditions will increase with the anticipated temperature change. If so, it will certainly provide considerable challenges to the practice of reindeer husbandry, particularly in the use of winter pastures. The largest temperature increase is projected for inland Finnmark (Figure 14.2). These findings demonstrate the complex and diverse nature of the environment in which reindeer herding has evolved. Inland temperatures in a hundred years in Finnmark, Norway, may resemble those of the coastal area of Finnmark (Nordreisa) today. This will then constrain Sami reindeer husbandry in a manner similar to constraints experienced in the coastal regions today (Magga et al., 2011).

More detailed scenarios for Finnmark (Engen-Skaugen, 2007) show that the annual precipitation may increase by 5 to 30 per cent; the snow season may be one to three months shorter than today; and annual maximum snow depth may decrease by 5 to 60 per cent. The largest reductions in snow season and snow depth are projected in the coastal areas of northern Norway. Comparisons of reindeer herders' reports and climate data from the area indicate that temperature and precipitation conditions alone are not critical for the reindeer. However, various combinations of these variables lead to different snow structures, which may make the pastures more or less accessible to the reindeer (Vikhamar-Schuler et al., 2012). Likewise, recently in the high Arctic Svalbard archipelago, an extreme winter warming event in 2012 caused ground ice layers to form, creating problems for local

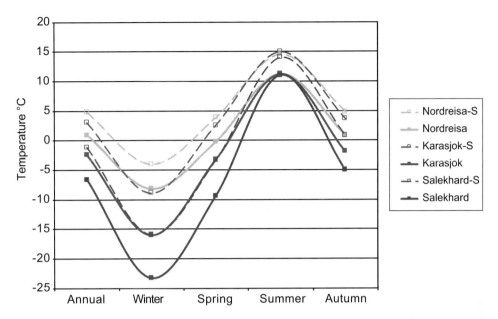

Figure 14.2 Annual and seasonal mean temperatures in coastal Finnmark, Norway (Nordreisa, yellow), inland Finnmark, Norway (Karasjok, red) and Yamal-Nenets AO, Russia (Salekhard, blue). Unbroken lines show 1961–1990 observed values. Dotted lines show similar average calculated scenarios (S) from fifty downscaled climate models for the year 2085 (Magga et al., 2011). (A black and white version of this figure will appear in some formats. For the colour version, please refer to the plate section.)

population of wild reindeer (Hansen et al., 2014). Increases in precipitation with swift changes of temperature in winter are expected to create future climate challenges for reindeer herding in some areas, although mobility and migration with the herds in the past usually helped mitigate the negative effect of adverse grazing conditions.

Land-use Change

It is important to recognize that reindeer nomadism requires the use of extensive areas of land. Degradation of pasture lands combined with the consequences of a changing climate presents substantial challenges to the future of reindeer husbandry (Vistnes et al., 2009). For herders, the principle issue is the security of habitats and landscapes where reindeer are grazed in various seasons and under different conditions.

Only recently have traditional areas of reindeer husbandry in Finnmark (as well as in Yamal and other northern regions) become of interest to others, such as oil and gas, and mining industries. With a changing climate and increased industrial development in the North, the ability of herders to adapt using their fine-tuned survival skills has been jeopardized, as development can delay, block or terminate critical migrations between winter and summer ranges (Oskal et al., 2009; Vistnes et al., 2009; Maynard et al., 2010;

Oskal et al., 2010; Degteva and Nellemann, 2013). The IPCC AR5 chapter on polar regions (Larsen et al., 2014) argues that climate warming is accelerating access to northern lands for oil and gas, and other industrial development (Forbes et al., 2009). Ensuring access to these resources by these new industries, such as the build-up of road and other infrastructure, inevitably shrinks the size of grazing areas used by reindeer herders. It has been characterized as one of the major human-induced factors in the Arctic contributing to the loss of 'available room for adaptation' for indigenous reindeer husbandry (Nuttall et al., 2005; Forbes et al., 2009; Oskal et al., 2009).

The 'High North' strategy of the state of Norway envisions the development of new industrial projects such as mining, oil and gas extraction, green energy, and tourism in the same regions as those traditionally used by the Sami people. As these industrial projects are being planned, it has been officially decided to decrease the number of reindeer and reindeer herders in the Finnmark County of northern Norway, in order to maintain a 'sustainable' reindeer husbandry in the future (Turi and Keskitalo, 2014).

According to Mattias Åhrén, President of the Sami Council, while speaking at the Sixth Ministerial Meeting of the Arctic Council in Tromsø 29 April 2009:

We face many challenges related to the climate change. First, increased access to non-renewable resources in our homelands has created a 'race to the Arctic' and a change in land use. We humbly ask the states that participate in the race to the Arctic to be mindful that you all base your claims to the resources in the High North on claims to rights to indigenous territories.

Loss of reindeer pastures occurs principally in two ways: (1) through physical destruction; and (2) through the effective, though non-destructive, removal of habitat or through a reduction in its value as a resource (Vistnes et al., 2009). Physical destruction of reindeer habitat chiefly results from the development of infrastructure, including the construction of pipelines, roads, hydroelectric facilities, buildings and even artillery ranges. Of far greater concern is the gradual abandonment by reindeer of previous high-use areas due to their avoidance of areas disturbed by human activities. A series of studies have documented a reduction in the use of rangelands by reindeer varying between 48 and 96 per cent compared with the pre-development distribution within a band of 2.5–5 km from cabins, dams, power lines and roads (Vistnes and Nellemann, 2001; Nellemann et al., 2003, Vistnes et al., 2004). Approximately 25 per cent of former reindeer pastures in the Barents Euro-Arctic region have effectively been lost owing to disturbance from infrastructure development; in some of the productive coastal ranges of Finnmark the figure is as high as 35 per cent. As much as 1 per cent of the summer grazing areas used traditionally by Sami reindeer herders along the coast of northern Norway are lost every year, which is similar to the amount of grazing land used by one nomadic family in summer (Magga et al., 2011).

Snow Change and Adaptation

Indigenous knowledge is a critical element in building future climate adaptation strategies for reindeer herding communities (Eira et al., 2010). The basic components of reindeer herding are reindeer, land and people with their unique knowledge base. The reindeer must have enough food, water and shelter, and they must be able to move freely in order to, among other things, escape from insects and run free. Furthermore, the reindeer should be in the right place at the right time to fit the seasonal and environmental conditions of its natural surroundings. The Sami technical term *guodohit* ('to herd') comes from the word *guohtun* ('access of pasture plants through the snow') which transforms into *guoduhit* ('to let or put to graze'). Across northern Finnmark, reindeer grazing land is covered with snow for as many as 250 days per year. Weather and snow conditions are the two factors that almost completely control the survival of the reindeer (Eira et al., 2010; Eira, 2012).

The EALÁT project has, therefore, recommended that national adaptation strategies to climate change must recognize minorities' and indigenous peoples' traditional knowledge, and their cultural and linguistic rights. As Johan Mathis Turi, Chair of the International Centre for Reindeer Husbandry (ICR), stated on the UN World Environmental Day in Tromsø in 2007, the concept of adaptation, rather than stability, is inherent in reindeer herding societies:

We have some knowledge about how to live in a changing environment. The term 'stability' is a foreign word in our language. Our search for adaptation strategies is therefore not connected to 'stability' in any form, but is instead focused on constant adaptation to changing conditions.

The Sami reindeer herders' knowledge base is therefore a prerequisite for adaptation to the new hazards due to change in the Arctic. Sami researcher Inger Marie Gaup Eira (Eira, 2012; Eira et al., 2013) has documented that herders' knowledge about adaptation is encapsulated in their language and professional vocabulary. She furthermore reports that Sami reindeer herders in the Kautokeino area of northern Norway use some 318 noun stems designating various types of snow and snow conditions, which provide a unique knowledge base for adaptation to snow change. With the rich derivation morphology of the Sami language, it seems safe to concur with Krupnik and Müller-Wille (2010) that Sami have the richest terminology on snow of all languages of the world. While part of the content may be based on objective physical properties, there are also semantic components that refer to the use, time, locality and processes of reindeer husbandry (Eira et al., 2010).

Eira concludes that:

Humans describe the natural environment on the basis of their local experience and their interactions with nature in terms of its relevance to their daily lives. These descriptions are incorporated into traditional local languages and form a specialized terminology that is unique and specifically applicable to local needs and practices. Therefore, snow defines most of the conditions which must be met to support reindeer pastoralism across Eurasia. Snow is a prerequisite for mobility, tracking, visibility and availability of pasture plants like lichens and grasses.

(Eira et al., 2013)

A major finding is that Sami herders' snow knowledge is more holistic and better integrated with the ecology of the herds and pastures than international nomenclatures for snow terms (Eira et al., 2013). The richness and relevance to reindeer herders of Sami traditional snow terms reveal a distinctly different view of snow compared to that of the purely physics-based international classifications. In addition, the study illustrates the importance of using traditional Sami terminology when developing climate change adaptation strategies for Sami reindeer husbandry, thus stressing the importance of the two ways of knowing. Based on this understanding, the team developed a novel observational system and a specially designed herding diary, which focused on snow conditions and their impact on herding practices in winter (Eira, 2012; Eira et al., 2013). Future understanding of adaptation to snow change in the Arctic and new monitoring systems should include herders' knowledge, since the continuing warming of the snow in reindeer pastures will place growing pressures on reindeer herding.

Social and Economic Adaptation

Reindeer herding is based on the sequential and flexible use of a wide variety of ecological niches under differing climate conditions. This flexibility allows herders to adapt to climate variation and to increase their social and ecological resilience (Mathiesen et al., 2013). Our surveys in combination with other studies of the social organization of reindeer pastoralists (Turi, 2008; Sara, 2009) confirm that indigenous reindeer herding has developed an integrated resilience for coping with climatic uncertainty based on traditional ecological knowledge. As a place-based, coupled socioecological system, reindeer pastoralism is primarily a local practice (Sara, 2009), and strategies for building and maintaining resilience based on traditional knowledge are implemented at this level. Among the Sami, the social organization through family-based reindeer herding *siidas* (i.e. groups of households that keep their animals together in a joint herd and cooperate on tasks associated with herding) is an important framework for climate change resilience (Turi and Keskitalo, 2014). Government regulations at the regional, national and international scales, however, have a profound impact on reindeer herding and often reduce, directly or indirectly, the resilience and adaptive capacity by affecting the herders' ability to maintain flexibility, mobility and socioecological diversity (Turi, 2008; Turi and Keskitalo, 2014).

Yamal-Nenets Area, Russia: Projected Climate Variability and Change

The mean seasonal temperatures in 1961–1990 for Salekhard, the capital of the Yamal-Nenets Autonomous Area were: –22.9°C (winter), –8.9°C (spring), +11.1°C (summer) and –4.9°C (autumn). Climate scenarios indicate that summer temperatures in Yamal may increase by 2 to 4°C in the next hundred years, while winter temperatures may increase by as much as 7 to 8°C (Benestad, 2011). In summer, the average temperature in Salekhard is projected to be higher than the present summer average temperature in Kautokeino (Oskal et al., 2009). The large warming that is calculated for the Yamal Peninsula may be caused by the changes in ice conditions in the Barents and Kara Seas surrounding the peninsula. Future scenarios indicate that Yamal winter temperatures in 2070–2100 may be comparable to inland Finnmark winter temperatures in Norway today. Therefore, by 2099, the range of mean autumn, winter and spring temperatures in Salekhard are projected to closely

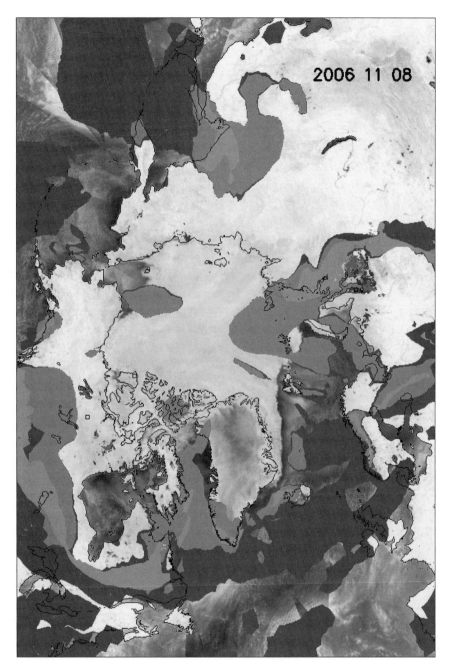

Figure 14.4 The 'rain-on-snow' circumpolar map for 8 November 2006 built on the Polarstereographic 12.5 km AMSR-E grid. A grey scale image of the AMSR-E 89 GHz polarization ratio is used for the background. The coincident ECMWF precipitation is shown in blue, with precipitation that occurs in the vicinity of subfreezing temperatures shown in cyan, and ECMWF snowfall shown in grey. The ROS categories are as follows: liquid layer (red), recent refreeze (purple), longer-term refreeze (pink), and temporal classification (beige). The extensive ROS event that was observed near the Yamal Peninsula in November 2006 can be clearly seen on the map. (A black and white version of this figure will appear in some formats. For the colour version, please refer to the plate section.)

immediate adaptive solutions and new societal opportunities. Loss of pasture lands exacerbated by a changing climate will challenge the future of reindeer husbandry across northern Eurasia. It is important to support knowledge-sharing on impacts and adaptation measures connected to climate change and to the loss of grazing land, while also recognizing the value of traditional knowledge as a foundation for adaptation. We are concerned about the explosion of human activities linked to climate change and to the destruction of grazing land for reindeer and caribou in the Arctic. Grazing land used for reindeer must be protected as an adaptive measure in the face of climate change and as an asset to sustainable Arctic societies. It is important to identify institutional mechanisms, which constrain indigenous peoples' original socioecological resilience and their ability to adapt to climate variability and change (Mathiesen et al., 2013).

Furthermore, in the face of new industrial development in the Arctic, there is a real danger that the benefits from increased exploitation of Arctic resources will flow to mainstream society, while indigenous and local people will bear the costs. Several Arctic nation states have not yet developed strategies for the adaptation of indigenous reindeer husbandry and the lack of such plans may increase herders' vulnerability to future changes. Reindeer herders' cultures and their traditional knowledge are inevitably affected by institutional governance, economic conditions, and other regulatory practices across the Circumpolar North (Reinert, 2006; Reinert et al., 2009; Sara, 2009; Turi and Keskitalo 2014). It is important to increase the information and insights about circumpolar reindeer herding and how resource conflicts could be avoided by including representatives of local Arctic communities and land-rights holders at an early stage of planning for all industrial development projects.

Indigenous traditional knowledge, cultures and languages provide a central foundation for adaptation and building socioecological resilience in the face of these rapid changes. Education based upon traditional knowledge, cultures and languages should be provided locally. There is a need for a new type of education in the North that incorporates multidisciplinarity, multicultural and holistic approaches for sustainable development, including reindeer herders' traditional knowledge and worldviews. Adaptation to rapid climate change and new hazards in the North requires the building of local competencies. Reindeer herders must have the essential tools to meet new global challenges, at a time when various national and global actors are able to coordinate their goals and strategies in the home areas of the herders. Both scientific and traditional experience-based knowledge, knowledge transformation, education and training of future Arctic leaders are the keys to the future sustainability of reindeer herders' societies and their adaptation.

Engaging reindeer herders' youth directly in herding practices and providing enhanced education are also critical factors in the future sustainability of indigenous reindeer husbandry and its cultural foundations.

Adaptation to climate change also requires the training of local Arctic leaders in long-term sustainability thinking. Such educational goals need to rely on the best available knowledge about adaptation and the herders' knowledge base should be included. It is imperative that all available forms of knowledge are considered in developing adaptive strategies to climate change. Renewable industrial activities aimed at combating climate change and the new strategies for ecosystem services in the Arctic must respect the rights of indigenous peoples. At the same time, Arctic residents must be given the possibility to adapt to the changed environment and to seek sustainable societal opportunities through education and understanding. One of the main reasons for the enhanced adaptive capacity of reindeer husbandry in Yamal, as opposed to Finnmark, might be the importance of family-based reindeer husbandry in Yamal with its traditional knowledge nested within the culture of Nenets reindeer nomadism.

Acknowledgements

The data used in this chapter were collected during the IPY 2007–2008 EALÁT-Reindeer Herders Vulnerability Network Study (IPY #399), supported by the Research Council of Norway (Project IPY EALÁT–RESEARCH: Reindeer Herders Vulnerability Network Study: Reindeer pastoralism in a changing climate, grant number 176078/S30, MFA).

References

Bartsch, A., Kumpula, T., Forbes, B. C. and Stammler, F., 2010. Detection of snow surface thawing and refreezing in the Eurasian Arctic with QuikSCAT: Implications for reindeer herding. *Ecological Applications*, 20(8), 2346–58.

Benestad, R. E. 2011. A new global set of downscaled temperature scenarios. *Journal of Climate*, 24(8): 2080–98, http://dx.doi.org/10.1175/2010JCLI3687.1

Degteva, A. and Nellemann, C. 2013. Nenets migration in the landscape: Impacts of industrial development in Yamal Peninsula, Russia. *Pastoralism: Research, Policy and Practice*, 3:15, www.pastoralismjournal.com/content/3/1/15

Eira, I. M. G. 2012. *Muohttaga jávohis giella Sámi árbevirolaš máhttu muohttaga birra dálkkádat Rievdamis* [The Silent Language of Snow Sámi traditional knowledge of snow in a time of climate change]. Ph.D. Thesis, University of Tromsø.

Eira, I. M. G., Jaedicke C., Magga O. H. et al. 2013. Traditional Sámi snow terminology and modern physical snow classification: Two ways of knowing. *Cold Regions Science and Technology*, 85: 117–30.

Eira I. M. G., Magga O. H and Eira, N. I. 2010. Muohtatearpmaid sisdoallu ja geavahus. *Sámi dieđalaš áigečála* 2/2011 [Sámi Snow Terminology: Meaning and Usage], pp. 3–24.

Engen-Skaugen, T. 2007. Refinement of dynamically downscaled precipitation and temperature scenarios. *Climate Change*, 84: 365–82.

Federova, N.V. 2003. Каслание длиной в две тысячи лет: человек и олень на севере Западной Сибири (A Migration Lasting 2000 Years: Humans and reindeer in the north of West Siberia). Available at: http://yamalarchaeology.ru/index.php/texts/archeol/84-fedorova-n-v-2003b

Forbes, B.C., Stammler, F., Kumpula, T. et al. 2009. High resilience in the Yamal-Nenets social-ecological system, West Siberian Arctic, Russia. *Proceedings of the National Academy of Sciences*, 106(52): 22041–8.

Grenfell, T. and J. Putkonen. 2008. A method for the detection of the severe rain-on-snow event on Banks Island, October 2003, using passive microwave remote sensing. *Water Resources Research*, 44, W03425.

Hansen, B. B., Isaksen, K., Benestad, R. E. et al. 2014. Warmer and wetter winters: Characteristics and implications of an extreme weather event in the High Arctic. *Environmental Research Letters*, 9(11): 114021, http://iopscience.iop.org/1748–9326/9/11/114021/article

Krupnik, I. and Müller-Wille, L. 2010. Franz Boas and Inuktitut terminology for ice and snow: From the emergence of the field to the "Great Eskimo Vocabulary Hoax." In Krupnik, I., Aporta, C., Gearheard, S., Laidler, G. and Holm, L. Kielsen (eds.) *SIKU: Knowing Our Ice. Documenting Inuit Sea Ice Knowledge and Use* Dordrecht: Springer, pp. 371–400.

Larsen, J. N., Anisimov, O. A., Constable, A. et al. 2014. Polar regions. In Barros, V. R., Field, C. B., Dokken, D. J. et al. (eds.) *Climate Change 2014: Impacts, Adaptation, and Vulnerability. Part B: Regional Aspects. Contribution of Working Group II to the Fifht Assessment Report of the Intergovernmental Panel on Climate Change*. Cambridge, UK and New York: Cambridge University Press, pp. 1567–612.

McCarthy, J. J., Martello, M. L., Corell, R. W. et al. 2005. Climate change in the context of multiple stressors and resilience Arctic. *Arctic Climate Impact Assessment*. Cambridge, UK and New York: Cambridge University Press, pp. 945–88.

Magga, O. H, Mathiesen, S. D., Corell, R. W. and Oskal, A. (eds.) 2011. *Reindeer Herding, Traditional Knowledge and Adaptation to Climate Change and Loss of Grazing Land. A project led by Norway and Association of World Reindeer Herders (WRH) in Arctic Council, Sustainable Development Working Group (SDWG)*. International Centre for Reindeer Husbandry Report 1: 2011. Alta, Norway: Fagtrykk Idé AS.

Mathiesen, S. D., Alftan, B., Corell, R. W. et al. 2013. Strategies to enhance the resilience of Sámi reindeer husbandry to rapid changes in the Arctic. *Arctic Council Arctic Resilience Report (ARR). Interim report to the Arctic*. Stockholm: Stockholm Environmental Institute (SEI) and Stockholm Resilience Centre (SRC), pp. 109–12.

Maynard, N. G., Oskal, A., Turi J. M. et al. 2010. Impacts of Arctic climate and land use changes on reindeer pastoralism: Indigenous knowledge and remote sensing. In Gutman, G. and Reissell, A. (eds.) *Eurasian Arctic Land Cover and Land Use in a Changing Climate*. Dordrecht: Springer, pp. 177–205.

Nuttall, M., Berkes, F., Forbes, B. et al. 2005. Hunting, herding, fishing and gathering: Indigenous peoples and renewable resource use in the Arctic. *Arctic Climate Impact Assessment*, pp. 649–90.

Oskal A., Maynard N., Degteva A. et al. 2010. *Oil and Gas Development in Reindeer Pastures in Northern Eurasia: Impacts and Solutions*. 2010 State of the Arctic Conference IPY, 16–19 March 2010, Miami, Florida.

Oskal, A., Turi, J. M., Mathiesen S. D. and Burgess P. (eds.) 2009. *EALÁT Reindeer Herders' Voice: Reindeer Herding, Traditional Knowledge and Adaptation to Climate Change and Changed Use of the Arctic*. Information Ministerial book, International Centre for Reindeer Husbandry and Association of World Reindeer Herders. International Centre for Reindeer Husbandry Report 2: 2009.

Oskal, A., 2008. Old livelihoods in new weather: Arctic indigenous reindeer herders face the challenges of climate change. *Development Outreach*, 10(1): 22–5.

Nellemann, C., Vistnes, I., Jordhøy, P., Strand, O. and Newton, A., 2003. Progressive impact of piecemeal infrastructure development on wild reindeer. *Biological Conservation*, 113(2), 307–17.

Pogodaev, M., Oskal, A., Avelova, S. et al. 2015. *Youth. The Future of Reindeer Herding Peoples. Executive Summary. Arctic Council Sustainable Development Working Group. EALLIN Reindeer Herding Youth Project 2012–2015*. Kautokeino, Norway: International Centre for Reindeer Husbandry (ICRH).

Reinert, E. S. 2006. The economics of reindeer herding: Sámi entrepreneurship between cyclical sustainability and the power of state and oligopolies. *British Food Journal*, 108(7), 522–40.

Reinert, E. S., Aslaksen, I., Eira, I. M. G. et al. 2009. Adapting to climate change in Sámi reindeer herding: The nation-state as problem and solution. In Adger, N., Lorenzoni, I. and O'Brien, K. L. (eds.) *Adapting to Climate Change: Thresholds, Values, Governance*. Cambridge, UK and New York: Cambridge University Press, pp. 417–32.

Sara, M. N. 2009. Siida and traditional Sámi reindeer herding knowledge. *The Northern Review*, 30: 153–78.

Turi, E. I. 2008. Living with climate variation and change. A comparative study of resilience embedded in the social organization of reindeer pastoralism in Western Finnmark and Yamal Peninsula. Ph.D. Thesis, Institute of Political Science, University of Oslo, Norway.

Turi, E. I. and Keskitalo, E. C. H. 2014. Governing reindeer husbandry in western Finnmark: Barriers for using traditional knowledge in local-level policy implementation. *Polar Geography*, 37(3): 234–51.

Turi, J. M. 2002. The world reindeer livelihood: Current situation, threats and possibilities. In Kankaanpää, S., Müller-Wille, L., Susiluoto, P. and Sutinen, M.-L. (eds.) *Northern Timberline Forests: Environmental and Socio-economic Issues and Concerns*. Kolari, Finland: The Finnish Forest Research Institute, pp. 70–75.

Vikhamar-Schuler, D., Hanssen-Bauer I., Schuler T. V., Mathiesen S. D. and Lehning M. 2012. Use of a multilayer snow model as a tool to assess grazing conditions for reindeer. *Annals of Glaciology*, 54(62): 214–26.

Vistnes, I., Burgess, P., Mathiesen, S. D. et al. 2009. *Reindeer Husbandry and Barents 2030. Impacts of Future Petroleum Development on Reindeer Husbandry in the Barents Region. Report for StatoilHydro Barents 2030 Scenario Programme*. International Centre for Reindeer Husbandry Report 1: 2009.

Vistnes, I., Nellemann, C., Jordhøy, P. and Strand, O., 2004. Effects of infrastructure on migration and range use of wild reindeer. *Journal of Wildlife Management*, 68(1): 101–8.

Vistnes, I. and Nellemann, C., 2001. Avoidance of cabins, roads, and power lines by reindeer during calving. *The Journal of Wildlife Management*, 65(4): 915–25.

15

'Everything That Is Happening Now Is Beyond Our Capacity' – Nyangatom Livelihoods Under Threat

Sabine Troeger

> Pastoralism is the finely-honed symbiotic relationship between local ecology, domesticated livestock and people in resource-scarce, climatically marginal and highly variable conditions. It represents a complex form of natural resource management, involving a continuous ecological balance between pastures, livestock and people.
>
> (Nori and Davies, 2007: 7)

Individual and social adaptation to climate is nothing new, neither as an empirical nor as a theoretical construct. Within the range of many recent studies, pastoralism is viewed as a system that is highly adaptive to harsh and, at the same time, fragile environments. Would these flexible and environmentally alert societal systems not be predestined to adapt to environmental transformations triggered by climate change – especially, as climate variability has always been a feature accompanying pastoralists' lives?

This chapter investigates why, and due to what factors, pastoralists declare: 'everything has become beyond our capacity'. Following Adger et al. (2009), we contend that limits to adaptation are endogenous to society. They are contingent on attitudes to risk, on ethics and culture. Correspondingly, the argument relates the observed landscape of risk – with the risk of climate change impacts being one element of the given risk scenario, to the way it emerges through practice, 'for these landscapes are eventually navigated by agents who both experience and constitute larger formations, in part from their own sense of what these landscapes offer' (Appadurai, 2011: 285). A closer examination of the transformation of what was formerly a culturally, firmly established and resilient livelihood system into one that is now fluid and subject to new interpretations and definitions due to adverse forces and negative impacts, of which climate change is but one, explains the title's negative message.

The following arguments will focus on two sets of parameters in interplay that are at the centre of this chapter. The first is the modes and the patterns of climate change across; whereas the second is contingent upon Nyangatom society itself – a pastoralist group in south Ethiopia of about 10,000 people. Together, they represent the background against which the pastoralists navigate a landscape of climate change risks. The argument then turns towards the perspective of further threats to the pastoralist society and its quest for successful adaptation.

Climate Change in Ethiopia

The accuracy of climate prediction is limited by fundamental, irreducible uncertainties. Some can be quantified, but many simply cannot. A certain level of uncertainty therefore persists in our understanding of the future climate (Dessai and Hulme, 2004).

Meteorological data on annual rainfall patterns in Ethiopia cast doubts about the severity, even the overall reality, of climate change impacts with regard to patterns of rainfall. While there is solid evidence for the increase in temperature by 1.3°C between 1960 and 2006, and estimates point to further increases between 1.1° and 3.1°C by 2060, annual rainfall data cannot offer such clear predictions. In particular, the highland zones, considered to be comparatively productive and the 'bread basket' of Ethiopia, are sometimes assumed to be sheltered from impacts. A frequently encountered argument is that measurements of total annual rainfall are, with acceptable variation, more or less the same as before or even slightly increasing in some areas, for example, in the western parts of the country (NMA, 2007). This contradicts statements from the World Bank that rank Ethiopia among the world's twelve most-impacted countries (as of February 2009) and as the second most vulnerable when the focus is on drought (IRIN News, 2009).

In recent climate change research, increased attention has been given to the knowledge and experience of individuals and communities about past and recent climate, and the way this shapes peoples' perceptions of the future climate. The relevance and validity of local knowledge have been widely demonstrated in climate change studies (Orlove et al., 2000; Riedlinger and Berkes, 2001; Berman and Kofinas, 2004). Although some critical voices exist, especially against integration of local knowledge in scientific assessments, our attention here focuses on the people on the ground, the farmers and pastoralists, who experience the changing climate patterns from day to day and from season to season. This choice was especially relevant, as even today some regions of Ethiopia have no weather stations to supply scientific observational data.

For thirteen months, from February 2009 until March 2010, we conducted climate change and adaptation research at thirteen sites in four regions of Ethiopia – Tigray, Amhara, Oromia, and what is officially called 'Southern Nations, Nationalities and Peoples' Region. The surveys were done by intercultural teams of eight master's students from Germany, together with eighteen international experts and three Ethiopian students. As several teams worked partly in parallel, we accumulated a total of twenty-four months of fieldwork. This provides a comprehensive insight into the climate change impacts experienced by people in different regions of Ethiopia. The research was jointly supported by the Sustainable Land Management Programme of the Ethiopian Ministry of Agriculture (SLM-Ethiopia-MoARD) and the German Technical Cooperation Agency (Gesellshaft für Technische Zusammenarbeit – GTZ-SLM).

The research made use of various assessment tools from ethnographic research, such as participatory observation, open narratives and semi-structured interviews. Research activities and the methodological approach were adjusted to the nature of target groups, including a clear distinction between gender and generational cohorts. The portrait drawn from the data is highly differentiated and takes into account various perspectives within the societies. The validity of the data was ensured by methodological triangulation.

This triangulation was especially relevant because we were well aware that farmers, pastoralists, meteorologists and climate scientists 'measure' different things when they discuss 'climate change'. The measured 'amount' of rainfall, whether annually or by season, is not taken as an isolated arithmetic mean by local farmers and pastoralists. Rather they consider it in direct relation to what it is supposed to produce, namely, as water requirements for particular crops or for livestock. Farmers and pastoralists, in particular, record seasonal and intra-seasonal changes and variation that they consider as especially destructive, notably dry spells, rain causing floods and, in certain geographic localities, hailstorms (see also Jennings and Magrath, 2009). Analysis of the collected data emphasizes a twofold pattern of climate change impacts in Ethiopia as depicted in Figure 15.1.

These localized assessments leave no room for doubt about the severity of current climate change impacts. Climate change is having an impact everywhere across the country, even in places that might register the same annual rainfall amounts as before. The face of climate change changes from place to place. In the western highland region of Ethiopia, it is characterized by more irregularities and weather extremes, even though the same amount of rainfall may be recorded by the end of the year. In the country's eastern-southern lowland, on the other hand, the increased frequency of droughts is clearly evident. The Nyangatom territory, located in the southern lowlands, is impacted by the latter pattern, which is especially characterized by considerably shortened *Belg* rains. The *Belg* rains are part of the customary Ethiopian rainfall pattern: they usually start end February–early March and last until end May–early June. The eastern and southern regions of Ethiopia, demarcated by the Rift Valley, are highly dependent on this specific rainfall for their annual agricultural production. Failure of this rainfall pattern may hit hard a community that was formerly well-adapted to this fragile environment. A key uncertainty in this scenario is the role of Indian Ocean sea surface temperature in driving rainfall patterns. The absence of a clear period of rainfall recovery points to ocean warming as the dominant negative driver for rainfall, and thus to a continued and accelerated drying over the coming decades.

The Nyangatom people currently discuss the spatial and quantitative changes of temperature, precipitation and floods. In their perception, all of these deviations started around the year 1989 and since then have shown a gradual decline in precipitation and floods. More severe changes are perceived to have started after 1998. They include an increase in temperature, as well as the increased variability of precipitation. Comparing the two rain periods, the *Belg* rains (i.e. *Akoporo* rains) are the most affected (Figure 15.2).

In 2011, the disaster scenarios projected by the IPCC AR4 (2007) showed up earlier than they were predicted. Large parts of the Horn of Africa were struck by one of the worst droughts in sixty years that translated into a severe food crisis. In July 2011, the United

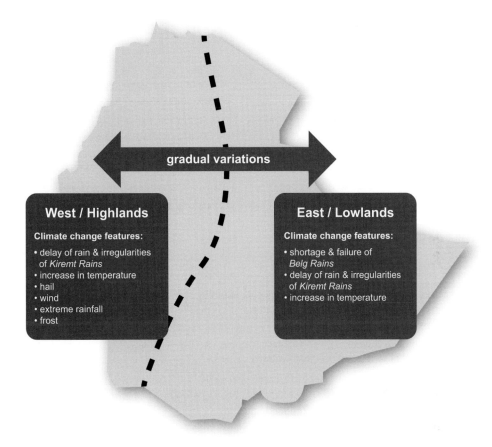

Figure 15.1 Map showing climate change patterns in Ethiopia.

Nations declared a famine. But already as early as January 2011, considerable quantities of Nyangatom livestock had died. In the past, severe droughts had often produced much harm to the people and had left some pastoralist communities without any livestock to speak of. But the drought that started in 2011, whether it was caused by global climate change factors or the effect of a very harsh La Niña event as suggested by IRIN News (as of 15 July 2011), took place in the wake of four to five years of more or less failed *Belg* rains.

A 'Culture of Sharing' – Constituents of Nyangatom Society

Today, the finely honed symbiotic relationship between local ecology, domesticated livestock and the Nyangatom people is obviously disturbed. In order to understand the opening statement by one of the Nyangatom elders, the main elements of their agro-pastoralist society need to be highlighted.

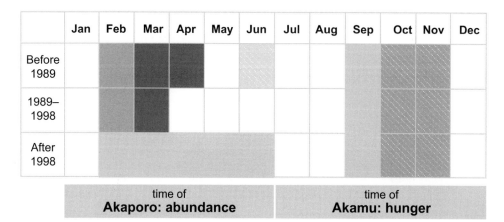

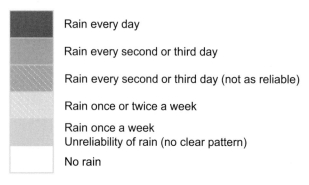

Figure 15.2 Calendar of observed changes in rain patterns. (A black and white version of this figure will appear in some formats. For the colour version, please refer to the plate section.)
Source: S. Troeger *et al.* 2011:34.

Streams of Life 1: Livestock

Livestock in the Nyangatom life, and especially cattle, are more than just animals for the production of milk, meat and leather. Prosperity and a person's position in society are defined by the number of cattle owned by a household. Cattle are the means of paying a dowry and therefore a prerequisite to marriage. The deliberate choice of a wife and the possession of cattle are the very basis of a man's concept of self. Besides their cattle, the Nyangatom also keep smaller ruminants (sheep and goats), which are especially important for barter exchange in times of hardship.

Animal products play an important role in daily consumption arrangements. Smaller ruminants are used for slaughter in the time of *Akamu* (the dry season and usually the time of food shortage, even hunger). While the cattle are moved to distant pastures in line with the rules of a transhumance system, the smaller ruminants usually remain in the settlement and serve as a means of survival for elders, women and children. During periods of drought, the goats are vital for milking. Additionally, they are very important for ceremonial use, especially for prayers for rain. Besides meat, milk products such as butter have a high social value. They are consumed on a daily basis, but primarily they are proudly offered to invited elders and well-respected people. Adding to their economic importance, all animals and especially cattle have a sociopsychological meaning. The possession of livestock, in and of itself, as well as the consumption and use of livestock products, are major elements of self-esteem and social understanding.

Streams of Life 2: Crop Cultivation

The Nyangatom base their crop-growing activities on two different agricultural cycles in the course of the year: the rain; and the river flood. Rainfed cultivation is practised with the help of the *Belg* rains (*Akoporo*) between February and May. The river-based cultivation depends on the *Omo* flood, which enriches the riverbanks with moisture and sediment in the course of the *Kiremt* rains (July–September) in the Ethiopian highlands. Sorghum is the most precious crop for local farmers, as compared to maize and beans. It is especially liked for cultural reasons; the Nyangatom describe sorghum as sacred and use it for many ceremonies. The existence and consumption of sorghum is fundamental for self-esteem and identity.

The Confluence of Streams of Life: The Nyangatom 'Culture of Sharing'

Livestock and crop production are pillars of Nyangatom society, and they play an important role in terms of food security, culture and sense of self-worth. Furthermore, livestock and sorghum are a major means to build friendship. The Nyangatom create strong social networks around their territory by endowing and receiving livestock and sorghum. These reciprocal relations represent an important support in times of hunger and hardship.

Akamu, the hunger period, does not affect all Nyangatom to the same degree and at the same time. While some Nyangatom depend on flood-based (river) agriculture, others depend on livestock and rainfed cultivation. During the dry season, exchange relations represent a major means of survival.

Within Nyangatom settlements, all exchanged goods and presents – including food aid today – are shared with friends and relatives. This culture of sharing is a major element of the society's adaptability to harsh and uncertain conditions and livelihood security. It characterizes the Nyangatom community as a whole, as well as its daily life. Reciprocity at all levels is a coping measure and a major force in this agropastoral society.

Vulnerability Turning into a Story of Extinction

In general, the risk scenario that the Nyangatom are facing, along with neighbouring pastoral communities in South Omo, is characterized by a continuous decline in access to pastures and, accordingly, in herd mobility. It is caused by processes of encroachment, motivated by various interests – from the global to the local level – as well as by natural

Figure 15.3 The Nyangatom culture of sharing. © Julia Pfitzner. (A black and white version of this figure will appear in some formats. For the colour version, please refer to the plate section.)

forces, such as invasive species as illustrated below. People are 'displaced' in the words of David Turton (2014), considering their current loss of access to vital resources like land, pasture and water, even if physical displacement is not involved.

Adaptation Means Navigating Through Landscapes of Risk

Today, the Nyangatom are confronted with many adverse stressors that define the landscape of risk facing pastoralists and other local actors (Figure 15.4). Field observations unveil their combined efforts to adapt to harsh environments, of which recurrent failure of rains represents only one element. Land fragmentation is one of the key reasons why the ability of pastoralists to overcome drought has been severely reduced. Flintan (2011: 5) highlights the following causes of rangeland fragmentation in East Africa and across the Horn of Africa: the establishment of extensive protected area systems; alien invasive 'toxic' plants; and conflicts among neighbouring pastoralist groups triggered by the ever-decreasing access to vital resources, pastures and water. All of the factors of land fragmentation that Flintan names are relevant for Nyangatom society and neighbouring pastoral communities.

This process of land fragmentation was politically supported by the official definition of 'free land' as pastoral land. The Federal Constitution of Ethiopia (1995, Article 40–5) states that: 'Ethiopian pastoralists have the right to free land for grazing and cultivation as well as the right not to be expelled from their own lands.' But the moment the government chooses to claim grazing land and declares it no longer 'free', pastoralists lose any right to use it for their herds. In Ethiopia, peasants obtain land, which becomes their legal possession, whereas pastoralists merely exploit 'free land' (Pavenello and Levine, 2011: 14).

As a consequence, the Nyangatom face new challenges that heighten their vulnerability: nature conservation in the form of two national parks established in South Omo – the Omo National Park and the Mago National Park. Both were created during the era of Haile Selassi in the 1960s, but nowadays their presence has been reinforced by globally condoned measures against biodiversity loss, and they are increasingly enforced and safeguarded by park rangers (Turton, 2011: 159–64).

Furthermore, *Kibish*, the former cultural centre for the Nyangatom community, has been lost to the invasive tree *Prosopis juliflora*. Former grazing and cropping grounds are overgrown by *Prosopis*, and the seasonal river *Kibish* has fallen dry, as the deep-rooting plant has consumed all the river water. *Kibish* nowadays is deserted.

Additionally, increasing conflicts over the land and resources with the neighbouring Turkana group moving from Kenya threaten the Nyangatom, as they directly border the Turkana area on the Ethiopian side. Fights over cattle have become fiercer during the past years, as nowadays almost all herders carry automatic rifles adding a deadly edge to old fights that not so many years ago fulfilled a traditional social role.

Large portions of the pastoralists' territory have already been assigned for large-scale agricultural initiatives. A huge sugar-growing project with plantations extending over 175,000 ha is already being implemented by the state-owned Ethiopian Sugar Cooperation. This development was announced to the pastoralists in South Omo by the late Meles Zenawi in his speech on 25 January 2011 (Zenawi, 2011).

It is difficult to foresee how Nyangatom community members will cope with their current livelihood crises themselves. Responses may also vary from year to year, or season to season. The entire risk-scape must be taken into consideration. The chapter's title 'everything that is happening now is beyond our capacity' evokes this multidimensional risk landscape. The interim conclusion may be that it is up to the people who will, based upon their cultural values, knowledge, risk perception and awareness, define the options for adaptation, their 'adaptive capacity', but also the limits. The impacts and related new risks of climate change factors will vary accordingly.

Despite this variation, the impact of certain climate change factors on the cultural framework of the Nyangatom pastoralist society can be unambiguously stated, as argued below.

Nature is not giving any signals any more. How can we know? Our calendar is not working anymore!
An elderly woman of about 65 years

Successful adaptation to particularly harsh environments commonly finds expression in precise and elaborate culturally transmitted description of local landscapes and their

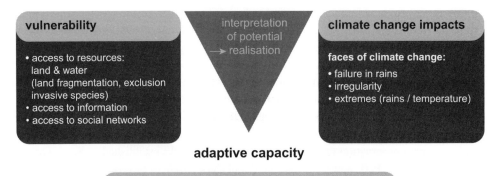

Figure 15.4 Schematic showing adaptive capacity of the Nyangatom.
Source: S. Troeger.

seasonal variations. The annual calendar with explicit directives for seasonal activities serves as a cultural backbone for the society. It encompasses an understanding of the entire universe, not just a way of living. On a particular date, people plant seeds in the ground. It involves a lot of physical energy, hope for the future and also the certainty of a new harvest, as well as various beliefs and magical actions.

Failing seasons and the resulting disorientation cause an existential shock to individuals and to the society. The climate change impact, as interpreted and constructed by people, threatens their belief systems, cultural practices and, as a result, their social relationships, as it questions centuries-old rules and certainties about agency and life. When people nowadays discuss how to 'redefine' their calendar, the fundamental cultural impact of environmental change becomes obvious. The system has moved beyond a threshold and there is no return – a state of societal anxiety has been reached which dictates that innovation and new cultural guidelines need to be defined.

But not all members of the community agree that there is a need to 'redefine'. While young people evaluate the changes as irreversible, the elders hesitate and do not want to comment on future projections. The generations are taking a different stance in relation to (re)conceptualizing nature and environmental shifts. Younger people interpret the change as more profound than their parents. Yet all Nyangatom unanimously discuss the need to re-name the months, a proactive measure to achieve a new equilibrium with nature and thus increase resilience.

Everything that is happening now is beyond our capacity

Moro, woman pastoralist leader, 58 years

New landscapes of risk emerge. They are highlighted, interpreted and strategically incorporated into the lives of people. If we try to observe 'adaptation', it will not be possible to differentiate between strategies in the context of one particular risk that may be easily isolated. We can only observe how people enact their risk landscapes, and in which way they make sense of new signposts, as they navigate through them. In relation to Nyangatom society, some new strategies, calculations, alliances and procedures can be identified. They led to the adoption of preventive measures with special effects in obvious relationship to climate change impacts (see also November, 2008: 1526). These are illustrated below and each perspective is introduced by a personal statement recorded from people in the field.

We do not count these fields any more, they are from the past!

woman, about 50 years

In responding to the failing *Belg* rains, the Nyangatom have almost given up cultivating sorghum, as this crop needs a longer maturing period than maize and beans. They have given up rainfed agriculture and now increasingly rely on flood-based agriculture along the *Omo* river banks. As a consequence, and due to the above-mentioned loss of land to the invasive *Prosopis juliflora,* land pressure along the Omo river has increased considerably. But even the river-based agriculture does not fulfil its promises anymore! Due to the ever-increasing unpredictability of rainfall in the Ethiopian highlands, the productivity of the riverbanks has gone down and is further threatened by unexpected second or third flooding during the time of cultivation that may wash away the newly planted seeds.

We move from place to place because we think the rain might come!

experienced herder, about 35 years

Due to the failing rains, pastures have deteriorated and animals, especially cattle and sheep, are weak and unproductive. Furthermore, as pastures are degraded, the cattle have to be driven to the distant mountainous areas, like *Tirga* in the Omo National Park, with better grazing grounds. Whereas before this transhumance only occurred during the dry period, nowadays the cattle are away more or less throughout the year, leaving the more vulnerable family members back in the settlement without access to much-needed cattle products. This implies not only a complete shift in the migratory pattern, but also new conflict scenarios in the ever-increasing contest for scarce resources.

Today I have to walk a long distance to find a friend ready to give me a goat!

woman, about 50 years

By these developments, the old culture of sharing is fading away – and the entire Nyangatom safety network is threatened. The Nyangatom do not have the elements required to maintain their exchange relationships and thus refrain from the culturally established ceremonies. Many individuals are left alone in their struggle for survival – and are increasingly dependent on infrequent deliveries of food aid.

With regard to the gender dimension, these developments are especially severe for women. Until recently, in addition to their marriage obligations, they could rely on friendships among women, mostly based on exchange of cattle or goat products, especially butter and sorghum. These items that were formerly at the centre of friendship relations are no longer available in sufficient quantity and as a result, women are increasingly dependent on their husbands. In the months when their husbands move away with the family herds, women have no culturally legitimate power to sell a goat for crops, and they become, together with their children, increasingly food-insecure.

We have forgotten what it means to be a Nyangatom

elderly woman, about 60 years

Consumption and sharing, pride, ceremonies, self-identification and self-esteem – all of these categories in Nyangatom life are, in one way or another, connected to livestock, livestock products and sorghum. Today, the Nyangatom life lacks its most decisive elements, as cattle are not accessible as before and are hardly productive. Even animal blood cannot be tapped excessively, as this will further weaken the livestock. Sheep suffer from the heat, do not want to eat, but rather stay in the shade of the few trees, and are not productive any more or even die. Goats are less productive than before and can only be milked to a limited extent to safeguard survival of the offspring. Sorghum is grown less and less, and its yields are insufficient. Also, the people are not sure about the culturally 'right' time for their ceremonies, which used to stabilize their social identification, and often opt to postpone and even refrain from such cultural events.

Pastoralists, once stable in pastoral production, increasingly find themselves in danger of losing their livelihood. They may try to restock their herds to start the cycle of livestock husbandry again, but time and again see their efforts wiped out by drought exacerbated by the weakening reciprocal and social ties that once offered coping mechanisms via sharing and exchange (HPG, 2009).

Nowadays, people hardly meet for social events and have nothing to exchange with their friends and neighbours. As declared by a male pastoralist of about 45 years: 'There is nothing to exchange. We only meet at food-aid distribution stations.'

Adaptation to Landscapes of Risk – Future Perspectives and Conceptual Outlines

Adaptation is commonly perceived as a re-stabilization of livelihood and the restoration of a people's security and resilience (Adger et al., 2009). But from the evidence collected among the Nyangatom people, 'adaptation' in the sense of adopting more elaborate technologies or improved seeds will not guide the Nyanatom towards higher resilience. What is needed are new strategies that have never been tried and experienced before. As a consequence, these strategies will imply, in many cases, new fields of action to be explored, as well as new institutions of governance and management to support the required innovations.

While listening to the people back in January 2011, when the drought had already taken shape across Nyangatom land, the true meaning of the statement put in the title of this chapter, 'everything that is happening now is beyond our capacity', was quite evident. Agro-pastoralist groups in south Ethiopia, like the Nyangatom and their neighbours, are at the threshold of giving in after 'an unrelenting battle with failed rains over eight years (that) has left them with no sense of a future nor any hope for a better life' (IRIN News, 2011). As stated by the people themselves, traditional community-based adaptation measures are not working anymore due to the intensity of environmental changes, together with the processes of land fragmentation and the paradigms of pastoral land rights and natural resources management that are dominant in Ethiopian society.

Vulnerability is a function of three somewhat independent variables: 1) ecosystem resilience – the degree to which the system can tolerate environmental shocks and remain productive; 2) socioeconomic affluence – access to assets for livelihood security in cases of environmental shock; and 3) institutional capacity for climate change adaptation (Fraser, 2007; Twyman et al., 2011). In view of this definition, one path towards successful adaptation for the Nyangatom, that in turn will foster enhanced livelihoods and adaptive capacity, will be security in land access and land holding. Policies aimed at promoting pastoralism must take the pastoralists' cyclical movement of livestock and people into account. Migratory routes need to be mapped and protected, and pastoralist communities have to be provided with basic services. Altogether, more attention needs to be given to how people, including many local actors, such as women and men, elders and youth, define and interpret their navigation through their homeland that is becoming a landscape of increasing risk.

References

Adger, N., Dessai, S., Goulden, M. et al. 2009. Are there social limits to adaptation to climate change? *Climatic Change*, 93(3–4): 335–54.

Appadurai, A. 2011. Disjuncture and difference in the global cultural economy. In Szeman, I. and Kaposy, T. (eds.) *Cultural Theory: An Anthology*. New York: John Wiley & Sons, pp. 282–295.

Berman, M. and Kofinas, G. 2004. Hunting for models: Grounded and rational choice approaches to analyzing climate effects on subsistence hunting in an Arctic community. *Ecological Economics*, 49: 31–46.

Dessai, S. and Hulme, M. 2004. Does climate adaptation policy need probabilities? *Climate Policy*, 4: 107–28.

Flintan, F. 2011. The causes, processes and impacts of land fragmentation in the rangelands of Ethiopia, Kenya and Uganda. Regional Learning & Advocacy programme for Vulnerable Dryland Communities (REGLAP).

Fraser, E. D. G. 2007. Travelling in antique lands: Using past famines to develop an adaptability/resilience framework to identify food systems vulnerable to climate change. *Climate Change*, 83: 495–514.

HGP (Humanitarian Policy Group) 2009. *Pastoralism, Policies and Practice in the Horn and East Africa: A Review of Current Trends*. London: Overseas Development Institute.

IPCC. 2007. *Climate Change 2007: Impacts, Adaptation and Vulnerability. Contribution of Working Group II to the Fourth Assessment Report of the Intergovernmental Panel on Climate Change*, Parry, M. L., Canziani, O. F., Palutikof, J. P., van der Linden P. J. and Hanson, C. E. (eds.) Cambridge, UK and New York: Cambridge University Press, 976 pp.

IRIN News. 2009. Global: Twelve countries on climate change list, www.irinnews.org/report/85179/global-twelve-countries-on-climate-change-hit-list

IRIN News. 2011. Eastern Africa: Too soon to blame climate change for drought, www.irinnews.org/report/93204/eastern-africa-too-soon-to-blame-climate-change-for-drought

Jennings, S. and Magrath, J. 2009. *What happened to the seasons? A paper for the Future Agricultures Consortium International Conference on Seasonality*. Seasonality Revisited International Conference, Institute of Development Studies, 8–10 July 2009, Brighton, UK.

NMA (National Meteorological Agency) Ethiopia (2007), Addis Ababa.

Nori, M. and Davis J. 2007. *Change of wind or wind of change? Climate Change, adaptation and pastoralism*. Working paper prepared for the World Initiative for Sustainable Pastoralism, https://cmsdata.iucn.org/downloads/c__documents_and_settings_hps_local_settings_application_data_mozilla_firefox_profile.pdf

November, V. 2008. Spatiality of risk. *Environment and Planning A*, 40: 1523–27.

Orlove, B.S., Chiang, J. C. H. and Cane, M. A. 2000. Forecasting Andean rainfall and crop yield from the influence of El Ninõ on Pleiades visibility. *Nature*, 403: 68–71.

Pavanello, S. and Levine, S. 2011. *Rules of the Range – Natural Resources Management in Kenya-Ethiopia Border Areas*. HPG Working Paper, September 2011, www.odi.org/sites/odi.org.uk/files/odi-assets/publications-opinion-files/7307.pdf

Riedlinger, D. and Berkes, F. 2001. Contributions of traditional knowledge to understanding change in the Canadian Arctic. *Polar Record*, 37: 315–28.

Tröger, S., Grenzebach, H., zur-Heide, F. et al. 2011. *Failing Seasons, Ailing Societies: Climate Change and the Meaning of Adaptation in Ethiopia*. Addis Ababa: Deutsche Gesellschaft für Technische Zusammenarbeit (GTZ).

Turton, D. 2011. Wilderness, wasteland or home? Three ways of imagining the Lower Omo Valley. *Journal of Eastern African Studies*, 5(1): 158–76.

Turton, D. 2014. *'Development' by Dispossession in Ethiopia's Omo Valley*. Presentation to the Anglo-Ethiopian Society, 17 September 2014. Unpublished.

Twyman, C., Fraser E. D. G., Stringer, L. C. et al. 2011. Climate science, development practice, and policy interactions in dryland agroecological systems. *Ecology and Society*, 16(3): 14–23.

Zenawi, M. 2011. Speech by Meles Zenawi during the 13th Annual Pastoralists' Day celebrations, 25 January 2011, Jinka, South Omo, www.mursi.org/pdf/Meles%20Jinka%20speech.pdf

Part IV
Sources of Indigenous Strength and Resilience

16

'Normal' Catastrophes or Harbinger of Climate Change? Reindeer-herding Sami Facing Dire Winters in Northern Sweden

Marie Roué

Sapmi is the name that the Sami people give to their territory, which extends across the northern parts of four countries: Norway, Sweden, Finland and Russia. The Sami have occupied these lands since the retreat of the inland ice, and their territories extended even further south before they were displaced by colonization. Of a total of 70,000 Sami, 20,000 live in Sweden, where reindeer-herding activities extend across one-third of the country's land area. The present research has been carried out in the northernmost part of Swedish Sapmi, in the county of Norrbotten, a region of almost 100,000 km^2 that constitutes a quarter of the country.

In Norrbotten, 4,000 Sami own reindeer (*Rangifer tarandus*) and maintain a distinctive way of life, as well as a living culture and language. There are also a number of Sami who are not engaged in reindeer husbandry, although many still have family ties to reindeer herders. For all Sami, natural resources continue to be vital, for herding but also for hunting, fishing and other activities, including tourism. While this reliance on natural resources makes them vulnerable to climate change, at the same time, the Sami have a great adaptive capacity that is rooted in their knowledge of the environment and their practical know-how.

This chapter investigates the specific nature of climate change impacts that may affect the Sami in Norrbotten. We will then consider if and how they may be able to adapt to these changes, taking into consideration the many other challenges that reindeer herders face in Sapmi today. As expressed by a young Sami herder:

There are different problems. For winter grazing, the problem is that we have so little space ... and we have more problems due to forestry and maybe soon, mining, and also snowmobile tourism. Because the winter pasture is so important, climate change will probably, in my opinion, be a big problem. Because when the climate changes, traditionally we have always moved around. I mean, in ten thousand years, the climate has changed a lot ... but then, we had places to go. Now we don't.

Accordingly, this chapter will consider climate change impacts against the backdrop of other socioenvironmental impacts of relatively recent origin, whose cumulative effects severely restrict the adaptation opportunities available to Sami herders.

Ice Crusts Over Winter Pastures: A Catastrophe for Reindeer Herding

There is general agreement that environmental change due to global climate change is already well under way in the Arctic and Subarctic (ACIA, 2005; IPCC, 2007, 2014). This said, forecasting remains a difficult enterprise and the various scenarios put forth by scientists remain highly speculative. The future impacts of climate change on reindeer herding are difficult to ascertain. Moen (2008), for example, points out that effects on reindeer husbandry could be positive in summer due to higher plant productivity and a longer growing season. Shorter winters could also be viewed as a positive development. However, my own fieldwork and discussions with herders provide support for a more pessimistic assessment.

Indigenous hunters and pastoralists with whom I have worked during forty years of fieldwork in Canada and Scandinavia have repeatedly emphasized that, for wild caribou and domestic reindeer alike, the critical factor for herd survival is access to winter forage. In winter, both caribou and reindeer feed almost exclusively on lichens that they access by digging through a layer of snow. Under specific meteorological conditions, however, one or more layers of ice may form on or within the snowpack. In some cases, this ice layer may be so thick that animals cannot break through it to reach their food. Such conditions sometimes occur in spring, when a warm spell that melts the snow surface is followed by a period of severe cold. It may also occur early in winter, when freezing rain on the snow cover is transformed into ice when temperatures drop. If pastures are locked away under an impenetrable ice crust for an extended period, then animals may perish from hunger and a catastrophe may be imminent.

In the 1970s, a Sami from Kautokeino, Norway, explained to me that these conditions can devastate entire herds while the herder stands by helpless. Having lived through this ordeal himself, he described the overwhelming sense of despair that led him to the difficult decision, a few years later, to abandon herding. His decision was motivated by the tough work conditions that ruin one's health and the high risks associated with a livelihood where, regardless of one's talent or hard work, one can lose everything from one day to the next and be forced to start again from nothing.

Nielsen's dictionary defines the Sami term for this phenomenon *čuokke* as 'ice-crust on pasture' and the verb *cuokkot* as 'to get covered with a crust of ice'. (Nielsen, 1979). Other definitions offer additional details, such as *čuokke-boazo*, 'a reindeer that has suffered from a lack of pasture due to *čuokke*'. These brief definitions hardly do justice, however, to the condition's potentially dramatic consequences.

Another Sami term illustrates the singular importance of access to pasture. *Guohton* has often been translated into Swedish as 'bete' or pasture. However, our research in northern Sweden shows that this transposition from Sami to Swedish loses an essential element (Roturier and Roué, 2009). When Sami say that there is good *guohton*, it means not only that there are lichen pastures for reindeer to graze, but also that these lichen are accessible. Lichen may be present, even in abundance, but if they are not accessible to the herds, then these pastures are not referred to as *guohton*.

To the untrained eye, one area of pasture may resemble another. But herders have different terms for a pristine, untrodden field of snow (*oppas*) that reindeer can graze by digging craters to reach their food, and a pasture covered with old craters and trampled snow (*čiegar*) where access to ground lichens is obstructed by snow compacted by an earlier passage of the herd (Roturier and Roué, 2009). The herders have different ways to check the state of the snow in winter: 'We usually check the snow with a stick. If it goes easily through it's usually good … This year it is not so bad, quite good pastures. The bottom part here is *seanas* [a granular snow that resembles sugar], it is quite soft' (Ole Isak Eira, in Oskal et al., 2009).

The disastrous consequences of pastures locked away by ice have been recorded throughout the Arctic, and are as devastating for wild caribou as they are for domestic reindeer. For example when Robert Flaherty arrived in the Belcher Islands in Arctic Canada in 1914 he reported seeing mounds of caribou bones on the tundra. The local Inuit explained to him that a few decades earlier, the entire Belcher Island herd starved to death due an episode of rain and a subsequent freeze that locked the entire archipelago in ice.

References to catastrophic years for reindeer herding appear in the earliest written records about the Sami. For the Sami of northern Sweden, the first eyewitness accounts are those of Petrus Laestadius, a priest of Sami and Swedish origin who is famous for his highly puritanical form of Lutheranism that he spread across Lapland at the beginning of the nineteenth century. In his travel journal (Laestadius, 1831), he writes on several occasions about snow and ice conditions that lock away the lichen pastures and cause the reindeer 'to die like flies'. To convey to other Swedes the devastating impact of such a disaster for the Sami (he uses the term Lapps), he compares it to losing one's property to fire. He was witness to two dramatic episodes of *tjuocke (čuokke)*, the first in 1828 and a second in 1832. The precise nature of his observations, perhaps due to his scientific background in botany, makes him a valuable eyewitness: 'It can be supposed that the Mountain-lapps' reindeer herds in Arjeplog, in this one year (1828), were diminished by at least one-third. Rich Lapps became poor, and those who earlier had small reindeer herds became unable to sustain as nomads, so that many left for Norway and settled there' (Laestadius, 1831). The year 1832 was equally devastating and led numerous families of forest Sami into ruin: 'Of the Lapps, who actually are my people, the best part, the Forest-lapps, have been almost totally ruined: their want is incomparable to that of any others. The reindeer-herds of the Mountain-lapps are at least halved' (Laestadius, 1831).

A more recent account for the same group of Sami from the Swedish mountains is provided by an ethnologist, Hugh Beach, who worked with Sami from the village of Tuorpon from the late 1970s. He recorded their memories of past winters that were catastrophic. While there were many bad winters in the early thirties, the winter of 1936 is remembered to this day as one of the worst in herding history. During the reindeer catastrophe of 1936, the village of Tuorpon lost almost half of its herd, from 8,167 reindeer in 1935 to 4,735 in 1938. These figures provide a precise measure of the village's loss (some 42 per cent of the herd), a proportion similar to those recorded by Laestadius a century earlier. The Sami herder, Isak Parffa, is quoted as saying: 'The grazing lands looked like a graveyard, and Tuorpon did not fully recover from this blow until the 1950s' (Beach, 1981: 174).

According to Beach, the same conditions occurred again in the 1970s. During the winter of 1973/74, the pastures were locked in ice. Herders were forced to gather their reindeer in a large corral and feed them with artificial fodder, partially subsidized by the state (Beach, 1981: 432). While increasingly common today, this was one of the first instances of artificial feeding and a further step towards the 'modernization' of reindeer herding. In the winter of 1976/77, conditions were similarly disastrous. However, one group of Sami herders, informed by telephone of the icing of the lowland winter pastures, opted for an unusual and risky plan of action. Their herds were just coming down from summer pastures in the mountains. Rather than have them continue towards the coast, as was the usual practice, the herders instead decided to stop the migration and have the herds over-winter in the mountains, on lands traditionally grazed in summer. Their audacity paid off, as their herds gave birth to a large number of calves the following spring.

My own discussions with a Sami from Kautokeino, Norway, in the 1970s, reveal other strategies that may save a herd if icing occurs in the spring, or at the end of winter when spring is approaching. When the interior plateau of Finnmark is covered with ice, one strategy is to accelerate the migration and rapidly move the herd to the sea coast in the hope that the ice and snow will have already melted there. Such a decision is not without risk, however, as the females are pregnant and ready to give birth. Alternatively, it may be on mountain tops that one can find places where strong winds have swept away the snow cover and the herds can find rock lichens to feed on.

These various episodes from earlier times give some indication of the challenging circumstances that are emerging today due to global climate change. The increasing frequency of difficult winter conditions requires herders to come up with new strategies to ensure the survival of their herds under extreme conditions (Mathiesen et al., 2013).

Čuokke Today: A 'Normal' Catastrophe or a Harbinger of Climate Change?

The Swedish government declared the winter of 2007 in Sapmi a 'climate disaster'. They committed 37 million kroner to the Sami parliament to assist reindeer herders in the face of this crisis. Was this year, albeit catastrophic, still within the range of 'normal' events? As we have seen, snow and ice conditions that devastate caribou and reindeer herds in winter have been recorded periodically across the Arctic since the early nineteenth century, and have probably always existed. But can the sequence of recent 'bad' winters in northern Sweden be considered 'normal', given their nature and frequency? Or are we witnessing the emergence of a new phenomenon, whereby frequent disastrous winters are becoming a new norm due to global climate change?

Although research into these questions is ongoing, the available data are as yet too few to be conclusive. We will nevertheless consider and compare two sources of information. The first is local – through our field research at the level of the *sii'da*, the basic unit of Sami social organization. The second source is the official report on climate change impacts published by the Swedish government (SOU, 2007).

In 2009, I worked in collaboration with Samuel Roturier and Per Sandström, who coordinates participatory mapping with reindeer herders as part of the Reindeer Herding Plan. Building upon previous work where herders identified 'key zones' for reindeer pasture in winter, we established more detailed maps, again in collaboration with herders, but at the level of the subgroup, the *sii'da*, rather than the broader level of the Sami village. These maps record, week by week, the successive stages of herd displacement and pasture use throughout the entire winter period. In this way, herder strategies for pasture use were meticulously recorded beginning with the year of the fieldwork – 2009, and tracing back over the four previous years.

The data show that herders did not follow the same pattern every year, due to variations in winter conditions from one year to the next (Figure 16.1). They also show that the 'key zones' for reindeer pasture that were designated at the level of the Sami village were, in fact, only suitable for use for half of these years. For the six winters between 2004 and 2010, three were catastrophic for reindeer herding: 2006/07, 2007/08, and 2009/10. In contrast, conditions during the winters of 2004/05, 2005/06 and 2008/09 were good for the herds. For each of the harsh winters, the icing phenomenon discussed above, *čuokke*, prevented the reindeer from gaining access to ground lichens, their main winter forage. *Čuokke* occurs when temperatures fluctuate to such an extent that warm spells melt the snow cover, which then freezes solid when sub-zero winter temperatures return. The ice crust that forms is so thick that reindeer cannot break through it to access their food.

During interviews (Samuel Roturier, unpublished data, 2009), one herder provided a very clear definition of *čuokke*: 'When it is *čuokke*, it means that the pastures are locked in. *Čuokke* … it is serious. When it is *čuokke*, it is impossible.' He contrasted this term with another Sami word that designates pastures blocked by ice, but only covering specific areas: '*Bodne-vihci*, it can be at one place only. Say at Gräsviken, there it is *bodne-vihci*. But here, maybe it is good. Then it is not *čuokke*. *Čuokke*, it covers everywhere.'

While pasture use was mapped over a five-year period to which additional interview data allowed us to also add the winter of 2009/10 (see Figure 16.1 for an overview), the time series nonetheless remains short. It is difficult to attribute the observed phenomena to climate change rather than a mere succession, albeit exceptional, of bad winters. Indeed, Sami herders themselves are very cautious in this regard and avoid speculative statements, even though they express grave concerns about the future of reindeer herding.

It is therefore useful to complement our field observations with data from another source and at another scale: the national and regional observations conducted by scientists under the auspices of the Swedish government. In 2007, the Swedish government published a report on global climate change in which impacts on the northern part of the country are addressed (SOU, 2007). The prospects for reindeer herding are not reassuring, and several potential negative impacts of climate change are emphasized:

The conditions for conducting reindeer herding in Sweden will be seriously affected by climate change. The growing season could be extended and plant production during the summer grazing is

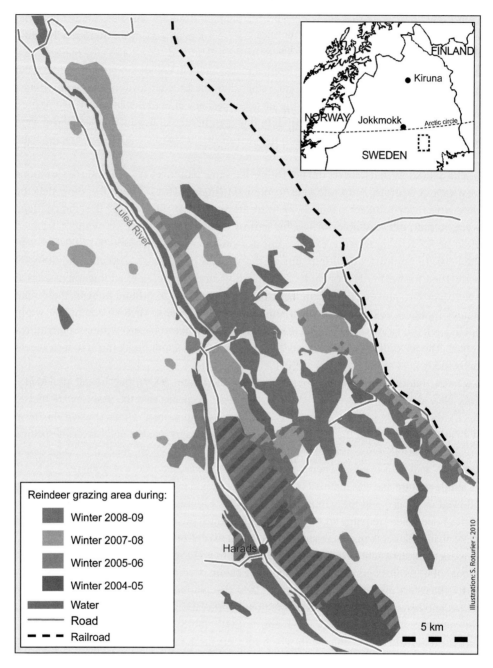

Figure 16.1 Pasture use by Sami reindeer herders over four winters. © Samuel Roturier. (A black and white version of this figure will appear in some formats. For the colour version, please refer to the plate section.)

expected to increase. Insect plagues could become worse and the snow conditions in the winter will become more difficult. The bare mountain areas above the tree line are expected to decrease, which could lead to more frequent conflicts of interest with other sectors.

(SOU, 2007: 375)

Moen (2008) provides a more nuanced assessment of the effects of climate change in northern Sweden, noting that an increase in vegetative growth due to warmer and longer summers could compensate for an increased density of mosquitoes and reindeer parasites. But he nevertheless expresses concern about a decrease in lichen growth rates, an essential food item for reindeer in winter.

Increased temperatures will likely lead to an increased productivity in boreal forests. Unless forestry responds with increased thinning, this will result in more closed-canopy forests with less light reaching the ground. Lichen cover is strongly negatively correlated with crown cover and the amount of light at ground level. This is probably related to competition from mosses, which have a lower light saturation point and are more adapted to moist growing conditions.

(Moen, 2008)

With respect to temperature, for the region of Norrbotten the SOU report (2007) predicts a temperature increase of one to two degrees between 1991 and 2010, two to three degrees between 2011 and 2040, three to four degrees between 2041 and 2070, and four to seven degrees between 2071 and 2100 (according to the RCA3-EA2 scenario). The assessment of changes in precipitation is equally disturbing. From the present to 2020, the amount of rain falling in winter is expected to increase in the north, and by 2080, the quantity of snow will drop significantly.

According to the climate scenarios, the winters will become warmer and wetter. There appears to be an increasing risk of difficult snow conditions, with ice and frozen crusts on snow that are very difficult for the reindeer to penetrate when looking for food, as the amount of rain in the winter will increase according to the scenarios. At the same time, the temperature will alternate more frequently above and below freezing point.

(SOU, 2007: 60)

Knowing that *čuokke* is caused by freezing rain and melting snow associated with fluctuating winter temperatures, it seems likely that the conditions that we are observing today may be the harbinger of a changing climate that will seriously threaten reindeer herding. Indeed, the official report of the Swedish authorities comes to a similar conclusion, suggesting that the difficult winter conditions experienced in 2006/07 may become a new norm:

An increase in the occurrence of ice and frozen crusts can result in the reindeer having poorer winter grazing, causing them to have to utilize the body fat reserves built up during summer grazing to a greater extent, with reduced fitness as a consequence (Moen, 2006). In other words, there is a risk that the problematic conditions that prevailed in large parts of the reindeer grazing area during the 2006/2007 winter could become more common.

(SOU, 2007: 379)

The Industrialization of Norrbotten: Cumulative Impacts

Access to winter pastures is a major challenge for reindeer herders. But meteorological conditions and access to pasture are not the only major changes facing reindeer herding. Today, there are numerous competing forms of land use that also render reindeer herding a precarious enterprise. Even though the county of Norrbotten is homeland to the Sami, this has not sheltered it from industrial development. Indeed, quite to the contrary, the rich natural resources of Norrbotten – hydroelectric potential, timber and minerals, especially iron – have played and continue to play a key role in the industrial development of Sweden as a whole (Magnusson, 2000). The people of this northern area, Sami and non-Sami alike, often express their bitterness about being exploited by southern Sweden.

Mining, Hydroelectricity and Transportation

In the 1980s, the advent of a new source of energy, namely hydroelectricity, triggered what has been called the 'second industrial revolution'. Through the construction of dams and reservoirs, hydroelectric power has completely transformed the landscape (Figure 16.2). The ensuing industrialization of the North has restricted lands available for reindeer herding. Today, herders must compete with hydroelectric development, mining, roads, railways and urbanization. The reduction of space available for winter grazing and the fragmentation of the territory make it difficult for them to move their herds from one area to another, and restricts their capacity to cope with 'bad winters'.

In Norrbotten, three industries are inextricably linked: mining in remote areas depends upon transportation of the minerals by rail to the sea for export, and both trains and mines depend upon electricity. Since the early eighteenth century, it was known that there were deposits of iron ore in northern Sweden. But large-scale exploitation only began at the end of the nineteenth century. In 1878, two English chemists invented a procedure that overcame the technical problem of producing high-quality steel from ore with high phosphorus content (Barck, 2003). In 1882, an English company began construction of a railway line to transport the ore from the mines of Malmberget to the harbour of Luleå. When finally completed in 1888, however, the railway proved to be unstable in summer when the ground thawed. The Swedish government had to take over the project and did not complete this second phase of work before 1892. The 'Ore Line' is still today one of the country's most important railways, accounting for half of Sweden's total rail freight tonnage. Ore trains run from the mines in Kiruna and Malmberget/Gällivare to harbours in Narvik and Luleå, and the same line transports tourists who come in summer to discover Lappland and hike the mountains. During the period of colonization, the state established a northern limit to agriculture, in order to limit conflicts between farmers and Sami. The Ore Line extends north of this agricultural line and creates a hazard where it traverses reindeer pastures. Herders try to prevent their herds from crossing the tracks, but every year reindeer are killed by the fast-moving trains.

Figure 16.2 Hydropower infrastructure and reservoir in the Laponia World Heritage site. (A black and white version of this figure will appear in some formats. For the colour version, please refer to the plate section.)

Just prior to the completion of the ore railway, the company LKAB (Luossavaara-Kiirunavaara AktieBolag) was formed to exploit the iron ore from the region around Gallivare and Kiruna. Since its creation, it has been the pride of the Swedish state, manufacturing iron ore pellets for the steel industries of Europe, North Africa and the Middle East. LKAB is the world's largest and most modern underground iron ore mine, and exploits an ore body 4 km long and 80 m thick that extends to a depth of 2 km. Although LKAB has produced over 950 Mt of ore in more than 100 years, only one-third of the ore body has been extracted. The region is also rich in many other minerals. Near Gällivare, the Boliden Company exploits one of Europe's largest copper mines and also produces gold and silver.

The mining industry has radically transformed the Sami territory since the early twentieth century. The mines themselves are environmentally destructive. They create whole new landscapes, as entire hills are literally consumed by the mining operation and replaced by mountains of rubble. Close to the mines, stock ponds hold waste waters containing toxic chemicals and acids used during the smelting and refining of the ore. These ponds contaminate water and soil, and consequently pose a threat to plants, animals and people. New towns, such as Kiruna and Gällivare, have been created where nothing existed previously. More and more space is required for housing, roads and other infrastructure, including those required for tourism. All of these industrial developments considerably reduce the area available to Sami reindeer herders.

Vattenfall, the national electricity company in Sweden, was founded in 1909 to exploit hydropower to produce electricity. The first plant, built in 1910 in Porjus, not far from the mines, supplied power for mining activities and the railway. During the twentieth century the Luleälv River was dammed at several locations along its course, and today continues to produce a quarter of Sweden's hydroelectric power. Not surprisingly, Norrbotten is self-sufficient in electricity, including the power that is supplied to all the processing industries. Half of the electricity produced is exported to other parts of the country.

This intensive use of northern rivers has deprived the Sami of precious calving pastures and ousted them from what were previously their main travel routes, the rivers themselves. In winter, frozen lakes and rivers were highways for the Sami and their migrating reindeer. But the ice that forms today on dammed rivers and on reservoirs is dangerously uncertain. While appearing to be 'normal' when viewed from above, the large volume of water released from reservoirs to generate electricity in winter erodes the ice from below. Several Sami have drowned when their skidoos have broken through the ice while crossing dammed waterways in winter. In addition, for the Sami who live in the mountains in summer, the migration of their herds along the edge of the reservoir has become much more precarious than in earlier days when they walked along flat valley bottoms, which are now flooded. As a young Sami herder explains: 'We have to move through this very very tricky terrain for about nine weeks in a row, and that's not so much fun for our families.'

Today the Sami have to live with these changes. They have heard from their parents or grandparents how the first dams, erected at the beginning of the twentieth century, permanently flooded fertile valleys that were traditional calving grounds. At that time, the Sami were not in a position to protest. They had no political power and it was unacceptable for anyone to stand in the way of 'modernity and progress'. Later, towards the end of the same century, they tried to oppose further dam construction on the same river, but they were again ignored. Viewed from Sweden's south, the lands of the Sami continue to be perceived first and foremost as a source of raw materials, largely empty of people, and just waiting to be developed by anyone able to extract its wealth.

The Forest Industry

Forests and their derived products – timber, tar and charcoal – are the first resources to have been exploited industrially in the north. The timber industry continues to make an important contribution to Swedish industrial production. During the first half of the twentieth century, selective logging created a more open forest cover that was suitable for the growth of ground lichens, and thus for reindeer husbandry (Berg et al., 2008). But the increasing industrialization of forestry has become in recent times a major obstacle for reindeer herding.

In winter, as soon as there is significant snow cover, the Sami move their herds from summer pastures in the mountains down to the forests. Here, the reindeer feed mainly on ground lichens. Arboreal lichens, which only grow in old-growth forests, are also of vital importance, particularly at the end of winter or when snow and ice conditions prevent access to ground-lichen pastures.

Since the 1960s, industrial exploitation of forests has become increasingly intensive. Clear-cutting has become the norm, combined with soil scarification and even artificial fertilization. According to the Swedish Forest Agency, some 72 per cent of forest lands in Sweden are occupied by stands younger than 100 years. These forest stands are too young to have developed arboreal lichens and old stands, where arboreal lichens occur, are fast disappearing. Clear-cutting has eliminated 50 per cent of lichen-rich pine heaths. This puts reindeer herders in a difficult situation, as described by a Sami herder: 'Soon we will no longer have any old-growth forest left. It's being cut down all the time. Every winter when we come down, a new area has disappeared. The forest we were in last year is gone the year after.'

After clear-cutting, the ground is prepared for a new plantation by harrowing, which speeds up forest regeneration and improves the soil for planting seeds and seedlings. Harrowing leaves the mineral soil surface exposed and destroys the lichen. Furthermore, logging residues left on the ground when the forest is clear-cut and when it is thinned have a negative effect on lichen growth and obstruct access for reindeer herds. As one Sami pointed out, forestry in Sweden is supposed to be sustainable because it cuts down trees and ensures the production of new ones through planting. But for the Sami, forestry makes reindeer herding unsustainable. Furthermore, biodiversity is diminished through the re-planting of only a few commercially important tree species and the reduction and increasing uniformity of forest age.

Protected Areas

Only the high mountain regions, which are not suitable for industrial development, have been sheltered from the appropriation of space by industries that slice through hilltops to extract minerals, dam rivers and flood plains to produce hydroelectric power, and transform forests into immense fields that are clear-cut, ploughed, replanted and then cut once more. Only on the high plateaus surrounded by mountain peaks can the Sami of north Sweden graze their reindeer in summer and fish in large and bountiful alpine lakes. Indeed, they have been called the 'Lapps of the Swedish mountains' because of this seasonal altitudinal migration. However, during the same period as the industrialization of the north, another form of appropriation of Sami homelands by the Swedish state began with the creation of national parks, which set aside the mountains for the entire nation as areas for nature conservation and recreation.

In 1909, Sweden became the first European country to create national parks. Two of these early parks, each of considerable size, are mountain areas in Norrbotten that are also the summer pastures of the Sami. Today, eight of Sweden's twenty-eight national parks are located in this one county, where they cover over 6,000 km^2. This represents 85 per cent of the area dedicated to national parks in Sweden. Muddus National Park, characterized by its untouched forests of giant firs, was established in 1942 and expanded in 1984. Padjelanta,

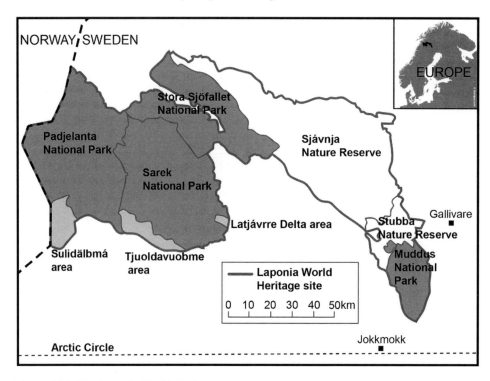

Figure 16.3 The Laponia World Heritage site, encompassing several national parks and nature reserves. (A black and white version of this figure will appear in some formats. For the colour version, please refer to the plate section.)

established in 1962, is the largest national park in Sweden and one of the largest in Europe, with a surface area of 1,984 km² and 800 km of hiking trails. In 1996, a UNESCO World Heritage site, called Laponia in acknowledgement of the Sami people, was created to protect the cultural and natural landscape (Figure 16.3). It encompasses several national parks and nature reserves that were already protected, including Sarek and Stora Sjöfallet National Parks, created in 1909 during the period of the industrialization of the north.

From its nomination in 1996 until 2006, the Laponia World Heritage site has been the arena of conflicts between the state, the Sami and the municipalities, due to their different aspirations for the area's use. Since 2007, the Laponia delegation, with a Sami majority on its committee, has collaborated with all parties to establish a new management plan, including a better zonation to prevent tourists from disturbing the herds at crucial times of the year. Since January 2013, a new body named Tuottjudus has taken over management responsibility for Laponia from Norrbotten County, with the possibility to become permanent after a six-year trial period. Tjuottjodit means in Lule Sami "to take care of, to administrate", a designation that signals a renewed relationship between the Swedish state and the Sami people.

In the past, predators have been a major point of contention. In 2002, the unexpected arrival of a wolf, which the Swedish Environmental Protection Agency advocated protecting at all costs, represented a major concern for the Sami. For the state, as well as for some environmentalists who continue to see Laponia as a 'wilderness', predator control is at odds with World Heritage status: 'in the case of a decimation of predators being allowed in Laponia, one of the bases for the area having been appointed a World Heritage site will fail' (Nilsson Dahlström, 2003: 6). The Sami, on the other hand, could not understand why their lands should become a research area for wolf specialists, who want to study wolf behaviour in the middle of their domestic herds. They are not satisfied with a state policy that argues that reindeer killed by wolves can be compensated financially. As for many other pastoral peoples, their attachment to the herd is not measurable in financial terms. A woman herder stated that she did not want money because money could not replace her lost reindeer that were the offspring of reindeer that she, her parents and grandparents had raised and selected.

Sami Adaptation Strategies

The Sami have always been confronted with an uncertain and risk-filled environment. Today, to be successful as a herder, one must be strategic, particularly during the challenging winter season when many forms of land use compete in a limited territory.

The Sami have developed various strategies to cope with and respond to the vagaries of the natural milieu. When winter begins, herders first drive their herds to good lichen pastures that cannot be used later in the season. These first pastures are in places where they know that snow will accumulate and pack as winter progresses, making access to forage later in the season difficult, if not impossible. Today, they also use pastures along railway lines early in the season, when fences along the rails contain the herds and prevent them from straying on to the tracks. Later in winter when drifting snow overtops the fences and renders them useless, they move their herds elsewhere to avoid the risk of reindeer being killed by passing trains.

Unfortunately, some strategies used in the past are no longer available to herders today. In winter, when there was *bodne-vihci* (ice crust over pastures but only in specific places) or even *čuokke* (widespread ice crust that blocks access to pastures over a large area), the herders could move their herds to old-growth forests where their animals could graze on arboreal lichens. This is no longer a possibility for most Sami villages, as almost all old-growth forest areas have been cut. By creating an increasingly uniform landscape with younger forest growth, industrial forestry is making herding more and more difficult. Herders need access to a mosaic of ecosystems which they can utilize in a selective and strategic manner according to their understandings of available forage, snow conditions and the requirements of their animals (Roturier and Roué, 2009).

The extreme snow and ice condition that prevent herds from feeding in winter over a large area, known as *čuokke*, is the most feared environmental challenge for herders. When *čuokke* occurs today, herders resort to two main strategies. The first is artificial feeding. During the two catastrophic winters of 2006/07 and 2007/08, the herders that we interviewed kept their animals alive by providing hay or pellets. While feeding ensures that the majority of the herd will not die from hunger, it nevertheless generates numerous new problems. Not only does it require a great deal of manual labour but it is also costly. Even though the state may subsidize up to half of the cost, artificial feeding remains very expensive. According to the official Swedish publication on climate change: 'it can cost around SEK 4 per day per reindeer, or SEK 2,000 per day for a herd of 500 reindeer' (SOU, 2007). If the problem begins early in the winter and persists, it could cost up to SEK 200,000 for the season, a sum that most herders cannot afford, even with state subsidies. It is not surprising that in 2008, after two successive bad years, many herders did not know what they would do if they had to face yet another bad winter. A third year of artificial feeding was well beyond their means.

Artificial feeding also brings other problems. Some reindeer suffer from eating artificial food. They become sick and, in some cases, even die. I observed women in summer collecting 'medicinal plants' that they store and use to relieve the stomach problems that reindeer encounter when fed artificial food in winter. This is yet another innovation, based on traditional knowledge, in response to environmental change.

Some herders have recently been adopting new strategies based on a reinvention of nomadism, but in modernized form. They have begun collecting lichen in sacks from locations where they are plentiful, for example along an airstrip in an urban area that is off-limits to their herds. Instead of using pastures the way their ancestors did by moving the reindeer herds, they now use modern technology to drive to places where lichen can be gathered and then bring the 'pastures' back to their reindeer. The feeding of reindeer on lichens gathered in other locations is not a completely new practice. It was used in the past, but only in special circumstances, such as to feed reindeer used for pulling sleds or carrying loads.

However, applying this practice today requires more skills than just driving a car, for environmentalists did not appreciate the ingenuity of the Sami and took them to court. According to the Sami who described this incident, they won the case by demonstrating that the area where the lichens were collected was a traditional pasture for which they had use rights. Today, therefore, to be innovative in the face of environmental change, herders need to couple their 'traditional' knowledge with 'new' sets of knowledge, including political and legal expertise.

When pastures are locked in ice and herders must move their herds from one location to the next in search of better conditions, they can either gather them in one big corral or provide artificial food at each place they stop. However, the extreme conditions make it difficult to control the animals and many reindeer go astray. Herders describe the situation as 'chaos'. Only when they re-unite the herd the following spring, in particular their females and calves, are they able to assess the extent of the losses.

Artificial feeding may also lead to the loss of cultural knowledge about how to manage 'free' animals in the natural environment, a skill that constitutes the very essence of this centuries-old culture. How can young herders acquire this essential knowledge when, as during the last several years, one winter in two is exceptionally bad and requires the herd to be fed artificially in corrals? Yet another problem is the conversion of free-range animals consuming lichens, mushrooms and grasses, and producing quality organic meat, into domesticated beasts fed with artificial, even processed, food. By compromising the quality

Figure 16.4 Sami from Tuorpon marking reindeer calves in the Laponia World Heritage site (July 2009). After a harsh winter, the Sami are particularly anxious to find out how many of their female reindeer are accompanied by calves. (A black and white version of this figure will appear in some formats. For the colour version, please refer to the plate section.)

of their product and diminishing its sustainability, artificial feeding may undermine the long-term economic and cultural viability of this way of life. This said, we should not lose sight of the fact that when conditions of *čuokke* occur, artificial feeding allows herders to save their herds from starvation, rather than helplessly watching them die.

After artificial feeding, a second strategy used in the event of *čuokke* is, as described by a herder, to keep the herds on the summer pastures in the mountains. As Hugh Beach had already observed in the 1970s (see above), this adaptation strategy may be used when herders are forewarned about iced-over pastures before they bring their herds down from the mountain. It means that another less fortunate herder has gone ahead, discovered the disastrous conditions and reported it to the others. One recent incident was described to me by a herder from Tuorpon. They were trying to bring the herd down from the summer pastures, but conditions were so bad that they lost control of the herd, which fled back to the mountains. They decided not to gather the herd again, but instead to take the risk of leaving the animals to fend for themselves in the mountains. As it turned out, this option worked well. They did not lose many animals and calf production the following spring was good (Figure 16.4). The herder commented that it takes experience, or at least a traditional way of thinking, to take this sort of risk as you have to trust the reindeer and trust nature. He felt that young herders would find this much more difficult. In his view, some young Sami have adopted a more Western way of thinking and are ill at ease when things get out of their control.

Even to these more 'modern' herders, however, it is evident that artificial feeding is not a long-term solution. A group of herders who provided feed for their herds during the two previous bad winters, decided this winter (2010) that they have to adopt a new strategy. They decided not to winter their herds on the traditional winter pastures in the forest, or on the high summer pastures, but instead in the halfway zone at the foot of the mountains. If such a strategy were to become widespread, it would be important to investigate whether it is sustainable for more than a small group, as these halfway pastures are transit areas for all herds when they are moved between winter and summer pastures. In Kautokeino, Norway, in the 1980s, I observed one case that triggered such a 'tragedy of the commons'. Instead of migrating between the interior in winter and the coastal area in summer (the typical seasonal pattern in Norway), a wealthy herder decided that it would be easier to leave his herds the whole year around on the halfway pastures. This area, however, served as the autumn pastures for all groups, and he was bitterly criticized when these pastures were soon depleted. When subsequently he lost control of his herd, nobody offered to help him locate his reindeer and in the end, he lost many animals.

Conclusion

In this chapter, I have considered the challenge that climate change represents for Sami reindeer herding. I have paid particular attention to the winter season, a critical bottleneck for herding, and more precisely the threat that the Sami refer to as *čuokke*, the formation of ice layers over large areas of winter pasture that prevent reindeer from accessing lichen, their principal winter forage. These dramatic events, which have occurred from time to time in the past and which were much feared, are happening with increasing frequency during recent years. According to climate scenarios developed by the Swedish government (SOU, 2007), there is a high risk that this phenomenon will increase in frequency and intensity during the coming decades.

If catastrophic winter conditions become common, or at least as frequent as during the period 2004 to 2010, when one out of two have been severe, then reindeer husbandry, as the livelihood and culture of an indigenous people, will be hard pressed to adapt and under serious threat. If the artificial feeding of reindeer becomes a necessity in winter rather than an exceptional measure, this would not only endanger the economic viability of reindeer husbandry, but it would have at least two other major consequences. First, it would impoverish Sami culture, and in particular the detailed knowledge possessed by the herders about snow and ice conditions in relation to vegetation, which is the foundation of their strategies for pasture use. Second, it would transform an animal raised on a wide variety of wild and thus organic plants – grasses, leaves, lichens and mushrooms – into a farm-raised animal dependent upon agricultural (hay) or industrial (pellets) products.

We have shown that due to competing uses of the territory for mining, hydropower, forestry, recreation, transport and urban development, some key adaptation strategies are no longer available to Sami herders. In the language of environmental impact assessment, the challenge is one of understanding cumulative impacts, where it is impossible to consider the most recent impact, climate change, without structurally considering the entire ensemble.

This said, herders are far from being passive victims and they are deploying numerous and innovative strategies in an attempt to face these threats, both new and old. Nevertheless, their future is not entirely in their own hands. They are familiar with the complexity of the challenges before them and understand the need to respond on a number of fronts. For this reason, they are active on the land through their constant innovation of herding practices and techniques, but also in the fields of communication, negotiation and indigenous politics at all levels. The first president of the UN Permanent Forum on Indigenous Issues in New York was Ole Henrik Magga, a Sami from Norway. The Sami are working with their Sami Parliaments, have their own political parties, and are also using their traditional organizations such as the Swedish Sami Herders Association (SSR). They are very active in participatory science (for example the Ealat Programme, see Oskal et al., 2009) and in the media. They have their own radio, communicate on blogs, and are developing skills in new domains such as tourism. In Sweden, after ten years of conflict, the Sami have also managed to convince the state to negotiate a new deal with respect to the Laponia World Heritage site.

The situation may be alarming, but it is far from hopeless. There are signs that progress is being made towards joint management with the state, for example in the cases of Laponia, forest management and Forest Stewardship Council labels. If their partners, and in particular the state, are willing to advance in this direction, then it may be possible for the Sami to work towards a more participatory management of their territory that will allow them to better position themselves to respond appropriately to the global changes that they are facing, including that of climate change.

References

ACIA (Arctic Climate Impact Assessment) 2005. *Arctic Climate Impact Assessment*. Cambridge: Cambridge University Press.

Barck, Å. 2003. *The History of LKAB*. Luleå, Sweden: LKAB.

Beach, H. 1981. *Reindeer-Herd Management in Transition: The Case of Tuorpon Saameby in Northern Sweden*. Uppsala: Acta Universitatis Upsaliensis.

Berg, A., Östlunda, L., Moen, J. and Olofsson, J. (2008). A century of logging and forestry in a reindeer herding area in northern Sweden. *Forest Ecology and Management* 256(5): 1009–20.

IPCC (Intergovernmental Panel on Climate Change) 2007. *Climate Change 2007: Synthesis Report. Contribution of Working Groups I, II and III to the Fourth Assessment Report of the Intergovernmental Panel on Climate Change*. [Core Writing Team: Pachauri, R. K. and Reisinger, A. (eds.)]. Geneva: IPCC.

IPCC. 2014. *Climate Change 2014: Impacts, Adaptation, and Vulnerability. Part A: Global and Sectoral Aspects. Contribution of Working Group II to the Fifth Assessment Report of the Intergovernmental Panel on Climate Change* [Field, C.B., Barros, V.R., Dokken, D.J. et al. (eds.)]. Cambridge, UK and New York: Cambridge University Press, 1132 pp

Laestadius, P. 1831. *Journal af P. Laestadius för första året af hans tjenstgöring såsom missionaire i Lappmarken*. Stockholm: Zacharias Haeggström.

Magnusson, L. 2000. *An Economic History of Sweden*. London: Routledge.

Mathiesen, S.D., Alfthan, B., Corell, R., Eira, R.B.M., Eira, I.M.G., Degteva, A., Johnsen, K.I., Oskal, A., Roué, M., Sara, M.N., Skum, E.R., Tury, E.I. and Turi, J.M. 2013: Strategies to enhance the resilience of Sámi reindeer husbandry to rapid changes in the Arctic. In: *Arctic Resilience Interim Report 2013*. Stockholm Environment Institute and Stockholm Resilience Centre, Stockholm, Sweden, pp. 109–112.

Moen, J. 2008. Climate change: Effects on the ecological basis for reindeer husbandry in Sweden. *Ambio*, 37: 304–11.

Nielsen, K. 1979. *Lapp Dictionary*. Oslo: Universitetsforlaget AS.

Nilsson Dahlström, A. 2003. Negotiating wilderness in a cultural landscape. Predators and Sami reindeer herding in the Laponian World Heritage area. *Uppsala Studies in Cultural Anthropology Serie, 32*. Uppsala: Acta Universitatis Upsaliensis.

Oskal, A., Turi, J. M., Mathiesen, S. D. and Burgess, P. (eds) 2009. *Ealat, Reindeer Herders Voice: Reindeer Herding, Traditional Knowledge and Adapting to Climate Change and Loss of GRAZING LAND*. Kautokeino: International Centre for Reindeer Husbandry.

Roturier, S. and Roué, M. 2009. Of forest, snow and lichen: Sami reindeer herders' knowledge of winter pastures in northern Sweden. *Forest Ecology and Management*, 258(9): 1960–7.

SOU (Swedish Commission on Climate and Vulnerability) 2007. *Sweden Facing Climate Change: Threats and Opportunities. Swedish Government Official Reports*. Stockholm: Edita Sverige AB, www.government.se/sb/d/574/a/96002

Tomasson, T. 1918. Renskötseln, dess utveckling och betingelser. In *Svenska Lapparnas Landsmöte I Östersund*. Uppsala: Almqvist & Wiksell.

17

Canaries of Civilization: Small Island Vulnerability, Past Adaptations and Sea-Level Rise

Marjorie V. C. Falanruw

The people of low-lying Pacific Islands are especially vulnerable to climate change, and are already experiencing impacts of sea-level rise (Fletcher and Richmond, 2010). Among the most immediately impacted are Caroline islanders who reside on low-lying islets and atolls in Yap State in the Federated States of Micronesia (FSM). The FSM is a young nation located in the north-west Pacific. The nation is composed of four states, the most western of which is Yap State (FSM, 2002; Falanruw, 2010). In addition to the high islands of mainland Yap, the state includes many inhabited and uninhabited atolls and low-lying islets that rise no more than a few metres above mean sea level. Over thousands of years, the *rematau*, 'people of the deep sea', from these atolls have developed technologies for food production and voyaging in small canoes in order to survive. Now rising sea levels are threatening supplies of fresh water and traditional food-production systems, including taro patches and atoll agroforests. In addition to sea-level rise, storm surges pose additional devastating threats to water supplies, food production and habitation.

As a result of these threats and already experienced impacts, outer islanders are migrating to mainland Yap, a high island with which they have strong cultural links. The ability of mainland Yap to sustain natural resources and food security in the face of climate change, while also accommodating a very large increase in population (over 40 per cent) (FSM, 2010), will be severely challenged. Most of the islands' more productive soils are very vulnerable to storm surge and important coastal taro patch habitat is expected to be inundated by sea-level rise. Meeting challenges of supporting increased populations on limited island resources will require wise planning, intensive preparation and hard work. During prehistoric times, the ancestors of present-day Yapese had successfully responded to the challenge of high population pressure through adaptive methods of food production and allocation of natural resources that avoided a tragedy of the commons. Today, however, these traditional technologies and natural resource management practices are fading. There is a need to document, evaluate, enhance and utilize traditional technologies of food production and natural resource management that could contribute to climate change adaptation.

This chapter shares some traditional technologies that could be used in adapting to climate change. In the past when Yap islands were densely populated and resources were at a premium, it appears that the entire island was utilized as a food-production system.

Remnants of these systems are still etched into the land and seascape in the form of ditched beds on land and large-stone fish weirs in the lagoon (Falanruw, 1994, 1995).

Traditional Agriculture

Adaptive features of traditional Yapese agriculture include the use of a wide variety of habitats and agrobiodiversity that enhance resilience in the face of erratic weather patterns. Food, building materials and fibres, medicines and other basic needs were derived from wild forests, shifting gardens, taro patch systems, agroforests and mangrove forests.

Wild forests were largely used for building materials, wild foods, medicines, collecting fruit bats and as sacred places. They also provided ecological watershed services. Closer to villages, forests were utilized for a mosaic of shifting gardens made by burn-girdling trees and small gardens that produced a variety of root crops, mixed with an array of other crops, in a multilayered garden. Main root crops included *Dioscorea* yams, taro and sweet potatoes; a variety of cucurbit vine crops; and a wide variety of bananas in the garden's second year. In today's gardens, harvests can begin in about three months, and one crop is followed by another. As each crop is harvested, it creates room for the remaining plants and the planting of others crops. After the second year, these gardens are visited less frequently and eventually the area is left fallow, returning to secondary forest or invasive weeds, especially if burned by wildfires.

Intensification of Shifting Gardens

As populations grew, the fallow periods were shortened so that the sites did not grow back to secondary forest. There is evidence that periods of drought and wildfires also contributed to deforestation. Without a forest canopy, soils were poorer and agricultural methods needed to be intensified.

One example of intensification is the construction of pyramidal trellises for the growing of *Dioscorea* yams, rather than large burn-girdled trees. This system also involved building mounds of enhanced soil and the training of the yam vines about the pyramid-shaped trellis as they grew. This handling of the vines seems to induce thigmomorphogenesis (i.e. the influence of mechanical stimuli on plant growth and development) that causes a thickening of the vine and shorter internodes resulting in greater leaf area and eventually a bigger yam harvest. When we compared the harvest of yams grown by this method as opposed to simply allowing plants to climb up sacrificed trees, we found that the traditional trellis method resulted in a harvest that was 2.5 times greater than the less intensive method that also used up more forest.

Other methods of agricultural intensification involved the construction of ditched garden beds. The deep ditches around the beds served to drain excess water while shallower closed ditches on the beds served to keep the garden moist. These garden beds did not develop a tree canopy, but when renovated, chunks of sod were laid upside down around the perimeter and nutrient rich soil that had accumulated in ditches was transferred on to garden beds. While ancient people could not have known the chemistry of how the management of aerobic and anaerobic conditions of waterlogged soils generates nitrogen and mobilizes phosphorus, they appear to have made use of the process. Some ditched bed systems were used for the cultivation of bananas, which were important in earlier days when they were accumulated in preparation for special events as mentioned in old chants. The banana beds were deeply mulched, and were sometimes followed by yam crops grown on pyramidal trellises in the phosphate-rich mulch left from the banana crops. After a succession of crops, sweet potatoes were often the last major crop grown in nutrient-depleted soils of shifting gardens and ditched beds before the areas were left fallow.

Taro Patches

In lower areas, water was directed into a series of excavated taro patches or flowed into marshes that were utilized for taro patches. Marshland taro patch systems included *Colocasia* taro grown in pits in deeply mulched peat soil. Measurements of the harvests from these taro gardens were made for comparison with commercial taro harvests in Hawaii. Results showed that these gardens, made by old women and young mothers using sticks as tools and no artificial fertilizers, produced yields per area as high as, and in some cases higher than, the yields of commercial taro farms that utilized prime manpower, machinery and chemicals (Falanruw, 1995).

In deeper areas, taro was grown within elevated baskets filled with mulch and soil, or on elevated beds. These elevated beds placed the taro between an upper aerobic zone and a lower anaerobic zone that contributes to the release of phosphates. In coastal areas, and on outer islands, these elevated growing sites also served to raise taro above saline soils. Outer islanders also grew more *Cyrtosperma chamissonis*, the hardier giant swamp taro, and *Alocasia macrorrhiza*, an especially hardy type of taro that can survive on nutrient-poor atoll soils.

In a few especially deep marshes in mainland Yap, a remarkable system of floating taro beds was used. Here the depth of the water precluded the construction of island beds, so beds of floating vegetative material and algae-rich mulch and flotsam were developed. Taro planted in these floating gardens grew hydroponically on mats that quaked when the gardener ventured onto the beds to harvest the taro.

Agroforests

In Yap's coastal villages, taro patch culture comes together with tree gardens. These areas, that must have once been rather swampy given their location at the bottom of the watershed, were drained by systems of ditched beds and series of taro patches that managed water. Soil from low areas was used to construct raised garden beds, house platforms and stone paths, and planted with a rich mix of food, fruit, condiment, fibre, and medicinal trees, shrubs and herbs. Especially important were a number of varieties of coconut trees, a great many varieties of breadfruit trees and Yapese chestnut trees (*Inocarpus fagifer*), bananas, garlic pear (*Crataeva speciose*), citrus and other fruits. Yap's agroforests served as a supermarket, hardware store and pharmacy. while providing the ecological services of a forest. Landscape architecture in the form of systems of ditches running through taro patches adjacent to raised areas made it possible to grow root crops in close proximity to tree crops.

Agrobiodiversity

The traditional agricultural system of Yap incorporated a wide range of species and subspecific varieties. There were about seven locally recognized varieties of coconut trees, each valued for different uses such as varieties with a thin husk and large nut with several litres of coconut water for drinking, and varieties with a thick husk that is especially useful for its fibres, to varieties with an edible husk. Likewise, some varieties were valued for their abundance of thick meat, or reserved for palm wine or medicinal purposes.

Breadfruit is a major staple and Yap has both seeded forms, cultivars without seeds and hybrids of the two. There are over forty locally named varieties of breadfruit valued for different attributes, and also for the different timing of their peak fruiting as this extends the breadfruit season. Breadfruit is a quintessential agroforest tree as it need only be planted once. Once established, it sprouts from its roots and spreads in this manner into favourable habitats and perpetuates itself.

Yapese chestnuts (*Inocarpus fagifer*), called *bu'oy*, provide kernels that yield a good mix of carbohydrates and protein that was especially important when fish resources were limited. Yap is the centre of *bu'oy* production in Micronesia, with at least seven varieties ranging from small nuts to fruits as large as a hand, with kernels that range from white to yellow and from mealy to oily. *Bu'oy* nuts can be stored for later use and when cooked in their husk they remain edible for some time making them a good disaster-relief food to send to other islands after a typhoon.

Four genera of taro are grown, each with its own habitat niches and production systems. *Cyrtosperma merkusii*, giant swamp taro, grows well in agroforests as it is more shade tolerant. It is hardy and can continue growing for some years and thus represents a major reservoir of food. *Colocasia esculenta*, true taro, requires more intensive cultivation and must be harvested at the end of its growing season. Nevertheless, it is especially high in energy and can be quickly multiplied to produce large yields. A bewildering number of varieties of this taro exist. *Xanthosoma sagittifolium* taro called *laiy* in Yap was probably brought to Guam on Spanish galleons, and then to Yap on the canoes that once sailed between Yap and Guam. It is a dryland taro grown in shifting gardens or about the house. It is hardy and can persist untended, but can produce greatly increased yields when well-tended. The main corm can also be left growing while side corms are harvested. White, yellow and red varieties occur. *Alocasia macrorrhiza* is a very hardy taro grown mainly in outer islands where it can survive on poor, rocky atoll soils. It requires special preparation, however, due to the high levels of oxalate crystals in the corms, although varieties have been selected to reduce this problem.

Some of Yap's most variable crops are old world yams in the genus *Dioscorea*. At least five species are present with a great many varieties of each species. Over fifty named cultivars exist, exhibiting an amazing range of colours, shapes, sizes and textures. There is also a remarkable genetic bank of bananas, including those that are best cooked, those that are best eaten raw when ripe, and those that are good eaten either way. Sizes vary widely from small rounded sweet fruit with thin skins likened to turtle eggs, to large very elongated fruit, to short oval fruit. Stalks of fruit are generally borne arched downward, but some are borne straight up. The flesh ranges from cream-coloured to bright orange, the latter being exceedingly high in vitamin A and carotenoids (Engleberger et al., 2006). Today unfortunately, the rich genetic heritage of agrobiodiversity is eroding – at a time when it will be needed to provide resilience in the face of climate change.

Adaptation to Climatic Variables

Yap islands lie in a zone of erratic weather, near the edge of the Asian monsoon. On average, half of the years exhibit a dry season followed by monsoon rains, but in other years the pattern is erratic. Weather and storm patterns are influenced by El Niño Southern Oscillation (ENSO) events that also do not follow a definite pattern. The Yapese system of agriculture seems to be adapted to this erratic weather pattern with the considerable range of systems and crops providing food throughout the year. *Dioscorea* yams are harvested during the winter and spring while another major staple, breadfruit, produces a counterpoint harvest during summer months. The third main staple, *Cyrtosperma* taro, can be harvested throughout the year, but taro patches are generally given a rest during peak harvests of other main crops. Intraspecific varieties extend the harvest period and a number of crops can be managed so that they produce when needed (Falanruw, 1995).

Vulnerability of Agricultural Systems to Sea-Level Rise

While the Yapese system of agriculture is fairly resilient, climate change is already taking a toll. In the outer islands especially, taro patches are suffering from saltwater intrusion and even coconut trees are dying. Storm surges are especially devastating for islets that lie no more than a few metres above sea level. In addition, outer islands have experienced a number of king tides – tides that just keep rising until the sea comes ashore and floods villages and taro patches. Even taro grown on raised beds has been ruined by these events. As a result of the salinization of freshwater and impacts on food production, along with coastal erosion, many outer islanders are migrating to mainland Yap. Mainland Yap, however, is also vulnerable to sea-level rise: most of the best soils are alluvial, which are vulnerable to sea-level rise and storm surge.

Mangroves and Marine Resources

Most of mainland Yap is rimmed by mangrove forests. These remarkable forests are very important as a buffer between sea and land, and for marine productivity. In addition, recent studies by US Forest Service scientists and Yap State foresters have revealed that they are a major carbon sink. While comprising only about 12 per cent of island vegetation, mangroves sequester some 34 per cent of the carbon held by Yap vegetation (Donato et al., 2011, 2012). Much of this carbon is held in deep mangrove muds at and below water level where it cannot be oxidized.

A variety of fishing methods are employed within mangroves, with minimal impact on them. Women collect a variety of crabs and shellfish, including large mangrove clams, *Anodontia edentula*, living in mud tunnels. Small areas within the trees are enhanced using traditional techniques that greatly increase the production of these clams. Considering the amount of support that has gone into culturing heavy-shelled bivalves, the traditional cultivation of mangrove clams with their thin shells and large proportion of meat has considerable potential.

Among the threats to mangroves are modern roads that pass through the interface between freshwater marshes and mangroves. In the past, Yapese constructed porous rock causeways with log bridges that allowed water circulation. Today, heavy machinery is available to make it possible to build non-permeable roads. Such roads interfere with the flow of freshwater needed by mangroves. In addition, such roads together with landfills will prevent mangroves from moving inland as sea levels rise compromising their survival and service as a living buffer against the sea. Observations of Yap's mangroves over the last ten years shows widespread thinning and serious dieback in many areas.

Seaward of most mangroves lie productive seagrass meadows. Yapese of old managed the mangrove-seagrass interface in order to enhance their preferred fishery of largely herbivorous species. A wide range of fishing methods were used throughout Yap's lagoon and beyond the reef. We have collected local names and descriptions of a huge repertoire of fishing methods targeting a wide variety of species. One of the most spectacular remnants of marine technologies are large-stone fish weirs that were used to concentrate fish for harvest when needed (Falanruw, 1994).

A Social System that Managed Exploitation

Resources cannot be managed if there is no way to manage the activities of people that use the resources. In the past, this management was provided by the hierarchical organization of a dense population. Utilization of land, and especially access to marine resources, was managed by this social organization that placed controls on exploitation in the hands of specific estates. In this way, the societies avoided the tragedy of the commons wherein resources available to all are more likely to be overexploited and depleted or damaged. A great deal of Yapese culture centred on the allocation, sharing and exchanging of natural resources.

Today's Situation

Traditional technologies and management of exploitation are fading at a critical time, as they will be needed to adapt to climate change. Some efforts are under way to enhance local management of marine areas and a proposal is being developed to survey traditional food-production technologies and to support community efforts to demonstrate the best of these technologies so that they can contribute to the development of food security in the face of climate change. This would contribute to 'Pacific Solutions' to address climate change: technologies for food production that do not rely on fossil fuels, big machinery and chemicals and thus both mitigate the causes of climate change while also adapting to climate change.

We portray our efforts towards Pacific Solutions as a world adorned by a *lubuw*, the traditional dance adornment, made from a leaflet of a young coconut frond that is placed around the neck of a dancer to protect them and give them grace. Today, the world needs this protection (Falanruw, 2000).

References

Donato, D. C., Kauffman, J. B., Murdiyaso, D. et al. 2011. Mangroves among the most carbon-rick forests in the tropics. *Nature Geoscience*, 4: 293–7.
Donato, D. C., Kauffman, J. B., Mackenzie, R. A., Ainsworth, A. and Pfleeger, A. Z. 2012. Whole-island carbon stocks in the tropical Pacific: Implications for mangrove conservation and upland restoration. *Journal of Environmental Management*, 97: 89–96.
Englberger, L., Schierle, J., Aalbersberg, W. G. L. et al. 2006. Carotenoid and vitamin content of Karat and other Micronesian banana cultivars. *International Journal of Food Sciences and Nutrition*, 57(5–6): 399–418.
Falanruw, M. C. 1974. Island life in the balance. *The Guam Recorder*, 4(1): 48–50.
Falanruw, M. C. 1989. Nature intensive agriculture: The food production system of Yap islands. In Johannes, R. E. (ed.) *Traditional Ecological Knowledge: A Collection of Essays*. Gland: International Union for Conservation of Nature (IUCN).
Falanruw, M. C. 1993. Taro growing on Yap. In *Proceedings of the Sustainable Taro Culture for the Pacific Conference*. Conference on Sustainable Taro Culture for the Pacific, 24–25 September 1992, Honolulu, Hawaii, pp. 105–9.
Falanruw, M. C. 1994a. Food production and ecosystem management on Yap. *Isla, A Journal of Micronesian Studies*, 2(1): 5–22.
Falanruw, M. C. 1994b. Traditional fishing on Yap. In Morrison, J., Gerharty, P. and Crowl, L. (eds.) *Science of Pacific Island Peoples: Ocean and Coastal Studies, vol. I*. Suva, Fiji: Institute of Pacific Studies, The University of the South Pacific, pp. 41–58.
Falanruw, M. C. 1995. *The Traditional Agricultural System of Yap*. Ph.D. Thesis, Suva, Fiji, The University of the South Pacific.
Falanruw, M. C. 2000. Towards a Pacific alternative. *The Yap Almanac 2000*. The Yap Institute of Natural Science, pp. 1–20.
Falanruw, M. C. 2004. Canaries of civilization and a Pacific alternative. *Our Planet*, 15(1), June 2004.
Falanruw, M. C. 2010. Yap State Natural Heritage: A Terrestrial Ecological Assessment, report on field research on mainland Yap and Outer Islands of Yap State, 120 page report accompanying a 42 minute film.
Fletcher, C. H. and Richmond, B. M. 2010. *Climate Change in the Federated States of Micronesia: Food and Water Security, Climate Risk Management, and Adaptive Strategies*. University of Hawaii Sea Grant Report, www.sids2014.org/content/documents/128Food%20Securiry%20Strategy.pdf
FSM (Federated States of Micronesia) 2002. *The Federated States of Micronesia National Biodiversity Strategy and Action Plan*.
FSM (Federated States of Micronesia) 2010. *The Federated States of Micronesia Census of Population and Housing*.

18

Peasants of the Amazonian-Andes and their Conversations with Climate Change in the San Martín Region

Rider Panduro

Located within South America's Altoandina region, Peru encompasses a large number of climate zones, a diversity of ecosystems and a high concentration of the Earth's species and genetic resources. Peru is also culturally diverse, with an indigenous population of 340,000 inhabitants (Chambi and Rengifo, 2011) and forty-two Amazonian ethnicities. For at least 20,000 years, different peoples have lived in the Andean-Amazonian landscape, moving across it and reshaping it through agriculture (Cardich, 1958; Thomson, 1960; MacNeish et al., 1970). The men and women of this landscape have developed a particular cosmovision of their world – one populated with equal, living beings, in which the climate itself is a person with whom you can converse through ritual (see Box 18.1).

This chapter addresses the worldviews of these communities that 'converse' with the climate. We focus on the region of San Martín, which is part of the north-east Andean-Amazonian watershed and home to some 250,000 indigenous and non-indigenous peasants, mainly small-scale agriculturists.

In this area lies a system of very old paths through which peoples connect via networks of interculturality and seed exchange. Called '*los caminos de las semillas*' (lit. the routes of the seeds), these paths that wind through a diversity of ecological zones allow the communities to enhance, disseminate and increase the diversity of their crops and, in this manner, to adjust to and synchronize with variable and diverse climatic conditions. To cope with the increasing competition for land and the increasing variability of climate, some peasant communities are taking the lead in re-creating their traditional agriculture to intensify their use of the land. By using a diversity of native crops and practices compatible with the conservation of nature, they are improving the microclimate and facilitating adaptation to the climatic crisis. These agrosilvicultural spaces are currently being converted into Centres for Mutual Learning, *in situ* laboratories to intensify the conversation with the changing climate and for the exchange of knowledge between numerous peasant families.

The Andean-Amazonian Peoples and the Development of their Cosmovision

In the cosmovision of the Amazonian-Andean peoples, the cosmos is enlivened as an animated world (Kush, 1962), as a living organism. When Luis Paredes Ramírez, wise

Box 18.1 Key concepts of Andean-Amazonian peoples

Many Andean-Amazonian peoples share a number of key concepts that are crucial to the understanding of our case study.

Acompañamiento: the act of living with communities as they manage the *chacra* and local landscape. Performed by technicians, the focus is on allowing the communities to lead as they experiment with managing their local environment (Ishizawa and Arnillas, 2010).

Ayllu: the extended family living in the *pacha* or local world. *Ayllu* includes not only the human communities but also the collectivity of all natural beings and community of deities (Ishizawa and Arnillas, 2010).

Ceremonial centres: These centres are found throughout the length and breadth of the Central Andes, including the countries of Peru, Colombia, Ecuador, Bolivia, part of Chile, Venezuela and Argentina. They were constructed during the major pan-Andean-Amazonian migrations, as well as during smaller or regional displacements. These displacements took place in accordance with climatic changes. It is from these ceremonial centres that the dissemination, adaptation, multiplication and genetic enrichment of the native seeds took place. Otherwise described as archaeological remnants constructed by pre-Colombian populations, these are both grandiose and small works constructed using stone or mud and bricks. These centres are places with high concentrations of energy, constructed under the guidance of the *Yachaq*.

Chacra: The *chacra* is understood not only as an agricultural space but as the scene of creation and strengthening of all forms of life, including forests (Rengifo Vásquez, 2010).

Cosmovision: the concept of life … the manner in which a person or group perceives the basic principles of how the natural, supernatural and human worlds are united. It includes philosophical and scientific assumptions, as well as the ethical principles by which people relate to nature and to the spiritual world (COMPAS and AGRUCO, 1998).

Converse: To converse is to synchronize, to beat the same rhythm, to achieve a mutual empathy in which each one perceives and moves differently in accordance with one's own nature. This corresponds to the characteristic of incompleteness that is typical of all beings of the living Andean world. In the living Andean world, every being has its own wisdom that corresponds with its capacities of perception and emotion, as well as to its own livelihood, and with its life experience. It is in this way that the toad – by its way of living – knows things about the Andean climate that humans – because of their own way of living – will not be able to know. But if humans converse with the toads, they can be enriched by the toad's wisdom, and vice versa. We therefore see that the conversation allows people to access the wisdom of others, notwithstanding the sharing with them of particular capacities of perception and emotion. The conversation is not limited to dialogue, to the flow of words, but comprises all of our abilities to synchronize ourselves, to pulsate to the same rhythm of those with whom we converse (Grillo, 1996).

Pacha: Local world where the Amazonian-Andean collectivities live.

Re-creation: Is the returning to the same as always, but in a different manner, in a reinvented manner and even almost unexpected manner. No 'archetype' is repeated. The same plants and their arrangements will be different. Memory, in this vision of the world, is only a reference. It guides you, but does not oblige you to repeat. The regeneration is synchronizing

> with the cyclical-climatic and agro-astronomic circumstances in a local space (Rengifo Vasquez, 2003: 189).
>
> **Yachaq:** Much of the rich traditional knowledge in the area resides in *Yachaq* – wise, curious elders of the communities. *Yachaq* are recognized for their capacity to 'converse' with nature – in other words, to read the signs of the non-human entities in the *pacha* and make the appropriate offerings to the deities. *Yachaq* provide seeds and knowledge that they share with other families, helping to stimulate and maintain a high genetic diversity in their environment.

elder (*Yachaq*) and Amazonian healer uses the potion from the sap of medicinal trees and enters into a trance, his songs or *takina* or *ikaros*, call the spirits or gods of the waters, hills and of all inhabitants of the Amazonian *ayllu*. His song calls to the *yana rumi* (black stone) to come from the Jalka, a high Andean region, saying: 'the *yana rumi*, his little voice is to come from the Jalka, to come with his covering poncho'. The stone in this ritual moment is alive and sacred.

All who live in the cosmos are at certain moments perceived as natural or spiritual beings; in other words, it is a world filled with living beings of equal status.

> The light rain is female, it is the weeping of the woman; the rain with thunder and lightning is male, it is the crying out of the man, or rather, the rains are also persons. In the same way, there exist many mermaids (deities of the water) in the deeps of the river Mishquiyacu, all of these are thus the mothers of the water, or rather, better said, they are its spirits or souls. Similarly, before one went to hunt and fish in the Laguna del Mundo Perdido, they did not go in making noise, with much caution one would pass around its banks, because if not, it would become angry (infuriated) and would make heavy rains which could kill you by its coldness. It would make like big bubbles of water. So we would go into these places with a lot of respect because they were sacred places. Also the water is an authority, because it calls people together, unites them, as does the Mishquiyacu when it provides fish. Then the inhabitants of Pilluana, Mishquiyacu and Tres Unidos come together; this catch happens when it rains in the upper parts of the salt mine.
>
> Misael Pinchi Sangama, Amazonian peasant, 75 years

This microcosmos is known as *pacha* and the health of the *pacha* depends on the health of all, and in return the health of all depends on the health of the *pacha*. In the same way, we can understand through Misal Pinchi's testimony that the climate itself is experienced as a person with whom one converses ritually and in a holistic manner. Moreover, these disharmonies, such as the current climate crisis, that occur in the *ayllu*, are understood by the peasant families as a result of the carelessness with which the climate or the weather has been treated. For these communities, this is why it is important to focus on re-establishing the health of our *ayllus* and *pachas* through a ritual conversation; in this case with climate change.

It is important to state that this ritual conversing is not limited to a spoken exchange of ideas. Through conversations, different signs or manifestations are exchanged between the three collectivities that inhabit this Amazonian-Andean *ayllu*, in accordance with the specific sensitivities that each being possesses for experiencing its environment, whether these be humans, deities or natural entities.

From this cosmovision, Andean-Amazonian peasant families synchronize their various agroforestry activities together with the circumstantial and spontaneous climate that occurs each year or each agricultural cycle; a synchronization that can be found in their communal calendars.

The North-Eastern Slope of the Amazonian-Andes and the Exchange of Seeds

The San Martín Region contains an enormous network of pathways through which people connect and exchange seeds. These routes are dubbed 'the pathways of the seeds and of the Amazonian-Andean wisdoms', alluding to the nomadic disposition of the peoples.

About 250,000 peasant inhabitants live in this region, of which 10 per cent are indigenous, including among others Kechwa, Awajun and Shawis. Moreover, over the past forty years, the region has been populated by peasants of more than 150 different Andean-Amazonian peoples, in search of new lands for agriculture. In this way, the region has become a space of high cultural diversity.

In this journey across the landscape, the peasant families occupy a diversity of ecological niches distributed across the verticality of ecological zones. They strengthen, disseminate and increase the diversity of their crops to harmonize with the diverse and variable climatic conditions. All these peasant families dedicate themselves to carrying out small-scale agriculture on mountainside soils, and are dependent upon the frequencies and unpredictability of the rains. They cultivate a diversity of native crops in association with various arboreal species, and in rotation with secondary forests, thus in harmony with the agroforestry, agrofestive and agroastronomic cycles.

These Andean-Amazonian pathways allow for the conservation of native germplasm in the habitat where it evolved, relying upon peasant strategies and knowledge about how to maintain and enhance native agricultural diversity. Through these pathways, families redistribute the seeds, exchanging germplasm and agronomic information, essential components for *in situ* conservation. From this point of view, the pathways of the seeds and knowledge are a window, not only on the varieties that a peasant family possesses, but also on the strategies that they use and the motivations that stimulate them to regenerate agricultural diversity.

Connected to these pathways are ceremonial centres where rituals for harmonizing with the agrofestive, agroforestry and astronomical cycles are conducted. In an environment where the relationship between human–nature–deity is central, these rituals help ensure harmonious continuity with nature. This indicates to us that intercultural relations with the landscape exist in a context of deep sacredness and respect.

These pathways are enriched over the years through intercultural exchanges, as new individuals come into the region. For an immigrant peasant like Uver Huatangarí Gamonal, connecting with the locals is important because:

We always help each other. When I recently arrived in this region, the natives helped me a lot. Don Daniel Pinchi and Don Darwin Pinchi, they always provided me with all kinds of seeds, *mallquis* (shoots) of plantain, yucca stems, cane, etc. We have always been friendly with each other, sometimes I went to help them on their plot and came back with some seeds. Now I share with other new friends who come to this region.

In the same way, for a peasant native to the place, intercultural relations bolstered by mutual help, revolves around seed exchange. Darwin José Pinchi Fasando says:

For friendship, mallquis (plantain shoots), seeds, some little thing is exchanged, given ... We always share the *chicha* (traditional drink made from maize). They always grind their mote (white corn), beans and ricachas [arracachas].

In turn, the immigrant peasant, Jesús Jiménez Córdova, states:

When we came for the first time, the natives didn't trust us but little by little we have learned how to act appropriately and have gained their friendship, often making contacts through the friendship between neighbours when we do paid work on their plots, or through the seeds that they give us to plant on our plots.

Meanwhile Edwin Gamonal Sarmiento indicates:

We and the natives have not had any problem since we arrived here. Rather, it is the opposite, as they know some seeds from here and know the region well, they know the time to sow. When we came here, we didn't know these things, these sowing seasons. They taught us. In that way, we avoided crop failure on a number of occasions. With the customs that are upheld here, we have come to understand each other, and we also share our customs with them. We also interact more in the *choba-choba* (collective work) at harvesting times.

Alterations of the Natural Cycles in the High Amazonía of the San Martín Region

These natural cycles with which the Amazonian-Andean peoples lived in harmony, are currently occurring in a much-accelerated manner and altered in their frequencies. This situation is seen to be exacerbated by the degradation of our 'agroecosystems' due to an increased rate of deforestation in Peru, calculated to be 57,000 ha per year. In the year 2000, in the Department of San Martín, more than 2,000,000 ha were deforested, which is 40 per cent of the total surface of the region (Reinders, 2003). Mrs Carmen Guerra Pinchi, an Amazonian peasant, tells us:

In the past, it was not as hot as it is now. The rains were more intense and continuous, back in the year 1950. The trees were higher. When I came to the Centre in the year 1970, there were not that many *chacreria* (an abundance of agricultural plots). Barbas was the Centre. It seemed far away to us and we were afraid to go there and run into jaguars.

Reinforcing this way of living, Mrs Julia Bayona Benítez, an immigrant peasant from the High Andes, states:

The forest is very important because it gives us oxygen, it attracts rain, it provides wood and many other things. For this reason, I always tell them that we have to work nicely, caring for our forests so the rains don't go away. Now, I can already see that the weather is changing because ten years ago the seasons were rainy. Today, it does not rain anymore like it did before. If we do not care for our forests, we will become like a New Lima; a dry land that does not produce well. Therefore, I conserve my forest because from there the water stream is born. I do not destroy it because I am causing damage only to myself. Because without forests my water can run out and my family will be the first to be affected. Moreover, when the forests grow in abundance the rains are constant and this benefits our crops.

Deforestation, and its replacement with monoculture plantations, promotes a high and disorderly immigration of Andean families and reduces spaces for regeneration of native diversity. In 1940, San Martín had 150,000 inhabitants, today it is estimated to be 800,000, and the region has the highest population growth rate in Peru. Added to this is the climate crisis that is altering the frequencies and reducing the volume of the rains upon which the native agroforestry and the various ancestral livelihoods of our rural populations depend.

These alterations in precipitation in the river basins and valleys, and the variations in temperature, are perceived and experienced in a very profound, but also very particular manner by the peasant families. In this regard, the fourteen- to sixteen-year-old peasants from the César Vallejo Education Institute in the Tres Unidos District, Picota province, San Martín Region, remark in remembering their childhood:

In the past, the rains were much stronger and abundant. That is because there were fewer farmers and there were more giant trees. Now it is hot and the harvests are small because migrants have come and they started to cut and burn the trees. Few original forests remain. Crops were more productive and the waterways were clean and supported living beings. Because of the breakdown of the ozone layer ... nature has changed: the waters are contaminated, there is no longer as much rain for our crops. Global warming has increased and the sun is much stronger nowadays. Also, rains are causing large material losses in many sectors and the majority of people are experiencing flooding because of the overflow of the rivers.

Nowadays, at the time of sowing, severe droughts occur which do not allow for the germination, growth and development of our crops. On the other hand, when these crops are maturing and ready for harvesting, there are unexpected rains that cause damage. As a consequence of these droughts and unexpected rains, diseases and epidemics are on the rise that drastically diminish our food production.

Crops such as dryland rice, beans, plantain, yucca and other cereal crops – and even the commercial coffee crops – are currently affected by this environmental problem. The alteration of the basic agroecological conditions required by these crops has additionally increased the occurrence of disease and epidemics. These changes and their impacts are described in the testimonies of peasants such as Edwin Gamonal Sarmiento:

The first years after my arrival, there were few diseases affecting the plantain, maize as well as yucca. But now it is different, because the summers which are more intense are harming the plants and more epidemics are occurring.

There is no doubt that peasant families are suffering a lot in this transitional phase of adapting to the current climate crisis. But Andean-Amazonian peoples are also taking action by intensifying their exchanges along the pathways of seeds and knowledge, occupying diverse spaces across ecological zones. To support them, we are re-creating old practices and strategies for conservation and sustainable use of biodiversity in family *chacras* and in communal landscapes.

Peasant Alternatives to the Climate Crisis: The Case of the Quechuas-Lamas Peoples

Intensifying Exchanges Along the Pathways of the Seed

In the area between the subbasin of Bajo Mayo and Alto Cumbaza, between 350 and 1,200 m above sea level in the Lamas province of San Martín Region, live 400 native Kechwa peasant families. Spread across eight communities, they are connected by a network of diverse meandering paths that crisscross over sixteen longitudinal and ten transversal kilometres. A diversity of plants are found along these paths, including fifty-five varieties of legumes, twenty-two varieties of peppers (*Capsicum* spp.), two types of cotton (*Gossypium barbadense*), three types of peanuts (*Arachis hypogea*), four types of maize (*Zea mays*), six types of greens and vegetables, thirty-seven types of plantains (*Musa* spp.), eleven types of gourds (Cucurbitaceae), nineteen types of roots and tubers, and seventeen types of fruits. Associated with over a hundred tree, shrub and liana species, these plants are moved and regenerated in small and multiple agroforest plots that average 2–3 ha in size. A typical family will manage four to five plots.

The paths cross three agroecological zones which can be divided into a lower (less than 500 m above sea level), middle (500–800 m) and upper zone (more than 800 m). In the lower zone, typical plant varieties found are those most adapted to limited precipitation including beans, plantains, coloured cotton, fruits and maize. The middle zone, which is the most fertile, gets over 800 mm of precipitation per year. In addition to the varieties cultivated in the lower zone, beans, plantains, peanuts, upland rice, cocoa, maize, greens, vegetables, roots and tubers are cultivated in the middle zone. Finally, in the upper zone, there are infertile soils and the climate is temperate with precipitation over 1,000 mm per year. Here, beans, roots and tubers are grown as well as coffee and cocoa. In contrast, plantains and maize are not often grown.

With the changing climate, many plant varieties found in the lower zones are already adapting to the middle and upper zones, just as the crops of the middle zone are growing very well in the upper zone. As the peasants find and plant crops in new zones, they begin to adapt to new associations and cycles, and their rituals and festivities are similarly adapted. The changes across the pathways of the seeds are thus reflected in the relationships between indigenous and non-indigenous families. For example, peasant families in different 'agroecosystems' may borrow plots from each other to sow and harvest different crops. Or they may help each other by providing manual labour for different agroforestry activities.

Intensifying Use of the Land

In the lower and middle portions of the paths, agroecosystems have considerably deteriorated due to the use of monocultures. As a result, families are obliged to move to the upper areas and towards the Low Forest region – areas that are likewise under high pressure from immigrant Andean populations. In following their new paths towards other regions of the Upper and Lower Amazon Basin, these indigenous families and communities link with other *campesino* families. While the different groups may hail from three distinct agroecological and socioeconomic conditions, they share similarities in their way of managing their agroecosystems.

Yachaq are taking the lead in 're-creating' their traditional forms of agriculture to render it more suited for a shrinking land base. From clearing, felling and burning, they are intensifying use of the land, using a diversity of native crops and experimenting with new practices that are compatible with the conservation of nature. The *chacra huerto*, home farmland, is thus refashioned into a more stable agroforest garden. Through this re-creation of their traditional agriculture, they are making these spaces highly productive, improving their soils and microclimates and increasing the resilience of their native diversity and its wild relatives. As they apply these new production techniques, they find new practices of association and rotation of agroforestry diversity, compressing in this way the agroforestry diversity and the existing ecosystems into spaces within the communal landscape.

In the *chacra huerto*, intensive use over long periods is applied to the small spaces of agricultural land that contain a high diversity and density of vegetable and animal species. In the communities' holistic vision, the forest forms part of the *chacra huerto* – the forest is within the plot and the plot is within the forest. Working the *chacra* is understood as the thinning, pruning and shearing of nature in search of the symbiosis between the natural diversity and the cultivated one.

In the *chacra huerto*, the distribution of vegetal diversity imitates and converses with the distribution of the adjacent natural diversity of the agricultural *chacra*. In other words, humans participating in this native agroforestry creation are imitating the architecture of the natural forest, which is heterogenic and diverse, and therefore each *chacra huerto* is also diverse in its composition and distribution.

With the *chacra huerto* as the peasant solution for the conservation of nature and native biodiversity, the peasant family conserves the native agrobiodiversity (including tree species), without excluding introduced species such as greens and vegetables. This diversity strengthens our family, community and intercommunal relations and our distinct traditional strategies for change; diversity always invigorates our seeds and, with that, our food-production sovereignty.

As the *chacra huerto* is rooted in the family, it is the family that develops the relevant expertise. Tending the home plot is done by the whole family with old and new knowledge transmitted across the generations. With the *chacra huerto*, the peasant families have a high productivity per species and a high production per unit area. While production is important, the principal focus is on food security, quality and sovereignty. Likewise, the *chacra huerto* is a space for permanent reciprocal care between human and nature, care that is given in accordance with the natural cycles. The *chacra huerto* is a highly dynamic space and provides a joint option with other different peasant strategies to establish *in situ* laboratories for testing agroforestry alternatives.

Campesino families estimate that each hectare of this type of land use, with respect to volume of production and income generated, corresponds with three hectares of monoculture. It also corresponds with three hectares of forests that are not slashed and burned, thus conserving the primary forests in this way and facilitating the natural regeneration of secondary forests (ARAA/CHOBA-CHOBA and PNCAZ, 2010).

One hectare of these secondary regenerated and conserved forests captures between 50 and 94 tonnes of C ha-1. One hectare of an agroforestry system similar to that of the *chacra huerto* holds between 20 and 50 tonnes of C ha-1 (Lapeyre et al., 2004). It is also estimated that each hectare of this type of land use captures 95 tonnes of carbon in five years with the herbal plants, leaf litter, organic material, trees, shrubs and permanent and annual crops. Additionally, the avoidance of slashing and burning these forests prevents the emission of 337 tonnes of carbon dioxide (CO_2) from one hectare of secondary forest (Reinders, 2003). In this way, the Amazonian-Andean *campesino* families in these diverse altitudes and ecological niches contribute to mitigating climate change with concrete, clean and energy-efficient alternative technologies.

These alternatives do not hinder the economic improvement of *campesino* families due to the significant production generated. While it is true that incomes are moderate and sufficient, they come with high indices of certainty against the climate unpredictability and the instability of markets. If we add to this the income that could be provided to them by carbon credits – when official policies become favourable to their traditional ways and their re-creation of land use – *campesino* incomes would increase, creating a better standard of living for them and their families.

Conclusion

The Andean-Amazonian *campesino* families of San Martín are facing new conditions that may be exacerbated by climate change. Led by the *Yachaq*, they are using their spaces to intensify the use of their land and increasing their own movements across *'los caminos de las semillas'* (the paths of the seeds). It is the *chacra huerto* that best exemplifies the diverse manifestations of the creative force of these wise and curious grandfathers and grandmothers. These include solidarity and reciprocity, dedication and delicacy, inclusion and tolerance, serenity and perseverance, dialogue and conversation, patience and care, empathy and symbiosis, dynamism and vitality, equity and equality, spontaneity and circumstantiality, order and obedience, knowledge and action, seriousness and responsibility, weaving and knitting, reason and sense, renovation and regeneration, hospitality and shelter, rigour and discipline, sensitivity and emotionality – all of these with culturally established limits and within a framework of profound respect for both the familiar and the foreign.

The appropriateness of these lifestyles is evident from the continuity of much of our cultural actions, such as those that are experienced by our peoples: for example, the traditional festivities, the ancestral rituals, the secrets and signs of creation by the crops and animals, the organic structure in its different levels, and the diverse and multiple family and community crafts.

Although these manifestations and many of their acts are conducted silently and may be sidelined, they nevertheless continue to be relevant for our places and for the well-being of the majority of our peoples.

Here the dedicated 'accompaniment' of professionals and of the regional institutions play an important role in supporting these lifestyles and making them visible in the outside world. We are working with the peasants agrosilvicultural spaces and converting them into Centres for Mutual Learning, *in situ* laboratories to intensify the conversation with the changing climate and for the sharing and exchange of knowledge among many peasant families.

Campesino life in this part of Upper Amazonia is lived in the *chacra*, the forest and the water. We accompany them, taking the collectivities as axis: human, natural and sacred, which together constitute the natural collectivity.

We accompany to strengthen and affirm their ancestral and re-created wisdoms and practices of conservation and sustainable use of biodiversity, sustained in their creational cosmovision, which is how the *campesino* families understand their development. Its way of being an 'imitation of nature' sustains and is sustainable in and of itself. To value the contributions that traditional cultures bring to biodiversity conservation and to the mitigation of and adaptation to climate change, institutions must be transformed into companions of these processes, respecting and strengthening a holistic vision of the natural, human and spiritual environment that permits the regeneration of biocultural diversity and the reestablishment of respect. For this to happen, it is likewise necessary for modern ecosystem science to advance towards a holistic science that permits the understanding, valuation and better interpretation, and with that, the respect and appreciation of other non-systemic lifestyles, recognizing our epistemologies between the modern and non-modern, that permits a friendly and healthy world for all. It is this relation that allows for the permanent regeneration of the biocultural diversity in such fragile, very diverse, highly heterogeneous and variable ecosystems as the Peruvian Andean-Amazon.

To collectively confront the uncertainties that characterize the modern world, our peoples are industrious, entrepreneurial and creative. Gold and silver, for example, are well treated in our ancestral way of living and moulded to give the form of corn cobs, peanut shells, potato plants, pepper fruits, the moon, the sun, the forest and domestic flora and fauna. They were offerings to the deities and to Mother Earth; that is to say they were there to strengthen our spirituality and relations of respect with the various expressions of life. Why not recover and strengthen our ways of living that contribute to the health of our ecologies and the well-being of the majority of people on the planet?

References

ARAA/CHOBA-CHOBA and PNCAZ. 2010. *Proyecto - Promoción de la diversidad agrosilvícola nativa para la estabilización de las familias campesinas en las comunidades de la zona de amortiguamiento del Parque Nacional Cordillera Azul (PNCAZ).* Tarapoto.

Cardich, A. 1958. *Los Yacimientos de Lauricocha. Nuevas Interpretaciones de la Prehistoria Peruana.* Acta Prehistórica I. Buenos Aires.

Chambi, W. and Rengifo, G. 2011. *Pequeña Agricultura Campesina en los Andes de Perú.* Lima: Proyecto Andino de Tecnologías Campesinas (PRATEC).

COMPAS (Comparando y Apoyando el Desarrollo Endógeno) and AGRUCO (Agroecología Universidad Cochabamba). 1998. *Plataforma para el Diálogo Intercultural sobre Cosmovisión y Agri-Cultura.* La Paz : AGRUCO/COMPAS.

Grillo, E. 1996. *Caminos Andinos de Siempre*. Lima : Proyecto Andino de Tecnologías Campesinas (PRATEC).

Ishizawa, J. and Arnillas, N. 2010. *Gestión de Proyectos Incrementales. La Experiencia del Fondo de Iniciativas de Afirmación Cultural (FIAC) 2002–2009.* Lima : Proyecto Andino de Tecnologías Campesinas (PRATEC), http://pratecnet.org/wpress/wp-content/uploads/2014/pdfs/Gestion.pdf

Lapeyre, T., Alegre, J. and Arévalo L. 2004. Determinación de las reservas de carbono de la biomasa aérea, en diferentes sistemas de uso de la tierra en San Martín, Perú. *Ecología Aplicada*, 3(1–2): 35–44.

MacNeish, R. S., Nelken-Terner, A. and García Cook, A. 1970. *Second Annual Report of the Ayacucho Archeological-Botanical Project.* Andover, MA: The Robert S. Peabody Foundation for Archaeology, Philips Academy.

Reinders, H. P. 2003. *Experiencias Agroforestales en el Cumbaza, San Martín.* Tarapoto, San Martín: CEDISA.

Rengifo Vásquez, G. 2010. *Los Caminos de la Sal: El Regreso al Territorio Excluido.* Lima: Proyecto Andino de Tecnologías Campesinas (PRATEC).

Rengifo Vásquez, G. 2003. *La Enseñanza es Estar Contento. Educación y Afirmación Cultural Andina.* Lima: Proyecto Andino de Tecnologías Campesinas (PRATEC).

Thompson, G. 1960. *The Foreseeable Future.* New York: Viking Press.

19

Climate Change, Whaling Tradition and Cultural Survival Among the Iñupiat of Arctic Alaska

Chie Sakakibara

Research on the human dimensions of global climate change should consider the way populations that are at risk confront uncertainty through cultural practices. This is a vital point for indigenous peoples around the world but particularly for those in the Arctic region where the effects of climate change are most dramatic (Krupnik and Jolly, 2002; Fox, 2003; Oozeva et al., 2004; Krupnik, 2010). The Iñupiat of Arctic Alaska are especially susceptible to climate and associated environmental changes because they rely on sea ice to hunt the bowhead whale (*Balaena mysticetus*). The Iñupiat identify themselves as the 'People of the Whales' because the bowhead whale has sustained their body and soul for many generations. The bowhead remains central to Iñupiaq life and culture through the hunting process, the communal distribution of meat and other body parts, associated ceremonials and other events that sustain well-being; what I call the 'Iñupiaq whaling cycle' (Sakakibara, 2009, 2010). For the Iñupiat, survival depends upon their right and ability to continue to hunt this culturally salient animal. For the People of the Whales, the bowhead whale integrates all elements of Arctic life: that is the sea, the land, the animals and humanity. Despite social change over the last century the bowhead remains central to Iñupiaq life (Worl 1980; Boeri 1983; Zumwalt 1988; Turner 1990, 1993; Lowenstein 1992, 1993; Hess 1999; Brewster 2004; Sakakibara 2008, 2009, 2010).

In this chapter, I use a humanistic approach to illustrate how collective uncertainty tied to the effects of climate change is expressed and managed in Iñupiaq practices. The chapter relates three case studies from my fieldwork carried out over the years 2004–2008 in Barrow and Point Hope in the North Slope Borough of Alaska (Figure 19.1). Together they show how the human dimensions of climate change, cultural resilience and identity politics are integrated in the Arctic. In so doing, I demonstrate how the Iñupiat reinforce their cultural relationship with the bowhead whale to better cope with an unpredictable future.

Climate change in the Arctic lays bare the essential interconnection of the climate, environment and society. Today the Arctic peoples are standing on the northernmost frontier of global climate change. Iñupiat, their indigenous neighbours in the Arctic and sub-Arctic, and most climate scientists agree that anthropogenic climate change is the major cause of recent alterations in physical, biological, cultural and social systems across the Arctic. The Arctic has warmed at nearly twice the rate of the rest of the world over the past century, and

Figure 19.1 The People of the Whales with a landed bowhead whale near Barrow, Alaska. © Jessica Jelacic. (A black and white version of this figure will appear in some formats. For the colour version, please refer to the plate section.)

scientists predict that warming trends in the Arctic will continue to outpace other global regions. Research predicts that sea ice extent on the Arctic Ocean will decrease dramatically by more than 50 per cent over the twenty-first century and snow and permafrost cover is projected to continue to decrease to at least 10–20 per cent of the current coverage (Arctic Climate Impact Assessment, 2005). Native peoples are cognizant of these probabilities and, thus, are concerned about the actual and potential impact of climate change on their cultural, spiritual and economic health. The Intergovernmental Panel on Climate Change's (IPCC) Fourth Assessment Report (2007) concurred that change in ice, snow and glaciers would have a tremendous impact upon Arctic people's subsistence. Updates to these reports suggest that changes are occurring even faster than anticipated (Arctic Climate Impact Assessment, 2005): in 2007, Arctic sea ice reached a record low (NASA Earth Observatory 2007a, 2007b), and in 2008, both the north-east and north-west passages were ice-free for the first time in recorded history (Revkin, 2008). In September 2010, the minimum level of sea ice is the third-lowest ever recorded in the Arctic Ocean (National Snow and Ice Data Center, 2010). As of 16 September 2012, the Arctic sea ice extent has fallen below its previous record low and the thickness of the ice cover is also in steady decline (Viñas, 2012).

Figure 19.2 Traditional Iñupiaq drum music performed at Nalukataq, the midsummer whale feast in Barrow. (A black and white version of this figure will appear in some formats. For the colour version, please refer to the plate section.)

Where I work in Arctic Alaska climate change increases environmental uncertainties and intensifies human–whale emotions. This passionate outpouring is evident in Iñupiaq efforts to foster traditional and newly invented events and performances that centre on the bowhead whale. My research suggests that vulnerable populations can and do confront environmental uncertainty by reaffirming their cultural identities and traditions, by strengthening bonds with the past to better adjust to an uncertain future. It is my hope that my studies reveal how the Iñupiat make conscious and unconscious efforts to secure their kinship with the whales through traditional expressive culture (Sakakibara, 2008, 2009, 2010, 2011). Embracing their values of tradition and innovation, the Iñupiat mobilize their cultural expressions as an elastic bond to preserve and develop their relationship with the whales (Figure 19.2).

Home is Where the Story Dwells: Storytelling and Drowning Home

The Iñupiat are masters at transforming their homes of snow, ice, tundra and gravel into humanized and inhabited places, and storytelling is the primary means of facilitating this process (Basso, 1996). For *Tikiġaqmiut* (Point Hope villagers), stories and the very act of narrating them help connect people to places, people to people, and ultimately people to the environment. In this way, storytelling remains and enhances its cultural ability to communicate meaning in a way that both produces and reproduces a sense of place. An examination of Iñupiaq storytelling clarifies how environmental change is culturally manifest through tales of the supernatural, particularly 'spirit-beings' or 'ghosts', which confirm human ties to place. With their environment becoming more unpredictable, the Iñupiat find themselves interacting with new spirit-beings in their drowning home. Storytelling, by renewing the kinship between humans and their land, has become a critical form of cultural adaptation that represents an enduring quality of Iñupiat lifeways. Although the rising sea may be eroding the ancestral land, storytelling weaves old and new homes into a viable place of cultural survival.

When *Tikiġaqmiut* lost their primordial homeland to the sea in the 1970s, the Village Council decided to move their settlement of *Tigara* (today known as Old Town) to its present location, New Town of Point Hope. Old Town used to contain sod houses made of whale jawbones. 'We felt like [we were] living in the belly of a whale,' said Pamiuq, now fifty-two (in order to preserve confidentiality and in view of potentially sensitive subject matter of this section, I use pseudonyms for my interviewees). Though only two miles away from the original location to the south-east, the impersonal rectilinear grid of the family houses in the New Town only serves to remind the villagers that their home lies elsewhere. Following continuous storms, flooding and erosion, the place that rooted the Iñupiat to their land is now underwater. As he pointed to the former shoreline, Pamiuq, barely audible over the howling wind, said, 'I hear the ocean getting closer day by day.' Born and raised in Old Town, Pamiuq is one of a handful of villagers who remains in his original family home: 'I don't have anywhere else to go … This [place] is all I am, and this is my home, nothing else.' As we spoke, waves heaved along the stony beach. As Old Town disappears into the sea, the Iñupiat spiritual relationship with the land and their cultural identity are threatened.

Embracing nearly 800 people, Point Hope is the north-westernmost settlement of North America and one of the longest continuously inhabited places in the Americas. Located approximately 125 miles north of the Arctic Circle, the Tikiġaq Peninsula has been occupied at least since 500 BCE (Rainey, 1947; VanStone, 1962; Giddings, 1967; Foote, 1992). Despite being the largest of the North Slope villages outside of Barrow (population approximately 4,200), *Tikiġaqmiut* maintain an active subsistence lifestyle. In addition to inland and coastal fish, animals and birds that they harvest, Point Hope is a traditional whaling community. Boys and girls go through family-based apprenticeships at an early age, and upon graduation from Tikiġaq High School, many youths specialize in daily subsistence activities in the village. Created in 1977, New Town of Point Hope clings to the tip of the peninsula on a long spit of gravel jutting west into the Chukchi Sea. Point Hope's other name, *Tikiġaq*, means the index finger (Lowenstein, 1992), and the peninsula points towards the migration path of the bowhead whale.

The Iñupiaq homeland has been experiencing a major transition since the 1977 relocation from their original settlement. With the loss of the majority of its original homeland,

Point Hope is one of the prime examples of changing communities, and it is just a matter of time before the remainder of Old Town vanishes into the Chukchi Sea. The shorefast ice retreats too soon and exposes more open water to the action of the wind, and brings waves towards the land. Normally, the ice provides a buffer for the coastline and helps prevent erosion (George et al., 2004), but now the land is exposed and vulnerable without sufficient sea ice to protect it. The resulting coastal erosion increases the impacts of sea-level rise, estimated to climb by nearly three inches every century due to warmer seawater and glacier melting (Weller, 2004; Overpeck et al., 2006). Worse still, the rate of sea-level rise is increasing (Meiser and Dyurgerov, 2002; Otto-Bliesner et al., 2006; Overpeck et al., 2006). A 2004 government Accountability Office report found that of Alaska's 213 native villages, 184 are battling floods and erosion (Knoth and Whitty, 2007). Another assessment foresees that in the coming decades, Alaska will require US$6.1 billion to relocate climate refugees and repair global warming's domino effect of fallen bridges, burst sewer pipes and disintegrating roads (Knoth and Whitty, 2007). As with coastal erosion and drowning homes, deteriorating sea ice also affects the emotional bonds to place among Arctic dwellers. The hunters in Alaska, Canada, Greenland and Chukotka struggle to find new and safe hunting routes, and they often encounter weather that is less predictable than in the past. Due to some of these changes, particularly fragile sea and lagoon ice, men and women have fallen through the ice and drowned in early spring or even in the winter when in the past such ice would have been safe for travel.

During my fieldwork and research on Old Town of Point Hope, I heard many villagers express their concerns and uncertainties in various ways: 'Overnight, everything was gone. Bang, a storm took our houses, one right after another,' one of my interviewees described. To him and to many other villagers, the disappearing Old Town still remains a special place which they call home. 'What's going on is [the] destruction of the oldest continuously known people and place. You're looking at them right now,' said Rayna, an elder of sixty-eight. Point Hope also contains some of the richest archaeological sites in Alaska that confirm the route that the ancestors of today's Native Americans took millennia ago (Foote, 1992). Because of rising seas, the coast is eroding, washing away forever precious artefacts and human memories along with history. Furthermore, continuous destruction and loss of *qalgis* (ceremonial space/buildings) and *siġluaqs* (underground caches where whale meat and other perishables are stored) have become a serious concern that influences their lifeways as well as their links to the land.

The 1977 Point Hope relocation marks a major shift in the direction of traditional supernatural tales. Ernest Burch (1971), who conducted fieldwork in several Iñupiaq communities in the coastal district between Point Hope and Deering during 1969–1970, found that the villagers considered non-residential areas of the tundra to be haunted. Burch's study illustrates how widespread social changes altered Iñupiaq spiritual experiences after the nineteenth century. He found that since many of the beliefs of early Christian missionaries were compatible with indigenous beliefs, the Iñupiat synthesized the two worlds. Burch emphasized how this synthesis happened without eliminating 'nonempirical beings' from the indigenous worldview. I found Burch's study valuable as there was substantial uniformity in the nature of composition of the 'nonempirical' environment during my fieldwork. My experience in Point Hope, however, convinced me that the Iñupiat do not necessarily perceive supernatural encounters as 'nonempirical', but rather as 'empirical' ones; in other words, they tell stories about actual experiences that reflect genuine perceptions and emotions.

Point Hope residents told me similar stories between 2005 and 2008 that they told Burch before the 1977 relocation. But the spatial distribution of humans and supernatural beings had been reversed since Burch's time. The place that was once 'haunted' is now the people's home (New Town was built on the tundra), and the people's old home (Old Town) is now inhabited by spirit-beings. Pamiuq, as well as eleven other elderly collaborators who now live in New Town but continue to call Old Town home, pointed out to me that the spirit-beings have themselves become 'restless'.

This observation is becoming increasingly common among the *Tikiġaqmiut* storytellers, especially those who were born before the relocation. In addition to the twelve elderly Old Town interviewees, many younger New Town villagers associated this shift with the changing environment. Moved by anxiety, villagers say the spirit-beings have begun to migrate from the tundra to Old Town. Four local hunters, the women at the parsonage and several youths recognized this geographic inversion. Although Old Town is mostly a non-residential area now, villagers witnessed spirit-beings, including the Little People, ancestral spirits, dragons, mermaids, spirit-beings shaped like animals, spirit-beings with no particular shape and spirit-beings shaped like humans. With the stories' emotional and sensational appeal to the villagers themselves as well as to outsiders, supernatural stories give high visibility to places and human sense of place; as ghosts haunt places, the increasing number of supernatural tales, perceptions and experiences set in the original settlement make the villagers' bond with Old Town more tangible.

Prior to the current environmental change, most introductions affecting the Iñupiaq culture and lifeways came from the lower forty-eight, such as Christianity, snowmobiles and canned food. Burch's study indicates that the Iñupiat had no trouble accepting these items into their lifestyles and making them their own. In other words, their lives have been built on change.

For the Iñupiat, the content of stories and the very act of narrating them helps connect people to place, people to people, and ultimately people to whales as the whale often serves as a metaphor of their homeland (Lowenstein, 1992). Stories serve to maintain a relationship and anchor the Iñupiat to their drowning home for old and young alike. Overall, storytelling is an expressive form of adaptation that reflects environmental uncertainties and the long-term well-being of the Iñupiaq homeland. There is a particularly strong reciprocal link between a human sense of place and the presence of spirit-beings in their storytelling. The types of stories and modes of telling them reveal the villagers' uncertainty about the future based on their knowledge of the immediate past. As their environment becomes more unpredictable, the Iñupiat engage with new spirit-beings in their drowning home. These stories clearly help the villagers process environmental changes as a way to retain and even grow a kinship with their drowning home, and also serves as an example of adaptation in studies of the human dimension of global climate change.

'No Whale, No Music': Whales, Performance and Climate Change

Another example that reveals the reflection of climate change upon traditional Iñupiaq culture involves musical practices, performances and emotional transformations among performers and community members (Sakakibara, 2009). To the Iñupiat, 'the whale is the drum and the drum is the whale' because the traditional drum membrane is often made from the linings of whale livers, stomachs or lungs (Sakakibara, 2009). Influencing the volume and seasonality of successful whale hunting, environmental change heightens Iñupiaq anxieties in three major ways: the availability of drum membranes; disruption of the whaling cycle because an unsuccessful whaling season implies no celebrations and/or no musical occasions; and social disharmony and mutual distrust as a result. Drum performances and their recent resurgence in community events indicate the increasing importance of whales, not their demise. Specifically, the Iñupiat newly endow their performance with an invitation for the whales to join their domain by reversing the human–whale relationship: traditionally, it was the whale that brought music and festivity to the people, but it is now the Iñupiat themselves who bring music to the whales to repair the broken whaling cycle as an act of collaborative reciprocity.

Abundant whale harvests bring the season of celebration with songs, dance and food. The productive 2005 spring whaling season in both Barrow and Point Hope brought a vibrant summer to these whaling villages. With sixteen whales landed, Barrow had four *Nalukataq* (midsummer whale feasts) days, each hosted by three to four successful crews. The villagers shared the whales, acknowledged the bounty and enjoyed the blanket toss. The drum dance in Barrow, however, had been cancelled due to continuous deaths in the village for the first three nights of *Nalukataq*. Julia and Jeslie Kaleak, Sr, one of the *umialik*s (whaling captain and his wife) whose crew gave up its dance, opined that this decision was out of respect to Christianity (20 July 2005). However, the joy and gratitude to the whales broke out on the very last night.

On the last *Nalukataq* night, a drum dance took place in the outdoor *qalgi* immediately following the blanket toss. It was so spontaneous that it seemed like the emotional flows of the whalers could never be contained. Although the execution of the drum dance was never announced to the public in advance, the drumbeat immediately lured many villagers into the circle. The drumbeat kept the village awake, excited, and enlivened until dawn raised the sun that had stayed shining low over the northern horizon throughout the night.

In the same year in Point Hope, five successful whaling crews that had landed seven whales participated in their own *qalgis* according to clan affiliation. Their *Nalukataq* (also called *Qagruk* in Point Hope) continued for three full days, involving all villagers and visitors in appreciating the whales through elaborated rituals and performance. The communal fulfilment was revealed in the blanket toss and eventually in the drum dance. In Point

Hope, sharing the whale and blanket tossing take place in the two separate *qalgis*. After the feast is over, the two clans from the two ceremonial grounds get together in one location for the drum dance. For this celebration, Tikiġaq High School gymnasium usually serves as a festive indoor *qalgi* to embrace both clans in one spot. The drum dances in both Barrow and Point Hope felt like villagers were nourishing another productive whaling season for the following year. This active whaling cycle, however, was broken rather abruptly in early 2006 with unexpected sea ice conditions.

The cultural continuity of the North Slope villages was disrupted in spring 2006 due to the difficult whaling season. 'It was one of the most impoverished seasons across the North Slope Borough coastal whaling communities for the past thirty-five years,' recalled an elderly whaling captain (23 June 2006). He continued: 'We have faced difficult seasons before and have had, of course, to adapt. But it had never been bad this much.' While this single statement cannot encompass all of local history, this situation sheds light on the three major problems that Iñupiaq performance confronts today: anticipated struggles in acquiring drum-membrane materials, disturbances in the timing of whale-related ceremonies and, more significantly, in the immediate continuity of ceremonies themselves. However, Iñupiaq attempts to cope with these new challenges with the continuity of tradition: drumming.

The whale drum definitely remains at the core of Iñupiaq performance, and climate change inevitably impacts the production of drums by influencing the level of whaling success. Pete Frankson of Point Hope noted: 'Whales get us dance. Whale skin is the best material for our drums' (19 June 2006). Frankson fears that difficulties in obtaining the correct material would directly influence Iñupiaq cultural survival and would eventually result in changes in the drumming tradition. Although the Arctic region has repeatedly gone through climate change historically and 'the death of hunting' has been predicted several times through oral history, one of the enduring aspects of Iñupiaq survival skills is flexibility (Bodenhorn, 2001). That said, the zero whale harvest in 2006 brought the Point Hopers no fresh supplies of whale membranes for the year, and drums need to be made annually. Without new supplies, the drummers were forced to use those from a previous time with uncertainties. As Ernie Frankson, Sr stated earlier, the whale-skin drums need meticulous care to maintain good condition. Dry Arctic air often makes the material fragile, and frequent repair or complete replacement of a membrane needs to be planned ahead of time. Additionally, in spite of its immense size, the portion of the whale stomach that can be used ideally for drum-making is limited. Considering the relatively large number of drums (five to twenty) required for ceremonial occasions, the decline in whale harvest may result in a serious loss of this crucial medium.

By influencing human–whale relations, drastic environmental changes may cause social disharmony in the whaling cycle. In 2006, the drum dance to celebrate the whales did not occur in either Barrow or Point Hope, and Point Hope was unable to host any *Nalukataq* at all. Barrow hosted only one *Nalukataq*, in honour of the three whales they had managed to land. At this feast, naturally, the quantity of whale meat in distribution was small compared to the year before. Additionally, with only three crews hosting the feast, despite their every effort and enthusiasm, the event itself appeared less vibrant. All crews related to the successful *umialiks* helped out to lead the feast to success, but undoubtedly the ratio of the servers to the served was out of balance. The drum dance was originally to follow the feast and blanket toss at night. In both Barrow and Point Hope, the 2006 devastation in whaling confirmed that the environment is a potent factor in determining the shape of Iñupiat well-being, musical performance and even their cultural identity.

While it is the most significant event in the whaling cycle, *Nalukataq* is not the only ceremonial occasion that is vulnerable to environmental change. As the oldest community of Iñupiat country, Point Hope retains a distinctive set of ceremonies and rituals that have been passed down in the village for centuries. In the early fall of a successful whaling year, the villagers traditionally enjoy *Qiñu*, the 'born of ice' ceremony at each *qalgi* in appreciation of the whale's tail (*akikkaq*). For *Qiñu*, when a whale is caught in the spring, its tail is cut off and then brought to a place near the captain's *siġluaq* and covered with *muktuk*, an Iñupiaq word that means the combination of the whale skin and blubber, or the underlying layer of fat below the skin. In particular, *muktuk* is indispensable for the people's survival, as it cleans blood and provides necessary vitamins A and C, thiamine, riboflavin and niacin (Freeman et al., 1998). Whale is a critical food among circumpolar peoples and is an essential element of their long-term survival. The ceremony is a renewal, deeply tied to subsistence activities and held in honour of the whale's spirit. Accompanied by an overnight drum performance, the *Qiñu* ceremony expresses gratitude to the whale and lets the whalers make a full circle into the following whaling season.

As its name and seasonality indicate, *Qiñu* is reliant upon the environmental cycle and can only be conducted when the first slush ice touches the shore of Point Hope. As the unpredictability of the climate increases and the formation of fall sea ice is delayed from late September to early November, many aspects of this sea ice-related ceremony now need to be delayed, modified or eliminated. Esther Frankson, elder of the community and a female whaling captain, shared her concern with me: 'We are not allowed to start our preparation for the following whaling season until the completion of *Qiñu*' (20 August 2005). More seriously, Suuyuk Lane, Sr, another elder, told me that the permafrost melting of *siġluaq* spoils the sacred *akikkaq* while it is being stored, which endangers the continuity of the ceremony itself (20 August 2005).

Another major facet of Iñupiaq lifeways threatened by unexpected change is mutual trust among community members. In fact, many villagers largely attributed the 2006 whaling collapse to taboo-breaking in the community during a ceremonial dance. Tensions arose because some villagers were concerned that someone revealed ceremony protocol to outsiders. The villagers' minds were dwelling on what they should or should not do to retain their cultural identity. An unsuccessful whaling venture sows seeds of distrust among community members. At Point Hope in 2006, it planted substantial uncertainties and mutual distrust that disrupted communal unity.

The Iñupiat of Barrow, in spite of their unsuccessful whaling, invited *Tikiġaqmiut*, other North Slope Borough villagers, and anyone who wished to attend to share the whales on *Nalukataq*. Many Iñupiat recognize and anticipate the link between climate change and their changing relationship to whales. George Ahmaogak, Sr, former mayor of the North Slope Borough, reminded me of this anticipated concern:

> Without whaling, our society will experience severe social disorder, nutritional problems ... We need whaling and the associated festivals to keep our connection with the whales. If [there is] no whale, [there will be] no festivals, and [the absence of whales and festivals] will cause a severe depression in this community.
>
> (20 September 2005)

Fannie Akpik also believes that it is time for the Iñupiat to make music to reconnect the human mind with the whale's because 'Now our community needs to be stronger than ever' (10 July 2005). Traditionally, the Iñupiat believe that without any whale harvest, they should refrain from conducting performances or festivities. Yet, according to Akpik, music now has a capacity to help people adapt to the changing environment:

> If we need to change things to survive, then we ought to. If we have no whales, then we really do need to have more music. It would help us and our souls tune into the mind of the whales. We also say the whale is always looking for a village with good music. Drum music may become our last hope to keep our relationship with the whale.
>
> (10 July 2005)

Although neither Barrow nor Point Hope was able to host the drum dance at Nalukataq in 2006, Barrow was privileged to host the Inuit Circumpolar Conference (ICC or *Inuit Issittormiut Siunnersuisoqatigiiffiat*) in July of the same year, welcoming international delegations from the six Arctic nations. The ICC is an international and inter-tribal organization that was originally founded by Eben Hopson Sr's vision of an Arctic home rule. The Iñupiat dancers organized four nights of drum dance and sharing of previously accumulated whales retrieved from their *sigluaqs*. Performance became a way to communicate with other pan-Arctic culture groups and a way to invite the guests to the Iñupiaq land. The Iñupiat-whale integrity was reactivated and reconfirmed through the circumpolar gathering around the drumbeat.

Muktuk Politics and Climate Change

The growing concerns over increasing unpredictability of climate, weather and environment have the People of the Whales connect cultural expression to political activism. This is where the ICC comes into the intersection of climate change, cultural adaptation and indigenous cultural rights. The ICC works to promote Iñupiaq and Inuit rights and to advocate for the preservation of their Arctic homeland. Current indigenous experiences of climate change and self-representation have been empowered by what I have called '*muktuk politics*', an Iñupiaq revitalization of their cultural identity through participation in international debates on climate change, whaling and human rights. A broader Arctic identity politics has its roots in the 1970s when Iñupiaq whalers successfully defended their whaling rights for the first time (Sakakibara, 2011). In this venue, politics, ethnic sentiments, identity and whaling rights have become entangled. Successful political struggles eventually contributed to effective self-representation in more recent global forums addressing climate change and its impacts (i.e. Indigenous Peoples' Global Summit on Climate Change in Anchorage; Amundsen Statement: 2012 ICC Climate Change Road Map).

The ICC has used international forums to remind the world that the Arctic is the home of many people and is a barometer of the world's environmental health. The ICC developed a Global Summit held in Anchorage, Alaska, in May 2009 where global indigenous peoples raised the visibility, participation, collaboration and role of themselves in local, national, regional and international processes. It was a five-day UN-affiliated conference with the involvement of approximately four hundred people from eighty nations. The organizer and participants' intention was to develop strategies and partnerships that engage local communities and formulate proposals for climate change mitigation and adaptation to be presented at the United Nations Climate Change Conference (COP 15) in Copenhagen, Denmark, in December 2009. Here, they discussed and promoted public awareness of programmes and proposals for climate change adaptation, and assessed proposed 'solutions' to climate change from their perspectives. It is also their goal to develop an Arctic-specific Circumpolar Climate Change Plan within each ICC country, reflecting traditional knowledge and the social and cultural impacts of climate change.

These voices from the Arctic have become part of the moral and ethical foundation that seeks strong leadership on climate change mitigation policy. In this way, the ICC has served as a vehicle for northern peoples to internationally represent indigenous perspectives at the COP 15 in addition to the meetings they hosted immediately before the international congregations in Barcelona, Bangkok, Bonn and other locations. The ICC began working with the UN Framework Convention on Climate Change (UNFCCC) and others to organize Arctic events. Prior to the COP 15 and the Global Summit meetings of 2009, the ICC called upon global leaders to listen to their northern voices through various meetings and by drafting an international position statement on climate change titled 'Inuit Call to Global Leaders: Act Now on Climate Change in the Arctic', co-signed by the ICC's chair and vice-chairs from Alaska, Canada, Chukotka and Greenland. These visions are also enhanced by the Amundsen Statement, which is also known as the 2012 ICC Climate Change Road Map. The climate change scientists and indigenous representatives at the Amundsen workshop pointed to the importance of the ICC developing a climate change strategy to build on its existing expertise and achievements in relation to environmental stewardship, international climate change negotiations and the protection of indigenous traditional and intellectual rights.

The environmental plight has facilitated a broader pan-Arctic unity among diverse peoples who share a common set of concerns and who share an interest in long-term survival strategies. Indigenous Arctic organizations are currently building on their knowledge of the past to confront environmental problems of the future, and this means active participation in international assemblies. Currently, the ICC puts climate change and human rights at the centre of its agenda. The Iñupiat realize that their present problems are different than those of the past, but they also believe that the new situation can be better mitigated by strengthening their ties with bowhead whales and their indigenous neighbours. This new form of *muktuk* politics also seeks to bring the cultural plight of the Iñupiaq and their circumpolar neighbours to a wider audience. The indigenous organizations are now ready to combat broader environmental problems, including climate change. As a symbol to unify the Iñupiat, *muktuk* can continue to nourish the People of the Whales for years to come.

Conclusion

The Iñupiat are revitalizing their cultural identity by reinforcing their cultural resilience and participating in international discussions dealing with climate change, whaling and human rights. To show this I explored how expressive culture plays a vital role in building strength among the People of the Whales today. For example, contemporary storytelling among the Iñupiat of Point Hope seeks to cope with an unpredictable future posed by climate change. Arctic climate change impacts Iñupiat lifeways on a cultural level by threatening their homeland, their sense of place and their respect for the whale that is the basis of their cultural identity. Among *Tikiġaqmiut*, environmental change is culturally manifest through tales of the supernatural, particularly 'spirit-beings' or 'ghosts'. The types of Iñupiaq stories and modes of telling them reveal people's uncertainty about the future, but their desire to get there with their past intact. Examining how people perceive the loss of their homeland, I suggest that Iñupiaq storytelling both reveals and is a response to a changing physical and spiritual landscape.

This chapter also described how Iñupiaq drumming and its continuity demonstrate an active adaptation process to environmentally induced cultural erosions. Drum performance and its resurgence in community events mark a strategic rebirth of Iñupiaq cultural identity. Some Iñupiat altered the rules governing their performances to better communicate with the whales and bring them back to them. In this way, the drumming that was originally brought to the Iñupiat by the bowhead whales is returned to the whales through collaborative reciprocity. Inevitably, the future holds further changes and challenges. Yet the Iñupiaq drumming for the whales continues to be an active agent in this process. My research participants' voices demonstrate how performance plays a significant role in expressing and reconfirming their ties with the whales and with their land. Iñupiat whalers now work on their own adaptation to survive environmental challenges via musicality to mend their disrupted whaling cycle. To the Iñupiat, the drumbeat is the heartbeat of the People of the Whales, and the instrument now has power to strengthen mutual exchange between the Iñupiat and the whales.

The Iñupiat-whale relations became an emotional filter that led to the active mobilization of cultural identity. Traditional Iñupiaq values are characterized by their strength in innovation and their faith in adaptability. As my Iñupiaq collaborators continually emphasized, to keep whaling as the environment transforms around them is a way to strengthen their identity and nurture their survival. Inevitably, the future holds further changes and challenges, but Iñupiaq relations with the whales continue to be an active agent in the process. Although the Arctic is the region experiencing climate change faster than any areas of the lower latitudes, we should note that this phenomenon is no longer solely a regional issue. Thus, learning from Iñupiaq cultural adaptation can shed light on future dislocations of human identity outside the Arctic. It is my hope that the voices from the Arctic have the potential to compel other researchers to explore the vital integrity of human–environment relations elsewhere.

Acknowledgements

The author extends her gratitude to the financial assistance provided by a United States National Science Foundation (NSF) Doctoral Dissertation Research Improvement Grant (Geography and Regional Science Program and Arctic Social Science Program, Award ID # 0526168) and NSF Early-Concept Grant for Exploratory Research (EAGER) (Arctic Social Sciences Program, Award ID #0939905), logistical support by the Barrow Arctic Science Consortium and the North Slope Borough Department of Wildlife Management, and grants from the Center for Ethnomusicology and the Earth Institute both at Columbia University. Jessica Jelacic at the Bureau of Land Management has graciously shared her map of the North Slope Borough for this work. The author is especially grateful for the support from the following individuals to complete this manuscript: Igor Krupnik, Arctic Studies Center at the Smithsonian Institution, Douglas Nakashima of UNESCO and Karl Offen at Oberlin College. The author is also appreciative of the friendship with Aaron Fox at Columbia University with whom she made numerous visits to Barrow for a collaborative project on intellectual property rights. Last but not least, her deepest gratitude goes to the people of Point Hope and Barrow, Alaska, for their continuous encouragement and friendship throughout my fieldwork and writing phases – *Quyanaqpak*. An earlier version of this paper appeared as "People of the Whales: Climate Change and Cultural Resilience among the Inupiat of Arctic Alaska" in *Geographical Review* 107 (1): 159–184 (January 2017).

References

ACIA (Arctic Climate Impact Assessment). 2005. *Arctic Climate Impact Assessment*. Cambridge: Cambridge University Press.
Basso, K. 1996. *Wisdom Sits in Places: Landscape and Language among the Western Apache*. Albuquerque: University of New Mexico Press.
Bodenhorn, B. 2001. It's traditional to change: A case study of strategic decision-making. *Cambridge Anthropology*, 22(1): 24–51.
Boeri, D. 1983. *People of the Ice Whale: Eskimos, White Men, and the Whale*. New York: Dutton.
Brewster, K. 2004. *The Whales, They Give Themselves: Conversations with Harry Brower, Sr.* Fairbanks, AK: University of Alaska Press.
Burch, E. S. Jr. 1971. The nonempirical environment of the Arctic Alaskan Eskimos. *Southwestern Journal of Anthropology*, 27(2): 148–65.
Foote, B. A. 1992. *The Tigara Eskimos and Their Environment*. North Slope Borough: Iñupiat History, Language and Culture.
Fox, S. 2003. *When the Weather is Uggianaqtuq: Inuit Observations of Environmental Change*. A multi-media, interactive CD-ROM. Produced at the Cartography Lab, Department of Geography, University of Colorado at Boulder. Distributed by the National Snow and Ice Data Center (NSIDC) and Arctic System Sciences (ARCSS), National Science Foundation.

Freeman, M. M., Bogoslovskaya, L., Caulfield, R. A., Krupnik, I. I. and Stevenson, M. G. 1998. *Inuit, Whaling, and Sustainability*. Walnut Creek, CA: Rowman Altamira.

George, J. C. C., Huntington, H. P., Brewster, K. et al. 2004. Observation on shorefast ice dynamics in Arctic Alaska and the responses of the Iñupiat hunting community. *Arctic*, 57(4): 363–74.

Giddings, J. L. 1967. *Ancient Men of the Arctic*. New York: Knoph.

Hess, B. 1999. The open lead – Iñupiat whale hunting in Alaska – brief unpublished article. *Natural History*. Barrow, AK: n.p.

Knoth, R. and Whitty, J. 2007. Sea change. *Mother Jones*. September/October: 52–9.

Krupnik, I. and Jolly D. (eds.) 2002. *The Earth is Faster Now: Indigenous Observations of Arctic Environmental Change*. Fairbanks, AK: Arctic Research Consortium of the United States.

Krupnik, I., Aporta, C., Gearheard, S., Laidler, G.J. and Kielsen Holm, L. (eds.). 2010. *SIKU: Knowing Our Ice Documenting Inuit Sea-Ice Knowledge and Use*. New York: Springer.

Lowenstein, T. 1992. *Things That Were Said of Them: Shaman Stories and Oral Histories of the Tikiġaq People*. Berkeley, CA: University of California Press.

Lowenstein, T. 1993. *Ancient Land: Sacred Whale: The Inuit Hunt and Its Rituals*. New York: Farrar, Straus and Giroux.

Meiser, M. F. and Dyurgerov, M. B. 2002. Sea level changes: How Alaska affects the world. *Science*, 297(5580): 350–1.

NASA Earth Observatory. 2007a. Record Arctic Sea Ice Loss in 2007, http://earthobservatory.nasa.gov/Newsroom/NewImages/images.php3?img_id=17782

NASA Earth Observatory. 2007b. Beaufort Sea Ice 2005 and 2006, http://earthobservatory.nasa.gov/Newsroom/NewImages/images.php3?img_id=17353

National Snow and Ice Data Center. 2010. Weather and feedbacks lead to third-lowest extent. *Arctic Sea Ice News & Analysis* (October 4, 2010), http://nsidc.org/arcticseaicenews/

Oozeva, C., Noongwook, C., Noongwook, G., Alowa, C. and Krupnik, I. 2004. *Watching Ice and Weather Our Way*. Washington, DC: Arctic Studies Center, Smithsonian Institution.

Otto-Bliesner, B. L., Marshall, S. J., Overpeck, J. T., Miller, G. H. and Hu, A. 2006. Simulating Arctic climate warmth and icefield retreat in the last interglaciation. *Science*, 311(5768): 1751–3.

Overpeck, J. T., Otto-Bliesner, B. L., Miller, G. H. et al. 2006. Paleoclimatic evidence for future ice-sheet instability and rapid sea-level rise. *Science*, 311(5768): 1747–50.

Rainey, F. G. 1947. *The Whale Hunters of Tigara*. New York: American Museum of Natural History Anthropological Papers.

Revkin, A., 2008. Experts confirm open water circling Arctic. *New York Times*, http://dotearth.blogs.nytimes.com/2008/09/06/confirmation-of-open-water-circling-north-pole/

Reynolds, J., Moore, S. and Rager, T. 2005. *Climate Change and Arctic Marine Mammals-An Uneasy Glance into the Future*. Unpublished presentation, Barrow Arctic Science Consortium, Barrow, AK.

Sakakibara, C. 2008. Our home is drowning: Climate change and Iñupiat storytelling in Point Hope, Alaska. *Geographical Review*, 98(4): 456–75.

Sakakibara, C. 2009. No whale, no music: Contemporary Iñupiaq drumming and global warming. *Polar Record*, 45(4): 289–303.

Sakakibara, C. 2010. Kiavallakkikput Aġviq (Into the Whaling Cycle): Cetaceousness and Global Warming among the Iñupiat of Arctic Alaska. *Annals of the Association of American Geographers*, 100(4): 1003–12.

Sakakibara, C. 2011. Climate change and cultural survival in the Arctic: Muktuk Politics and the people of the whales. *Weather, Climate and Society*, 3(2): 76–89.

Turner, E. 1990. The whale decides: Eskimos' and ethnographer's shared consciousness on the ice. *Études/Inuit/Studies*, 14(1–2): 39–52.

Turner, E. 1993. American Eskimo celebrate the whale: Structural dichotomies and spirit identities among the Iñupiat of Alaska. *The Drama Review*, 37(1), 98–114.

Vanstone, J. W. 1962. *Point Hope: An Eskimo Village in Transition*. Seattle: University of Washington Press.

Viñas, M.-J., NASA's Earth Science News Team (19 September 2012). 'Arctic Sea Ice Hits Smallerst Extent in Satellite Era', www.nasa.gov/topics/earth/features/2012-seaicemin.html

Weller, G. 2004. Impacts of climate change. In Nuttal, M. (ed.) *Encyclopedia of the Arctic*. New York: Routledge, pp. 945–6.

Worl, R. 1980. The North Slope Inupiat Whaling Complex. In Kotani, Y. and Workman, W. (eds.) *Alaska Native Culture and History. Senri Ethnological Studies* No. 4. Osaka: National Museum of Ethnology.

Zumwalt, R. L. 1988. The return of the whale: Nalukataq, the Point Hope whale festival. In Falassi, A. (ed.) *Time Out of Time: Essays on the Festival*. Albuquerque, NM: University of New Mexico Press, pp. 261–76.

20

Indigenous Knowledge for Climate Change Assessment and Adaptation: Epilogue

Igor Krupnik, Jennifer T. Rubis and Douglas Nakashima

This book has presented a global set of case studies – from the tropical islands, high mountains, the world's driest areas and polar regions – where indigenous peoples' priorities and issues intersect with the challenges presented by climate change. One question weaves its way through all the chapters: How can indigenous knowledge be brought together with science and technology to respond to climate impacts?

Preparation of this book coincided with the work of IPCC on its Fifth Assessment Report. During these same years, a short guide on indigenous knowledge and climate change was published with the specific audience of the AR5 authors in mind (Nakashima et al., 2012). Along with other efforts, this contributed to the inscription of indigenous knowledge in the Summary for Policy-makers of the AR5 Synthesis Report – a milestone of unquestionable importance (IPCC, 2014).

Building upon the momentum generated by the AR5 and contributing to the constitution of the knowledge platform outlined in the Paris Agreement of the UNFCCC (2016), chapters in this book cumulatively offer *four* messages coming from the world's indigenous peoples. These messages, in tandem with *two* specific challenges, are critical to the new emerging intellectual framework for the global climate change debate (see Rapley, 2012; Future Earth, 2014; Rapley et al., 2014; Nakashima 2015)

The Message of Local Scale

From the climate science perspective, current climate change is viewed as a global phenomenon. Yet people always experience its impacts locally, so that whatever response may come from humanity as a whole, it will be and historically has been a sum of local and regional actions. This effort to bridge from the global to the local is distinctly reflected in IPCC reports, where the section on the 'Physical Science Basis' is organized by environmental or disciplinary domains, whereas the 'Adaptation and Vulnerability' section has a strong focus by region.

The critical role of local and indigenous peoples in anchoring or 'scaling down' global change scenarios often remains overlooked and underappreciated. As a significant portion of the world's population participates in the new global village, indigenous people continue to live 'locally' – in the context of their home habitats, be it small islands, mountain valleys, traditional pasture areas or ancestral forest lands. Every chapter in this volume reinforces the message of the critical ties that indigenous people have built over generations of residence, management, and profound knowledge of the land (or sea) that they call 'home'. Through practice and knowledge, they keep the record of its minute changes and they share it within their communities and pass it among generations. Every study on indigenous peoples' knowledge confirms such a message that international organizations and government agencies are increasingly eager to accept.

Yet every time climate scientists and government managers listen to indigenous peoples (and to *all* people who possess local knowledge), they learn something valuable for their large-scale scenarios. For example, that not all small islands are born equal. Across Island Melanesia, the most important impacts over the coming years will arise from a new pattern of extreme drought and excessive rainfall, rather than from sea-level rise. Therefore, in certain places, climate impact assessment should focus on diminishing horticultural production, rather than on the drowning of local shorelines (Mondragón, Chapter 2). In the same vein, small atoll communities in Micronesia that are threatened by sea-level rise may find themselves facing a new set of problems when or if they relocate to larger, highly populated islands with limited resilience and land resources (Henry and Pam, Chapter 4; Falanruw, Chapter 17). To the surprise of many climate modellers, who anxiously focus on the shrinking of summer Arctic sea ice, warmer summers may be a mere inconvenience to Arctic indigenous reindeer herders; while the prospect of true disaster arises in winter through the threat of rain-on-snow events that can prevent herds from grazing and cause widespread starvation (Mathiesen et al., Chapter 14; Roué, Chapter 16). Almost every chapter in the volume contains similar 'gems' of people's local expertise about their homelands that provide unexpected revelations that overturn conventional global-scale wisdom about climate threats.

By the end of this decade, climate scientists and environmental agencies will be increasingly forced to translate their large-scale global and regional scenarios into real-life, high-confidence local models and plans. Whenever it is to happen, the knowledge and high-resolution 'lenses' of world's indigenous peoples will offer an authoritative template and a philosophy to follow. Such a philosophy will be progressively focused on intimate features of individual habitats, local adaptations, grass-root initiatives, attention to emotional and spiritual well-being and growing self-reliance – all trademarks of indigenous peoples' knowledge systems.

The Message of Self-reliance

Almost every chapter in the book presents a convincing story of indigenous peoples' interaction with their environment as a form of *integrated resource management* (Alangui et al., Chapter 7). Indigenous peoples historically viewed themselves responsible for the health of their habitats in a practical, social and spiritual sense, and in most cases continue to do so. Modern environmental scientists increasingly call such an approach *ecosystem stewardship*; and many argue that it should be promoted to the level of 'planetary' (or Earth) stewardship (Chapin et al., 2009, 2011a, 2011b). Indigenous peoples hardly ever have such an ambitious agenda, since their vision and actions are mostly local (see above). Yet the terms 'steward' and 'stewardship' are gradually becoming popular in indigenous public discourse (see 'Living Earth' Festivals organized by the Smithsonian National Museum of the American Indian) and may transform into a powerful political slogan for indigenous peoples' political movement by the end of this decade.

A clear message from many chapters in this volume is that of indigenous *self-reliance*, the conviction that indigenous peoples can keep their house (i.e. their habitats) 'in order'. Indigenous peoples rightly claim that they possess a thorough knowledge of their environment and that they have maintained enduring links with their habitats via technological and spiritual means over generations. The long-term sustainability of indigenous environmental management and its remarkable resilience is an overall rule rather than exception. Via such long-term practical experience, local peoples commonly know how to sustain their ecosystems: via sophisticated crop and plot rotation (Henry and Pam, Chapter 4; Briones, Chapter 6; Alangui et al., Chapter 7; Swiderska et al., Chapter 11), by following nomadic routes with their herds (Mathiesen et al., Chapter 14; Roué, Chapter 16), maintaining local networks of exchange of food resources, seeds, and crop varieties (Cabalzar, Chapter 3; Damon, Chapter 10; Panduro, Chapter 18) or even singing and drumming for the whales (Sakakibara, Chapter 19). People are not passive victims of change: they are actively observing, experimenting and evaluating alternative land-use and livelihood strategies – and they always did so (Salick et al., Chapter 9). Such individual stories coalesce into a strong common message of *self-reliance*, in sharp contrast to the public vision of indigenous peoples as 'canaries in the coalmine', mere casualties of global change, due to small numbers, vulnerable habitats or simple technologies. It may be just the opposite, as cited by Barber (Chapter 8: 119):

We, the Yolngu [Australian aboriginal peoples] have been here for 50,000 years and we have survived many changes in the past. It is going to affect you guys, not me. Because I've done it in the past. If the store runs out of food, that will simply make people go back to the bush and start eating healthy again.

In fact, as Mondragón (Chapter 2: 38) argues, 'these ... communities appear to be far more likely to successfully adapt to abrupt climate change than the encompassing, globalized societies and institutions that are seeking to help them in this process of environmental crisis and transition'.

We envision this message of self-reliance to be significantly strengthened in the coming years, as world attention shifts from planetary modelling to addressing specific environmental threats in particular geographic areas, and more effort will be necessarily put into local and grass-roots schemes. Here, the strength of indigenous management systems, as well as people's resolute message of self-reliance will feature prominently in global climate change discourse – of course, so long as indigenous voices are not sidelined for political reasons.

The Message of Indigenous Peoples' Rights

Indigenous peoples have been instrumental in adding yet another perspective to the climate change debate, namely on the relation between climate change and human rights, including the still to be fully defined collective and cultural rights. The earliest documents of the 1990s, including those coming from the 1992 UN Earth Summit, addressed this topic but indirectly, mainly via issues of economic inequality, colonialism and the opposition between world's industrial and developing nations. It was then another extension of the global North-South political negotiations and it left no space for the voice of indigenous peoples living within rich and poor nations alike. Such contexts actually left little room for the issue of human rights per se. The transition occurred but several years later, first via the concept of 'environmental justice' (see Perett, 1998; Tsosie, 2007) and later, by facing specific predicaments of indigenous and marginalized communities affected by climate change. It may be summarized in a simple statement: How can we respond to climate change in the ways that are just? (Wildcat, 2013: 511)

There are many reasons to look at the climate change and indigenous peoples' issues through the lens of justice (or rather injustice), as summarized by Margaret Hiza Redsteer and co-authors (Redsteer et al., 2013) and in this book (Chapter 12). Many of the world's indigenous groups, like the Native tribes in the south-western United States, have been disproportionally affected by climate change, because of the previous long history of unjust relations with the respective nation states. Their lands of today are often mere fractions of the former tribal habitats, as indigenous peoples were typically pushed into marginal and extreme environments where the sustainable lifeways were already a challenge. In almost every case, indigenous groups hold to the lands that are more vulnerable than those used by their non-indigenous neighbours (Maldonado et al., 2013). Indigenous groups commonly have a lower income level, less access to external resources for climate change adaptation and mitigation actions, and are often disproportionally subjected to the web of land-use regulations imposed by the management agencies of their respective nation states (Alangui et al., Chapter 7; Redsteer et al., Chapter 12; Roué, Chapter 16).

Of these, the factors of reduced space and strict land-use policies imposed by outside forces are most critical to indigenous peoples' efforts to adapt to the changing environmental and social conditions. In many cases, the prospects are bleak, as the onslaught on their ancestral lands continues:

During the last decades, the continuing advancement of the agricultural frontier has imposed a demographic pattern of instability that hinders the process of building communities, identity, and belonging ... Until recently, indigenous communities had been able to lessen the impact of those movements related to the agricultural frontier by strengthening their ancestral claims. The policy of appeasement and adjustment implemented from the 1990s, together with the processes of impoverishment and disinvestment that the farmers suffered in the zone of conflict ... unleashed a new, massive and aggressive cycle of movement into indigenous lands.

(Cunningham Kain, Chapter 13: 196)

We face many challenges related to the climate change. First, increased access to non-renewable resources in our homelands has created a 'race to the Arctic' and a change in land use. We humbly ask the states that participate in the race to the Arctic to be mindful of that you all base your claims to the resources in the high north on claims to rights to indigenous territories.
(Mattias Åhrén, President of the Sami Council, cited in Mathiesen et al., Chapter 14: 202).

In the worst known cases, like that of the Nyangatom people of southern Ethiopia (Troeger, Chapter 15), the most common threats, again, are discriminatory land legislations, ambitious development projects, and land reclamation for other needs, including internationally imposed biodiversity preservation programmes aimed to combat climate change.

This specific aspect of minority and human rights has been conspicuously absent in the IPCC assessment materials and at many high-level arenas of climate change discourse (Ford et al., 2016). Yet it is instrumental to indigenous peoples' participation in the joint actions to mitigate the impacts of a changing climate, including sharing their knowledge and observations for climate assessment and mitigation policies. Indigenous peoples' claim for environmental justice also helps articulate the plights of other disadvantaged (marginalized) groups within the world population, such as small sharecrop holders and fishermen, women and children, urban and rural poor, and discriminated communities – who have no representation in respective international bodies. We may say that indigenous voices strengthen our ethical and societal sensitivity by articulating key terms, like 'respect', 'rights' and 'recognition'. It helps restore a 'human face' to the climate change discourse that is so overwhelmingly loaded with politics, contested scholarly data and ideological claims.

This is why many chapters in this volume read like a sequel to the Summary of Indigenous Peoples' Proposals submitted to the twentieth UNFCCC Conference of the Parties (COP 20) in Lima, Peru in November 2014 (IIPFCC, 2014). That three-page document spells out requests for indigenous peoples' rights for full participation in all climate change agreements, UNFCCC institutions and related actions; respect for their rights to traditional lands and resources; the recognition of, and respect for, their traditional knowledge; and for support for indigenous community-based monitoring and information systems. It means that barely a few months after the release of the IPCC AR5 the struggle to expand the societal scope of the next IPCC AR6 has already begun.

The Message of Indigenous Innovation

As climate-triggered transitions in indigenous lands and seascapes continue unabated, indigenous peoples' self-reliance increasingly sounds like a model for common action. It is thus notable that almost every chapter in this book addresses indigenous peoples' strengths – be it their traditional knowledge, cultural resilience, kinship and regional networks, flexible nature of their economies or intimate spiritual bonding with Nature. Several chapters explore specific deliberate actions that indigenous communities have undertaken to illuminate their *proactive* response to environmental threats.

Some of those actions mirror strategies applied by other active players, such as local municipalities, town mayors, NGOs and citizen groups that are increasingly taking matters into their own hands. For indigenous groups, it includes the use of alternative resources, reliance on 'green' traditional technologies and age-old 'horizontal' networking in sharing resources and reducing pressure on most vulnerable ecosystems (Cabalzar, Chapter 3; Briones, Chapter 6; Salick et al., Chapter 9; Falanruw, Chapter 17; Panduro, Chapter 18). Others are more experimental and innovative, and rely on the means and technologies that were unheard of in ancestral time:

> (The Sami) are active on the land through their constant innovation of herding practices and techniques, but also in the fields of communication, negotiation and indigenous politics at all levels [...] The Sami are working with their Sami Parliaments, have their own political parties, and are also using their traditional organizations such as the Swedish Sami Herders Association (SSR). They are very active in participatory science ... and in the media. They have their own radio, communicate on blogs, and are developing skills in new domains such as tourism. In Sweden after ten years of conflict, the Sami have also managed to convince the state to negotiate a new deal with respect to the Laponia World Heritage site [that reinstates the Sami as the managers of a territory that they have occupied since millennia – Eds.].
>
> (Roué, Chapter 16: 245)

The Sami herders have succeeded in forging new alliances with scientists, international organizations, even foreign space agencies, like the US National Aeronautics and Space Administration (NASA) to collect and disseminate data on the impacts of climate change on their pasture lands (Mathiesen et al., Chapter 14). The Sami literally pioneered this path for other indigenous groups, who are now increasingly engaged in the documentation of their changing habitats using cutting-edge science technologies (see Krupnik, 2011). It is almost certain that building such connections among indigenous peoples, science groups, NGOs and national research programmes will provide novel insights and data for IPCC assessments, as well as viable alternatives to 'top-down' intergovernmental processes, including the UNFCCC.

Another strategy actively promoted by many indigenous groups is the revitalization of indigenous disaster preparedness and response. These practices that have withstood the test of time are better suited to local conditions than systems developed by international agencies. In certain cases, like small island states (Falanruw, Chapter 17) or on designated indigenous lands, such traditional practices may be easily incorporated into the official current policies and may eventually replace them. As the UNFCCC advances in its recognition of the enduring value of such practices, the politics of global climate assessment will be transformed, including the way recommendations on mitigation and adaptation strategies are elaborated for its Parties. As evidenced in this book, the role of indigenous voices in this process are hard to overestimate.

Challenges for Indigenous Peoples in a Climate-Changed World

The global community, including indigenous peoples, will face innumerable challenges as new uncertainties introduced by a climate-changed world continue to unfold. In closing this volume, we highlight two in particular. First, the need to construct an effective dialogue between indigenous peoples and scientists about knowledge, and between indigenous peoples and the state about decision-making. And second, the need to nurture indigenous knowledge within communities in order to ensure its continuing pertinence, vitality and intergenerational transmission.

The IPCC has provided prominent recognition of the role of indigenous knowledge in climate change assessment, adaptation and mitigation in the Summary for Policy-makers of the Fifth Assessment Synthesis Report (IPCC, 2014). This high-level recognition is a major and long-awaited achievement for indigenous peoples. It creates important opportunities while heightening expectations. The burning question is *how* to successfully bring together indigenous and scientific knowledge to reinforce collective efforts to combat climate change? In this regard, many of the chapters in this volume provide insights into the enduring value of indigenous knowledge, and the synergies that can be developed with science, including through the mobilization of cutting-edge technologies. It is clear, nonetheless, that we are at the beginning of a process, that many challenges remain, and that the objective of a broader-scale, systematic and institutional collaboration is only in the early stages of formulation.

Work in the circumpolar Arctic provides some insight into the long road still to be navigated. The Arctic Climate Impact Assessment (ACIA), published in 2005, clearly broke new ground in its inclusion of Arctic indigenous peoples' knowledge alongside scientific knowledge (ACIA 2005). It nevertheless acknowledged the lack of linkages between indigenous and scientific observations and interpretations, and underlined an absence of trust between the two communities (Huntington and Fox, 2005). Specific efforts were made to address this shortcoming during the International Polar Year (IPY) 2007–2008 through a collaborative research model that involved indigenous peoples not only in research, but also its design and the setting of the research agenda (Hovelstrud et al., 2011). The IPY provided an exceptional opportunity to explore and pilot a variety of approaches to bridge between indigenous and scientific knowledge, including the direct engagement of indigenous organizations and their networks, the development of community-based research and observing systems, the use of digital technologies for joint data mapping, sharing and archiving, and the trialling of 'match-making' between the research priorities of communities and scientists (Krupnik, 2011). Gofman and Dickson aptly portray both the enormous progress made, but also the long road still ahead:

The experience gained in IPY 2007–2008 by many indigenous groups and academic institutions can help them better understand the difficulties inherent to integrating non-academic and academic research [...] This experience opened doors to the next stage in collaborative polar research. The Arctic is a theatre where indigenous organizations are actors rather than props and it is time for them to play leading roles in polar research. IPY 2007–2008 was a baby step in that direction, but it was a giant baby step.

(Gofman and Dickson in Krupnik et al., 2011: 581)

At the global level, the UNFCCC is also struggling with the challenge of establishing a working dialogue between indigenous knowledge holders, climate scientists, policymakers and other key players. To this end, the Parties to UNFCCC are currently debating how to give life to the 'knowledge platform' inscribed in the COP 21 Paris Agreement (UNFCCC, 2015). To help bridge the ontological and epistemological gap between knowledge systems, UNESCO has been piloting Dialogue Workshops, which bring indigenous knowledge holders together with social scientists and climate scientists, in two world regions where climate change impacts are particularly severe: sub-Saharan Africa and the circumpolar Arctic (see, for example, www.climatefrontlines.org and www.arcticbrisk.org). These transdisciplinary dialogues allow for a sharing of observations, impacts and responses, while laying foundations for pooling knowledge and joint decision-making. Similar trials are under way in the biodiversity domain in the framework of the Intergovernmental Platform for Biodiversity and Ecosystem Services (IPBES). In its role as the technical support unit for the Task Force on Indigenous and Local Knowledge, UNESCO introduced Dialogue Workshops in the IPBES thematic assessment on Pollinators, Pollination and Food Production (Lyver et al., 2015). This model aimed at bringing indigenous knowledge holders and experts into dialogue with assessment authors has now been adopted by IPBES to reinforce the role of indigenous and local knowledge in its regional assessments of biodiversity and ecosystems services for Africa (Roué et al., 2017), Europe and Central Asia (Roué and Molnár, 2017), Asia-Pacific (Karki et al., 2017) and the Americas (Baptiste et al., 2017). These efforts to build synergies between knowledge systems are opening new avenues by which the growing recognition of indigenous knowledge can be translated into global action in partnership with indigenous peoples. Significant progress is being made, but it is only the beginning.

While the recognition of indigenous knowledge is rapidly expanding, at the same time, the knowledge systems themselves are under considerable pressure. In many parts of the world, indigenous peoples are witnessing a dramatic erosion of the wealth and dynamism of knowledge in their communities. Declining numbers of speakers of indigenous languages offers one indicator of this demise. This shift to the dominant languages of nation states, to the detriment of a multitude of indigenous languages, threatens the specialized terminologies and lexicons of indigenous peoples that conceptually structure entire knowledge domains (e.g. specialized knowledge of snow and ice regimes among Inuit hunters (Krupnik et al., 2010) and Sami reindeer herders (Eira et al., 2013)).

Perhaps unexpectedly, the expansion of literacy and formal education among indigenous peoples is accelerating the erosion of their knowledge and languages, especially in the many cases where school curricula are exclusively delivered in dominant national languages without the least indigenous content. Formal schooling disseminates the positivist and reductionist ontologies that underpin Western science, including the environmental and climate sciences, to the detriment of the diverse and holistic ontologies of indigenous peoples. Added to the above is the transformation of indigenous ways of life, livelihoods and knowledge systems due to displacement from traditional homelands, external competition for land and resources, and the relentless expansion of market-based economies to the most remote corners of the globe accompanied by Western patterns of consumption.

This book introduces individual stories of indigenous responses to climate change in many diverse areas of the planet. It is a tribute to the richness of indigenous peoples' knowledge and to its many applications. It is also a sign of the new era that ushers in the growing role of indigenous peoples and of their knowledge – in future integrative assessments at regional scale, in the ensuing drive for grass-roots self-reliance, in the value of peoples' rights, and in our collective actions to combat anthropogenic environmental threats.

It also makes clear that the challenge of collective action will require much more than just the will to act together, but also a determined, systematic and long-term effort on all sides to build mutual respect, trust and understanding as basis for knowledge-sharing and co-production. Finally, we cannot celebrate the contribution of indigenous knowledge without also supporting the efforts to keep this knowledge thriving in the face of threats to indigenous peoples' languages, ways of life and territories. These messages and challenges will frame the international climate change debate for the years to come.

Acknowledgements

Part of this Epilogue builds upon ideas first presented in an abridged popular format in Krupnik (2017). The authors are grateful to Margaret Hiza Redsteer and Julie Koppel Maldonado for many helpful comments on the original draft.

References

ACIA (Arctic Climate Impact Assessment). 2005. *Arctic Climate Impact Assessment*. Cambridge: Cambridge University Press.
Baptiste, B., Pacheco, D., Carneiro da Cunha, M. and Diaz, S. (eds.) 2017. *Knowing our Lands and Resources: Indigenous and Local Knowledge of Biodiversity and Ecosystem Services in the Americas*. Knowledges of Nature 11. Paris: UNESCO.
Chapin, F. S. III, Kofinas, G. P. and Folke, C. (eds.) 2009. *Principles of Ecosystem Stewardship: Resilience-Based Natural Resource Management in a Changing World*. New York: Springer.
Chapin, F. S. III, Pickett, S. T., Power, M. E. et al. 2011a. Earth stewardship: A strategy for social-ecological transformation to reverse planetary degradation. *Journal of Environmental Studies and Sciences*, 1(1): 44–53.
Chapin, F. S., Power, M. E., Pickett, S. T. et al. 2011b. Earth stewardship: Science for action to sustain the human-earth system. *Ecosphere*, 2(8): 1–20.
Eira, I. M. G., Jaedicke, C., Magga, O. H. et al. 2013. Traditional Sámi snow terminology and physical snow classification: Two ways of knowing. *Cold Regions Science and Technology*, 85: 117–30.
Ford, J. D., Cameron, L., Rubis, J. T. et al. 2016. Including indigenous knowledge and experience in IPCC assessment reports. *Nature Climate Change*, 6: 349–53.
Future Earth. 2014. *Strategic Research Agenda 2014. Priority for a Global Sustainability Research Strategy*. Paris: International Council for Science (ICSU), www.futureearth.org/sites/default/files/strategic_research_agenda_2014.pdf

Hovelsrud, G., Krupnik, I. and White, J. 2011. Human-based observing systems. In Krupnik, I., Allison, I., Bell, R. et al. (eds.) *Understanding Earth's Polar Challenges: International Polar Year 2007–2008. Summary by the IPY Joint Committee*. Rovaniemi, Finland/Edmonton, Canada: University of the Arctic/CCI Press/ICSU/WMO Joint Committee, pp. 435–56.

Huntington, H. P. and Fox, S. 2005. The changing Arctic: Indigenous perspectives. In ACIA, *Arctic Climate Impact Assessment*. Cambridge: Cambridge University Press, pp. 61–98.

IIPFCC (International Indigenous Peoples' Forum on Climate Change). 2014. *Executive Summary of Indigenous Peoples' Proposal to the UNFCCC COP 20 and COP 21. Lima, Peru. 30 November 2014*, www.iwgia.org/images/stories/int-processes-eng/UNFCCC/Executive SummaryIPpositionFINAL.pdf

IPCC. 2014. *Climate Change 2014: Impacts, Adaptation, and Vulnerability. Part A: Global and Sectoral Aspects. Contribution of Working Group II to the Fifth Assessment Report of the Intergovernmental Panel on Climate Change* [Field, C.B., Barros, V.R., Dokken, D.J. et al. (eds.)]. Cambridge, UK and New York: Cambridge University Press, 1132 pp.

Karki, M., Hill, R., Xue, D. et al. (eds.). 2017. *Knowing our Lands and Resources: Indigenous and Local Knowledge and Practices related to Biodiversity and Ecosystem Services in Asia*. Knowledges of Nature 10, Paris: UNESCO.

Krupnik, I., Aporta, C., Gearheard, S., Laidler, G. J. and Holm, L. K. (eds.) 2010. *SIKU: Knowing our Ice. Documenting Inuit Sea Ice Knowledge and Use*. Dordrecht: Springer.

Krupnik, I. 2011. Connecting to new stakeholders in polar research. In Krupnik, I., Allison, I., Bell, R. et al. (eds.) *Understanding Earth's Polar Challenges: International Polar Year 2007–2008. Summary by the IPY Joint Committee*. Rovaniemi, Finland/Edmonton, Canada: University of the Arctic/CCI Press/ICSU/WMO Joint Committee, pp. 575–92.

Krupnik, I., Allison, I., Bell, R. et al. (eds.) 2011. *Understanding Earth's Polar Challenges: International Polar Year 2007–2008. Summary by the IPY Joint Committee*. Rovaniemi, Finland/Edmonton, Canada: University of the Arctic/CCI Press/ICSU/WMO Joint Committee.

Krupnik, I. 2017. Living on a changing planet: Why indigenous voices matter. In Kress, W. J. and Stine, J. K. (eds.) *Living in the Anthropocene. Earth in the Age of Humans*. Washington, DC: Smithsonian Books, pp. 78–82.

Lyver, P., Perez, E., Carneiro da Cunha, M., and Roué, M. (eds.) 2015. *Indigenous and Local Knowledge about Pollination and Pollinators Associated with Food Production: Outcomes from the Global Dialogue Workshop*. Paris: UNESCO.

Maldonado, K. J., Colombi, B. and Pandya, R. (eds.) 2013. Climate change and indigenous peoples in the United States: Impacts, experiences, and actions. *Climatic Change*, 120 (3), special issue.

Maldonado, K. J., Shearer, C., Bronen, R., Petersen, K. and Lazrus, H. 2013. The impact of climate change on tribal communities in the US: Displacement, relocation, and human rights. *Climatic Change*, 120(3): 601–14.

Nakashima, D., Galloway McLean, K., Thulstrup, H. D., Ramos Castillo, A. and Rubis, J. T. 2012. *Weathering Uncertainty. Traditional Knowledge for Climate Change Assessment and Adaptation*. Paris and Darwin: UNESCO and UNU.

Nakashima, D. 2015. Local and indigenous knowledge at the science-policy interface. In Schlegel, F. (ed.), *UNESCO Science Report: Towards 2030*. Paris: UNESCO, pp. 15–17.

Perrett, R. W. 1998. Indigenous peoples and environmental justice. *Environmental Ethics*, 20(4): 377–91.

Rapley, C. G. 2012. Climate science: Time to raft up. *Nature*, 488: 583–5.

Rapley, C. G., De Meyer, K., Carney, J. et al. 2014. *Time for Change? Climate Science Reconsidered: Report of the UCL Policy Commission on Communicating Climate Science, 2014*. London: UCL.

Redsteer, M. H., Bemis, K., Chief, K. et al. 2013. Unique challenges facing southwestern tribes. In Garfin, G., Jardine, A., Merideth, R., Black M. and LeRoy, S. (eds.) *Assessment of Climate Change in the Southwest United States*. Washington, DC: Island Press/Center for Resource Economics, pp. 385–404

Roué, M., Césard, N., Adou Yao, Y. C. and Oteng-Yeboah, A. (eds.). 2017. *Knowing our Lands and Resources: Indigenous and Local Knowledge of Biodiversity and Ecosystem Services in Africa*. Knowledges of Nature 8. Paris: UNESCO.

Roué, M. and Molnár, Z. (eds.) 2017. *Knowing our Lands and Resources: Indigenous and Local Knowledge of Biodiversity and Ecosystem Services in Europe and Central Asia.* Knowledges of Nature 9. Paris: UNESCO.

Tsosie, R. A. 2007. Indigenous people and environmental justice: The impact of climate change. *University of Colorado Law Review*, 78:1625–77.

UNFCCC. 2015. *Adoption of the Paris Agreement.* (FCCC/CP/2015/L.9/Rev.1), 30 November–11 December. Paris, France: UNFCCC.

Wildcat, D. R. 2013. Introduction: Climate change and indigenous peoples of the USA. *Climatic Change*, 120: 509–15.

Index

adaptation, 1–3, 5–6, 8, 12, 15, 19–20, 23–25, 34, 37, 40, 42, 59, 68–69, 72, 74–75, 82, 84–87, 89, 93, 105–6, 108, 117, 121, 130, 132, 139–42, 145, 149, 152–56, 158–59, 163–64, 166–67, 185, 188–89, 194–97, 202–5, 210, 214–15, 221–22, 224–26, 229, 243, 245, 247, 254–55, 263, 268, 270, 274–77, 283, 285–86
   types of, 87
adaptive capacity, 10–11, 20, 23, 25, 28, 31, 37, 40, 60, 68–69, 119, 142, 162, 166, 171, 204, 221, 222, 225, 229
Africa, 5, 85, 102, 167, 186, 216, 220, 225, 237, 287, 290
agriculture, 3, 11, 42, 51, 54, 64, 76–77, 81, 83, 87–88, 98, 100, 123–24, 129–30, 135–36, 141, 152–55, 157–58, 160, 163–64, 166, 194, 219, 223, 236, 248, 251, 253–54, 257, 261
agrobiodiversity, 157, 248, 251, 262
agroecosystems, 158, 161, 258, 261
agroforestry, 26, 30, 32, 44, 136, 189, 257, 259, 261–62
agropastoral, 123, 125, 129, 141, 219
Akawas River, 95
Alaska, 11, 18, 20, 171, 265, 266–67, 269, 275, 277–79
Aleut, 2
Alocasia macrorrhiza, 30, 249–50
alpine, 125, 131, 136–37, 239
Amazon, 41–47, 50, 55–57, 261, 264
Amerindian, 42
ancestors, 46, 66, 98, 106, 111, 113, 118, 132, 161, 242, 247, 269
ancestral, 4, 26, 44–45, 54, 63, 106, 110–11, 113–14, 116–19, 156, 190, 195–97, 259, 263–64, 268, 270, 281, 283, 285
Andean, 12–13, 19, 83, 159, 161–62, 226, 254–55, 257–62, 264
Andes, 13, 15, 19, 83, 89, 152, 164, 254–55, 257, 259, 264
Apache, 182
Arawak, 44–45

arboriculture, 29–30
Arctic, 2, 9, 13, 15–20, 40, 138, 198, 200, 202–5, 207–8, 210–12, 225–26, 230–32, 245, 265, 267–69, 272, 274–79, 281, 284, 286–89
Arctic Climate Impact Assessment, 2, 16, 212, 245, 286
Arctic Council, 2, 200, 202, 212
Arctic Ocean, 266
arid, 95, 171, 174, 186
aridity, 181, 183–84
Atlantic, 2, 179, 189, 197
Atlantic Multidecadal Oscillation, 179
atoll, 61, 63–64, 68, 72, 144, 247, 249–50, 281
Australia, 5, 38–39, 106, 115, 117, 119–22, 145, 185
   Arnhem Land, 106
   Blue Mud Bay, *see* Blue Mud Bay
   Groote Island, 107
   Northern Territory, 118
   Numbulwar, 107
   Yirrkala, 107
autumn, 9, 127, 131, 204, 244
Awajun, 257

Balaena mysticetus, *see* whale, bowhead
banana, 51, 63, 193, 249, 253
Barents, 202, 204, 213
Barrow, 265–68, 271–74, 277–78
beans, 80, 87, 219, 223, 258–60
bees, 50, 131, 179
biodiversity, 29, 69, 100, 105, 119, 123–24, 129, 136, 141, 153–57, 160, 162–63, 165, 194, 221, 239, 260, 262–63, 284, 287
birds, 29, 45, 51–53, 79–80, 87, 160, 181, 183, 268
   migration of, 51, 87
Birgus latro, *see* coconut crab
Blue Mud Bay, 106–10, 114–15, 117–21
Bolivia, 152, 159, 162, 164, 255
   Cochabamba, 159
Brazil, 1, 50, 54
breadfruit, 63–64, 71, 249–51

breeding, 114, 152, 154, 156–58, 161–62, 166, 189
buckwheat, 129–30
buffalo, 115–16
burning, 48, 51, 54, 137, 146, 148, 261–62, 286

calendar, 33, 39, 41–42, 44–50, 78–80, 86, 88, 123, 125, 127, 132, 145, 150, 193, 221–22
    Tibetan, 124–25, 129, 132, 133
Canada, 5, 11, 13, 16–17, 19, 208, 230, 269, 275, 289
    Banks Island, 207
    Belcher Islands, 231
Cancun Adaptation Framework, 8, 152
canoe, 30, 79, 109, 114, 146
Caribbean, 2, 19, 82, 188–89
caribou, 6, 210, 230–32
cash crops, 134–35, 137
cash income, 85, 107, 131, 137
cassava, 51, 56, 158
cattle, 26, 97, 129, 174–75, 218, 221, 223–24
cereal, 99, 129, 259
ceremony, 46, 183–84, 273
    ceremonial exchange, 31, 66, 70
    ceremonial feasting, 31
    ceremonial songs, 177
Ch'ol, 84–89
children, 98, 114, 183, 188, 196, 218, 224, 284
Chile, 20, 255
China, 123, 127, 137–41, 152–57, 162–64, 166–67
    Guangxi, 137, 154, 156–57
    Guizhou, 154
    Hengduan Mountains, 124–25
    Hong Kong, 134–35, 137
    Shanghai, 134
    Tibet, *see* Tibetan Autonomous Prefecture
    Yunnan, 123–25, 127, 137
Christianity, 39, 270–71
Chukchi, 198
Chukchi Sea, 268–69
church, 98, 134
Chuuk, 58, 62–67, 69–73
Chuuk Lagoon, 58, 65, 67
Chuukese, 62, 64, 66–67, 71
Circumpolar North, 200, 210
climate change assessment, 1–3, 6, 286
coconut, 29, 35, 63, 249–51, 253
coconut crab, 29
Coffea arabica L., *see* coffee
coffee, 26, 76, 80–81, 87, 259–60
Colombian, 47, 255
colonial, 26, 86, 106, 108, 115, 143, 145, 149, 182, 194
Colorado River, 177, 181, 186
communal grazing, 160
community-based, 1, 23, 25, 94, 98, 101–5, 166, 224, 284, 286
Conference of the Parties, 8, 67, 73, 284
    UNFCCC COP 15, 275
    UNFCCC COP 20, 284
    UNFCCC COP 21, 8, 287
conflict, 182, 196, 223, 245, 283, 285
constellation, 13, 51–54, 86
Convention on Biological Diversity, 67, 138
Copoya, 75–81
coral, 58, 61, 63, 67, 70, 144, 161
corn, 81, 87, 159, 177, 179, 182, 184, 258, 264
crocodile, 112, 114
crop, 12, 19, 78, 86–87, 129–31, 136, 152–54, 156–59, 161–64, 167, 219, 223, 226, 248–49, 258, 282
cultivars, 56, 69, 250–51, 253
culture, 17, 30–31, 37–38, 42, 64, 72, 86, 105, 129, 131–32, 137, 148, 156, 163, 173, 182–83, 185, 193, 214, 219, 220, 223, 229, 242, 244, 249, 252, 265, 267, 270–71, 274, 276
customary, 53, 93, 102, 107–8, 145, 160–64, 166, 175, 216
cyclone, 28, 60, 63, 115
    Cyclone Pamela, 60, 63

dance, 88, 113, 253, 271–74
Dayak, 96, 98, 100–2, 104–5
deforestation, 19, 43, 93–94, 102–5, 132, 136, 192, 248, 258
degradation, 14, 48, 61, 93, 99, 105, 112, 137, 173, 192, 258, 288
Desana, 47–48, 51, 56–57
Dioscorea, 30, 248, 251
disaster, 16, 31, 40, 60–61, 73, 119, 216, 231–32, 250, 281, 285
displacement, 26, 32, 73, 119, 220, 233, 287
drought, 32–33, 37, 52, 56, 85, 87–88, 97, 143, 146–47, 149, 153, 155–56, 158–60, 164, 173–74, 176–79, 181–83, 185, 215, 217–18, 220, 224–25, 248, 281
dust storms, 173, 177, 178, 182

earthquake, 35–36
ecosystem services, 139, 153, 162, 211
Ecuador, 255
education, 121, 210, 259
ecosystem-based adaptation, 153
El Niño, 13, 19, 33, 142–43, 150, 251
elders, 11, 34, 46–47, 52, 94, 98–99, 104, 108, 158–61, 171, 173, 176–80, 182–84, 195–96, 217–18, 222, 225, 256
emigration, 70
employment, 70, 108–9, 173
encroachment, 137, 219
erosion, 35, 61, 63, 68, 81, 98, 136, 144, 154, 160–61, 175, 179, 194, 251, 268–69, 287
Ethiopia, 214–17, 221, 224–26, 284
    Amhara, 215
    Omo, 221, 223
    Oromia, 215
    Tigray, 215

ethnic minorities, 4, 137
Eurasia, 198, 199, 203, 208, 210, 212
Europe, 135, 237, 240, 287, 290
exchange, 8, 31, 33–34, 44, 47, 70–71, 145, 147–48, 162–64, 195, 218–19, 223–24, 254, 257–58, 263, 276, 282

fallow, 30, 51, 131, 147–48, 248–49
farmer, 14, 80–81, 152, 156, 161, 164–66
farming, 26, 80–81, 86–88, 99, 101, 125, 154, 159, 163, 175–77, 179, 182, 194, 196
Federated States of Micronesia, 58, 61, 67, 70, 72–74, 247, 253
   Chuuk, *see* Chuuk
   Kuttu, 63
   Moch, *see* Moch
   Mortlock Islands, *see* Mortlocks
   Namoluk, 63, 66, 73
   Satawan, 63
   Ta, 63, 69
   Tonoas, 66
   Weno, *see* Weno
   Yap State, *see* Yap State
Finland, 17, 213, 229, 289
Finnmark, 200–4, 212–13
fire, 85, 113, 143, 148, 231
firewood, 30, 99, 129
fish, 29, 32, 42, 44–47, 50, 52–55, 71, 112–13, 160–61, 192, 194, 239, 248, 250, 252, 256, 268
fisheries, 9, 19, 54, 158, 160–61
fishing, 3, 9, 32, 44, 47, 51, 99, 101, 109, 112, 116–17, 189–90, 194, 196, 212, 229, 252–53
   deep-water, 32
   overfishing, 54
   spear-fishing, 32, 109
floods, 36, 52, 53, 59, 127, 158, 178, 216, 251, 269
food, 3, 28–34, 45, 47, 51, 55, 63–64, 84, 86, 88–89, 97–98, 107, 114–15, 119, 145, 147–48, 154, 157–59, 164, 171, 175, 179, 182, 190, 193–95, 203, 216, 218–19, 223–25, 230–31, 233, 235, 242, 247, 249–51, 253, 259, 262, 270–71, 273, 282
   traditional, 179, 183, 195, 247, 253
food security, 3, 51, 64, 84, 86, 88–89, 158–59, 164, 193–94, 219, 247, 253, 262
forage, 6, 179, 207, 230, 233, 241, 244
   winter forage, 230, 233
forced displacement, 26
forest, 4, 26, 29, 32, 45–46, 51–53, 76–77, 85, 103, 93–105, 129, 136, 146, 148, 150, 156, 160, 192, 194–96, 231, 238–39, 241, 244–46, 248, 250, 259–64, 281
forest products, 129
   sustainable forest management, 98, 103
forestry, 11, 132, 160, 195, 229, 235, 238–39, 241, 245
free, prior and informed consent, 94

fruit, 51, 87, 101, 129, 146, 189, 194, 248–49, 251
full and effective participation, 8

garden, 30–31, 33, 129, 145, 147–48, 182, 248–49, 261
Garifuna, 189
gas, 18, 201–2, 205, 206
gathering, 3, 94, 134, 137, 175, 189, 212, 274
gender, 75, 148, 155, 194, 216, 223
giant swamp taro, 249–50
glacial retreat, 125, 127–28
glacier, 124, 127–28, 140, 269
globalization, 59
goat, 223
gold, 55, 237
grape, 131, 135
grass, 81, 160, 203, 242, 245
grassland, 137
grazing, 6, 99, 131, 137, 160, 175–76, 183, 201–3, 205, 207, 210, 213, 221, 223, 229, 231, 233, 235–36, 281
   grazing land, 131, 137, 202, 210, 221
   grazing permit, 175
   loss of grazing, 207, 210
   overgrazing, 173, 175
   range-sharing, 183
Greenland, 200, 269, 275
greens, 260, 262
Guam, 65, 70–71, 73, 250, 253
Guinea
   Kissidougou, 102
Gwich'in, 2

hail, 127, 130, 216, 261
handicraft, 44
harrowing, 239
harvest, 80–81, 87–88, 129, 140, 146, 148, 158, 193, 222, 248–49, 251–52, 261, 268, 272, 274
healer, 256
health, 3, 19, 26, 32–33, 48, 54, 62, 118–19, 131–32, 161, 164, 190, 194, 230, 256, 264, 266, 275, 281
herbs, 30, 80, 87, 171, 183, 249
herder, 14, 198, 223, 229–31, 233, 238–39, 241, 243–44
herding, 16, 129, 131, 139, 189, 198–205, 207, 210–12, 229–33, 235–36, 238–39, 241, 244–46, 285
high islands, 61, 64–66, 70, 247
high tide, 33, 63–64
Himalaya, 123–25, 127, 130, 136, 141
honey, 97, 99
Hopi, 176, 179, 183, 185–86
horticulture, 29
human–environment, 109, 277
hunter, 14, 56, 106, 108, 111–12, 114
hunting, 3, 13, 48, 99–101, 109, 111–13, 115–17, 119, 175, 189, 193, 225, 229, 265, 269, 271–72, 278
   subsistence, 119

hurricane, 32, 190, 194
  Hurricane Felix, 190
hybrid, 87, 115, 156–57, 164
hydrocarbon, 9
hydroelectricity, 236

ice, 2, 9, 17, 36, 71, 127, 135, 200, 204, 207–8, 212, 229–33, 235, 238–39, 241–42, 244, 265–67, 269, 272–73, 278, 281, 287
interdisciplinary, 39, 141, 185
Indian Ocean, 2, 216
Indians, 4, 56, 176
indigenous territories, 41, 75, 103, 190, 196, 202, 284
Indonesia, 94, 96, 98, 101, 105, 146–47, 150, 195
  Central Seram, 101
  West Kalimantan, 96, 98
inheritance, 28, 32, 81, 102
insect, 47, 52, 80, 87, 130–31, 183, 193, 203
integrated resource management, 96, 281
intellectual property rights, 152, 277
Intergovernmental Panel on Climate Change, *see* IPCC
Intergovernmental Platform for Biodiversity and Ecosystem Services, 287
International Indigenous Peoples Forum on Climate Change, 8
International Polar Year, 200, 286, 289
Inuit, 2, 13–14, 16–17, 19–20, 212, 231, 274–75, 277–78, 287
inundation
  saltwater, 64
  shoreline, 35
Iñupiat, 265, 267–71, 273–78
invasive, 220–21, 223, 248
IPCC, 1, 5–8, 10, 17–18, 39, 60, 68–69, 87, 89, 105, 126–27, 139, 144, 152, 154, 202, 207, 225, 245, 266, 280, 284–86, 288
irrigation, 130, 175
island, 19, 23, 37–40, 61, 62, 66, 68–70, 72–74, 142–43, 150, 211, 231, 247, 253, 281, 289

jaguar, 53

kava, 26, 34
kaya, 153, 158–60, 164
Kechwa, 257, 260
Kenya, 17, 19, 94–95, 105, 152–54, 158, 162, 164, 167, 221, 225–26
  Narok, 95, 103
knowledge
  transmission of, 84, 88, 188
Kuakuail II, 95, 98–99, 105
Kweywata, 146

La Niña, 33, 217
labour (manual), 84, 130, 134, 175, 193–95, 241, 261
lagoon, 29, 32, 66, 145, 192, 248, 252, 269

land
  land rights, 225
land tenure, 28, 66, 71, 81, 173
land use, 11, 31–32, 102, 121, 124, 128, 132, 134, 137, 140–41, 173, 177, 184–85, 202, 212, 236, 241, 262, 284
landraces, 153–57, 162, 164, 166
language, 2, 4, 26, 30–31, 38, 43, 46–47, 49, 63, 75–77, 99, 111, 121, 163, 183, 188, 190, 203, 210, 229, 245, 287–88
Lapland, 231
Laponia World Heritage site, 237, 240, 243, 245, 285
Lapps, 231, 239
lichen, 203, 230–33, 235, 238–39, 241–42, 244–46
livestock, 81, 85, 99, 129, 131, 141, 158, 160, 171, 174–77, 178–79, 183, 193, 196, 214, 216–17, 219, 224–25
logging, 39, 136, 193, 238–39, 245
Loita Maasai, *see* Maasai

Maasai, 95, 97
  Loita Maasai, 95–96, 99, 101, 103–4
magic, 35, 37–38
maize, 76–77, 79–80, 84, 86–88, 129, 154–58, 162–64, 219, 223, 258, 260
Maku, 44–45
malaria, 131, 148
mangrove, 29, 32, 35–36, 190, 248, 251–53
marginalization, 4, 76
Marshall Islands, 59, 61, 74
matrilineal, 32, 66, 148
Maya, 84
media, 36, 61, 117, 163, 245, 277, 285
medicinal, 30, 64, 87, 100–1, 129–31, 134, 137, 160, 179, 194, 242, 249–50, 256
Melanesia, 23, 37, 39, 142–43, 146, 281
Mesoamerican, 77, 80
mestizo, 86, 104, 196–97
Mexico, 6, 75, 83–85, 87, 94, 104–5, 186
  Atzacoaloya, 79
  Cancun, 87, 94
  Chapultenango, 78
  Copoya, *see* Copoya
  Ocotepec, 76, 78, 80
  San Pablo Huacanó, *see* San Pablo Huacanó
  Tila, 86
  Tuxtla Gutiérrez, 76, 77, 80
  Veracruz, 79
migration, 9, 28, 43, 47, 51, 54, 65–66, 70–73, 87, 141, 189, 196, 201, 205, 208, 211, 213, 225, 232, 238–39, 268
Mijikenda, 153–54, 158, 160
milk, 129, 132, 193, 218–19
minerals, 236–37, 239
mining, 11, 43, 104, 143–44, 148, 201–2, 229, 236–38, 245
Miriti-tapuya, 51

Miskitu, 95, 98–101, 104, 188–90, 192–94
mitigation, 2, 8, 12, 69, 84, 93–94, 105, 152, 263, 275, 283–86
Moch, 58, 62–67, 69, 69, 70, 72–74
Mochese, 62, 64, 67, 69, 71
Mongolia, 14–15, 18
monks, 125, 129, 132, 134–35
monoculture, 158, 259, 261–62
monsoon, 28, 125, 127, 174, 178, 181, 251
moon, 32, 33, 79, 88, 145–46, 264
Mortlock Islands, 60, 62, 66, 68, 73
Mortlocks, 58, 62–63, 74
mosquitoes, 131, 235
mountain, 19, 124–25, 130, 135–36, 181, 190, 232, 235, 239, 243, 281
mushrooms, 129, 131, 134, 242, 245
music, 267, 271, 274, 278
Muyuw, 143–47, 149–50
myths, 129, 188

NASA, 205, 207–8, 278–79, 285
national parks, 221, 239, 240
Native American, 171, 183, 269
Navajo, 171–87
Navajo-Hopi Land Settlement Act, 176
Nenets, 9, 198, 205, 207, 211
New Zealand, 5, 16, 120, 143, 166
Nicaragua, 94–95, 99, 104, 188–90, 193–95, 197
   Bocay, 190
   Indio Maíz, 190
   North Atlantic Autonomous Region, 189, 191, 194
   Puerto Cabezas, 194
nomadic, 9, 14, 18, 114, 139, 199, 202, 208, 257, 282
non-timber products, 129, 134
North America, 5, 186, 268
Norway, 198–200, 202–5, 211–12, 229, 231, 245
   Finnmark, *see* Finnmark
   Kautokeino, 198, 200, 204–5, 230, 232, 244
   Nordreisa, 200
Nualu, 101
Nueva Esperanza, 86
nuts, 30, 131, 250
Nyangatom, 214, 216–25, 284

Oceania, 25, 39, 72–73, 150
offerings, 53, 101, 134, 177, 182, 184, 256, 264
oil, 87, 201–2, 205, 207
oral traditions, 171, 188
orchards, 147–48

Pacific, 2, 13, 25–26, 28–29, 34–35, 38–40, 58–59, 61, 68, 72–74, 142–43, 148, 150–51, 179, 247, 253, 287
Paiute Tribe, 176
palm, 35, 51–53, 104, 147, 150, 250
Papua New Guinea, 38–40, 142–43, 148–50
   Gawa, 144–47

   Iwa, 144–46
   Kweywata, 144, 147
   Milne Bay Province, 142–45
   Muyuw, *see* Muyuw
   Nasikwabw, 144
   Trobriands, 145–47
   Vakuta, 146
   Yemga, 145
participatory, 61, 83, 86, 123–24, 138, 154, 162, 166, 216, 233, 245, 285
pastoralism, 3, 123, 139, 203–4, 211–12, 214, 225–26
pastoralist, 214, 217, 220–22, 224–25
pasture, 14, 85, 131, 160, 193, 203, 207–8, 220, 229–30, 233, 236, 242, 244, 281, 285
   degradation of, 201
   loss of, 210
   summer pastures, 199, 206, 232, 238–39, 243–44
   winter pastures, 200, 230, 232, 236, 244
peasant, 86, 254, 256–63
pepper, 51, 87, 260, 264
Peru, 153, 162, 254–55, 258–59, 284, 289
   Alto Cumbaza, 260
   Bajo Mayo, 260
   Cusco, 153
   Lama, 260
   Picota, 259
   San Martín, 254, 257–60, 262, 264
   Tres Unidos, 256, 259
pests, 115–16, 130–31, 134, 155, 153–56, 158–59, 161–62, 164
phenology, 47, 51
pig, 80, 116
Pirá-Paraná River, 46–47, 55
plantain, 193, 258–60
planting, 12, 28, 30–31, 48, 79–81, 85–88, 99, 129–30, 132, 146–47, 159, 164, 189, 193, 239, 248
plants, 30, 35–36, 45, 47, 51, 64, 78–80, 87, 99–102, 129–31, 134, 137, 146, 154, 158, 161, 164, 167, 173, 179, 181–83, 188–89, 192, 194, 203, 207, 220, 237, 242, 245, 248, 255, 260, 262, 264
Pleiades, 12–13, 19, 56, 146, 226
Point Hope, 265, 267–74, 276–79
polar, 2, 5, 9, 127, 198, 202, 207, 280, 286
policy, 15, 23, 25, 83, 107, 118, 137, 139, 145, 153, 196, 213, 225–26, 241, 275, 283, 287, 289
politics, 17, 23, 39, 59–60, 72, 89, 245, 265, 274, 276, 284–85
pollution, 48, 130, 132, 134
porcupines, 179, 183
potato, 12, 129, 159, 161
poverty, 55, 62, 76–77, 137, 155, 164
prayers, 101, 134, 137, 182–84, 219
precipitation, 12, 14, 125, 127, 130, 141, 161, 171, 173–75, 179–81, 198, 200, 207, 209, 211, 216, 235, 259–60
Prosopis juliflora, 221, 223

rain, 9, 12, 14, 34, 47, 50–55, 76, 78–80, 83, 88, 116, 131, 142–43, 146–48, 158, 160, 171, 177–79, 182–83, 190, 207–211, 216–20, 223–24, 230–31, 235, 251, 256–59, 281
   changes in, 178
   rainfall, 13–14, 19, 32–34, 37, 50, 52, 54, 76–80, 84–87, 119, 149, 155, 159, 161, 164, 173–74, 177, 182, 185, 208, 215–16, 223, 226, 281
rain-on-snow, 9, 207–211, 281
Rangifer tarandus, see reindeer
rapids, 44–45, 52
reciprocity, 59, 70, 162–63, 190, 193, 195–97, 263, 271, 276
REDD+, 93
reef, 28–29, 32, 58, 63–64, 66, 68, 113, 144, 161
reindeer, 9, 16, 19, 198–205, 207–8, 210–13, 229–36, 238–46, 281, 287
relocate, 35, 160, 176, 179, 269, 281
relocation, 18, 35–37, 176, 183, 268–70, 289
resilience, 1, 5, 9–11, 16–17, 32, 37, 58–60, 68, 71–72, 82, 85, 141–42, 148–49, 154, 156, 158, 163, 194, 204, 210–12, 222, 224–25, 227, 248, 251, 261, 265, 276, 281–82, 284
rice, 102, 130, 155–57, 193, 259–60
Rio Negro, 42, 43–44, 55–57
risk, 16, 26, 31–32, 34, 37, 58, 61, 65, 68, 71, 75, 84–85, 88–89, 112, 114, 119–20, 156, 158–59, 164, 183, 205, 214, 219–22, 225–26, 230, 232, 235, 241, 243–44, 265
ritual, 3, 12, 24, 30, 33–34, 41, 45–46, 48, 51, 54, 83–86, 88, 98, 102, 113, 148, 162, 188, 254, 256–57, 261, 263, 271, 273
river, 41, 45–50, 52–54, 101, 104, 109, 116, 123, 125, 160, 162, 177, 192, 205, 219, 221, 223, 238–39, 256, 259
root crops, 29–30, 129, 146–47, 248, 250
Russia, 2, 17–18, 198, 204–6, 211, 229
   Chukotka, 198, 269, 275
   Salekhard, 201, 204
   Siberia, *see* Siberia
   Yamal Peninsula, *see* Yamal
   Yamal-Nenets Autonomous Area, *see* Yamal-Nenets

Sapmi, 229, 232
sacred, 55, 98–99, 101, 113, 123–24, 127, 129, 134, 136, 138, 140–41, 153, 158, 160, 162, 164, 172, 175, 177, 219, 248, 256, 263, 273
   sacred forests, 153, 158, 160, 164
   sacred sites, 99, 101, 136, 138, 140–41, 162, 177
sago, 147–48, 150
Sami, 2, 9, 198–200, 202–4, 229–34, 236–46, 284–85, 287
San Pablo Huacanó, 76–81
sea ice, 17, 266, 269, 273
sea level, 26, 31–32, 35–36, 61, 119, 143–44, 150, 247, 251, 260
seagrass, 29, 32, 113, 252
seasons, 28, 41, 45–51, 54, 78, 123, 125, 127, 129, 131, 158, 174, 201, 222, 225, 258–59, 272
Secretariat of the Pacific Regional Environment Programme, 36, 40
seed, 86–87, 129, 147, 154–57, 159, 162–64, 166, 222–24, 239, 250, 254–58, 260–63, 282
   seed banks, 159, 164, 166
shaman, 44–45, 55, 86, 88
sheep, 129, 175–76, 178, 218, 223
Siberia, 198, 205, 206, 211
Small Island Developing States, 2, 9, 60, 62, 68, 151, 285
small islands, 5, 9, 23, 26, 60, 66, 68, 142, 144, 146–47, 281
snake, 45, 49, 52, 54, 100
snow, 16–17, 127, 131, 135, 137, 140, 171, 177, 178, 180–81, 198, 200, 203, 207–8, 211–13, 230–33, 235, 238, 241, 244, 246, 266–67, 287–88
   snowfall, 173, 177, 180, 184, 209
   types of snow, 203
soil, 14, 26, 29, 31–33, 35, 42, 51, 78, 80–81, 86–87, 129–30, 136, 140, 149, 153–54, 160, 174, 181, 192–93, 237, 239, 247–51, 257, 260–61
soil types, 26, 31
songs, 113, 116, 179, 188, 256, 271
sorghum, 219, 223–24
Spanish, 44, 76, 182, 250
species, 6, 9, 26, 29–30, 45, 51–54, 115–16, 129, 153, 160–62, 179, 181, 184, 220, 239, 250–52, 254, 257, 260–62
spirits, 35, 98, 100, 111, 113, 116, 134, 188, 190, 192, 256, 270
spirituality, 3–4, 264
spring, 9, 13, 15, 127, 131, 156, 174, 177, 179, 181–82, 200, 204, 230, 232, 244, 251, 269, 271–73
stars, 12–13, 146–48
stone, 144, 188, 248–49, 252, 255–56
storm, 18, 28, 78, 127, 158, 178, 190, 216, 251, 268–69
   surge, 26, 32, 61, 63, 247, 251
storytelling, 98, 267, 270, 276, 278
subsistence, 12, 26, 31, 34, 44, 64, 68, 77, 81, 84–89, 96, 119, 134, 160, 163, 175–76, 190, 194, 196, 225, 266, 268, 273
summer, 5, 32, 49, 51, 52–54, 76–78, 125, 127, 131, 139, 146, 148, 174–75, 178, 199–202, 204, 206, 230, 232–33, 235–36, 238–39, 242–44, 251, 260, 271, 281
sun, 51, 52, 79, 88, 97, 142–43, 145–46, 148, 188, 259, 264, 271
   sunshine, 155
   light, 79
supernatural, 28, 30, 35, 98, 255, 268–70, 276

swamp taro, 64, 68, *see also* giant swamp taro
swamps, 52, 98, 115, 160
Sweden, 199, 229–33, 235–36, 238–39, 245–46
  Arjeplog, 231
  Gällivare, 236–37
  Kiruna, 236–37
  Luleå, 236, 245
  Malmberget, 236
  Narvik, 236
  Norrbotten, 229, 235–36, 238–39
  Porjus, 238
  Tourpon, 231, 243, 245
sweet potato, 100, 248–49

taro, 30, 63–64, 67–70, 146, 148, 247–51
temperature, 32, 61, 76, 78, 80, 84–85, 87, 123, 125–27, 129–30, 132, 141, 154–55, 158–59, 161, 171, 173–75, 180–82, 184, 186, 200, 201, 204–5, 207–9, 211, 215–16, 230, 233, 235, 259
  changes in, 198
Tibetan Autonomous Prefecture, 123–24, 129
  Dechen, 124–25
  Mt Khawa Karpo, 123–24, 127, 129, 134, 136–37
  Shangri-la, 125
Tibetan Plateau, 123, 125, 127, 138–41
tides, 32–33, 63–64, 144–45, 251
  indigenous categories for, 33
timber, 100, 129, 134, 195–96, 236, 238
Tiquié River, 47, 51, 52, 55
tobacco, 107, 179
Torres Islands, 23, 24–29, 31–35, 37–40
tourism, 39, 134, 202, 229, 238, 245, 285
traditions, 11, 47, 67, 96, 98, 125, 152, 173, 182, 188, 267
transhumance, 129, 131, 134, 218, 223
tsunami, 26, 149
Tuapi, 194–95
tubers, 51, 193, 260
Tukano, 47–48, 51, 55–56
Tukanoan, 43–46, 52, 56
tundra, 205, 207, 231, 267, 269–70
Turkana, 85, 221
Tuyuka, 46–48, 51, 55–56
typhoon, 63, *see also* cyclone

unemployment, 62, 118
UNESCO, 6, 19, 39, 120, 140, 167, 240, 277, 287–90
  Laponia World Heritage, *see* Laponia World Heritage
  Local and Indigenous Knowledge Systems (LINKS) programme, 6, 120
  World Heritage, 237, 240–41, 246
United Nations Framework Convention on Climate Change, 2, 4, 8, 20, 94, 152, 189, 275, 280, 284–85, 287, 289–90

United States of America, 16, 18, 60, 65, 67, 70, 139–40, 171, 174, 186, 277–78, 283, 289
  Alaska, *see* Alaska
  Arizona, 150, 171, 173–74, 180–81, 185–86
  Barrow, *see* Barrow
  Black Mesa, 173–74
  Hawaii, 65, 70, 249, 253
  Los Angeles, 175
  New Mexico, 171, 173, 175, 182, 185
  North Slope Borough, 265, 268, 272–74, 277, 279
  Phoenix, 173, 186
  Point Hope, *see* Point Hope
  Utah, 171
urbanization, 63, 67, 236

values, 11, 15, 23, 31, 33, 59, 67, 98–99, 102, 104, 162–63, 166, 188, 192, 201, 221, 267, 276
Vanuatu, , 23–32, 35, 37–40, 143, 149–50
  Lo, 26, 31–32, 35–36
  Santo Island, 28
  Tegua, 26, 35–37, 40
  Toga, 26, 31
  Torres Islands, *see* Torres Islands
vegetable, 45, 129, 145, 156, 182, 260–62
vulnerability, 1, 8–11, 17–20, 26, 28, 59–61, 69, 75, 81, 84–85, 119, 141, 150–51, 162, 183, 194, 196, 200, 210, 221

war, 174, 176, 182, 196
  Second World War, 59, 171
water, 37, 47, 49–50, 52, 53, 61, 63–64, 77, 81, 96, 98, 101, 106, 112–14, 116, 123, 127, 129–30, 141, 144, 146, 148, 153–54, 159–60, 173–77, 179–84, 188, 192, 203, 216, 220–21, 237–38, 247–50, 252, 256, 259, 263, 269, 278
waterfall, 44–45
Weno, 58, 62–63, 65–67, 69
whale, 29, 267–74, 276, 278–79, 282
  bowhead, 265, 266–68, 275–76
whalers, 271, 273–74, 276
whaling, 265, 268, 271–74, 276
wind, 13, 17, 28, 32, 39, 52, 63, 78–79, 109, 112, 116, 144, 146, 155, 164, 174, 178, 178–79, 190, 193, 207, 226, 232, 254, 268–69
winter, 9, 18, 53, 87, 123, 126–27, 129, 131, 146, 174, 177, 181, 199–201, 204, 207–8, 211, 229–36, 238–39, 241–44, 246, 251, 269, 281
winter forage, 244
women, 3, 32, 44, 63, 76–77, 82, 94, 128, 144, 154–56, 163, 165, 177, 188–89, 194–95, 218, 221–25, 241–42, 249, 254, 256, 269–70, 284
wood, 30, 55, 99–101, 136, 195, 259

Xanthosoma sagittifolium, *see* taro

Yachaq, 256, 261–62, *see also* elders
yak, 129, 131–32
yam, 30, 32, 34, 146–48, 248–49, 251
Yamal, 17, 198–99, 201, 204–9, 211–12
Yamal-Nenets, 198, 201, 204–5
Yap State, 247, 249–53
Yolngu, 106–22, 282

youth, 46, 128, 161, 189, 196, 210, 225, 268, 270
yucca, 179, 258–60
Yup'ik, 11
Yuruparí, 54

Zoque, 75–78, 80–81, 83